XENOBIOTICS, ENVIRONMENT AND FUNCTIONAL BIOLOGY

About the Book

The xenobiotics compounds such as agro, pharmaceutical and industrial chemical substances are being used, they eventually reach the soil and water environments where they are utilized by the organisms such as soil fauna, microorganisms and plant root after their leaching and degradation. However, recent studies showed that the uptake of these xenobiotic substances lead to environmental toxicity, health hazards and degradation of biotic diversity. Therefore, the toxicity, biodegradability and efficacy of xenobiotics are considered to be great threat to human health, wildlife, flora and fauna. Many pharmaceutical and personal care products have become highly important emerging contaminant in the last decades because of their detection in virtually all-natural habitats which have toxic effects not only on human beings but also on flora and fauna. All categories of pharmaceuticals including pain killers (analgesics and anti-inflammatory), antibiotics (antibacterial), anticonvulsant drugs, beta-blockers, blood lipid regulators, x-ray contrast media, cytostatic drugs (chemotherapy), oral contraceptives and veterinary pharmaceuticals among many others have been found in the environment.

About the Author

Dr. M. Serajuddin is presently a Professor in the Department of Zoology, University of Lucknow, Lucknow. He is a Fellow of Zoological Society of India and Academy of Environmental Biology. Several awards of national and international reputes were given to him. Some important out of them are Life Time Achievement Award of Venus International Foundation, Shishak Shree Award of State Government of Uttar Pradesh, Adarsh Vidya Saraswati Rashtriya Puraskar from Gujrat, Indo Global Excellence Award for Zoology by the International Benevolent Research Foundation and Confederation of Indian Universities and Prof. Harswarup Gold Medal Award for his outstanding research and contribution in the field of Fish Biogenetics by Zoological Society of India.

Dr. Serajuddin has conferred upon the award of Distinguished Faculty for the contribution and achievement in the field of fish and fisheries by Venus International Foundation in the year 2016. He is teaching from 25 years with specialization in fish and fisheries. He has now engaged in doing research in fish biology and fish biodiversity. Dr. Serajuddin acted as Principal Investigator of major research projects funded by DST, ICAR, UGC and UPCST. He published 80 original research papers in the journals of national and international reputes, edited 05 books in his credit.

XENOBIOTICS, ENVIRONMENT AND FUNCTIONAL BIOLOGY

M. Serajuddin, M. Arshad,
Saima Amjad and Priyanka Agarwal

AMIGA PRESS INC.

DELHI (INDIA)

AMIGA PRESS INC.
H.O.: 28-A, Indra Park, Chander Nagar
Delhi-110051 (India)
Tel.: 011-22504028, +91-98736-57000
E-mail: amigapress@gmail.com
amigapub@gmail.com
Website : www.amigapress.co.in

This edition is published by Amiga Press Inc.

ISBN: 978-93-84533-51-9

First Published 2019

Printed and bound by Amiga Press Inc. New Delhi-India

DISCLAIMER

Preface

In recognition of the enormous advances in environmental sciences in recent years, we are pleased to present the book entitled "Xenobiotics, Environment and Functional Biology" which comprises of numerous publications, addressing the terminology of narrow disciplinary areas and review of research work in the field of medical sciences, biotechnology, toxicology, functional biology and newer application of environmental sciences. We appreciate the foresight and the constant diligent support of all the contributors whose valuable and appropriate response gave us the necessary impetus to publish the proceedings of the National Conference on "Impact of Environmental Xenobiotics on Human Health and Biodiversity" in the form of the present book in stipulated time period, and without whom this work could not have been brought forward. Several leading scientists have contributed their scientific articles of merit, making this book very meaningful. It is our hope that research articles published in this book will certainly bring about not only the strategies for healthy environment but also the conservation of biodiversity.

When the xenobiotics compounds such as agro, pharmaceutical and industrial chemical substances are being used, they eventually reach the soil and water environments where they are utilized by the organisms such as soil fauna, microorganisms and plant root after their leaching and degradation. However, recent studies showed that the uptake of these xenobiotic substances lead to environmental toxicity, health hazards and degradation of biotic diversity. Therefore, the toxicity, biodegradability and efficacy of xenobiotics are considered to be great threat to human health, wildlife, flora and fauna. Many pharmaceutical and personal care products have become highly important emerging contaminant in the last decades because of their detection in virtually all-natural habitats which have toxic effect not only on human beings but also on flora and fauna. All categories of pharmaceuticals including pain killers (analgesics and anti-inflammatory), antibiotics (antibacterial), anticonvulsant drugs, beta-blockers, blood lipid regulators, x-ray contrast media, cytostatic drugs (chemotherapy), oral contraceptives, and veterinary pharmaceuticals among many others have been found in the environment.

The overall aim of proposed national conference was to discuss the hazardous impact of xenobiotics on human health and biodiversity and to evaluate the strategies for their remedies in sustainable ways using modern techniques. Main focus was on the development of treatment process of industrial effluent, bioremediation of xenobiotic compounds, development of technologies for the improvement of environment, mitigation technologies for climate change and alternative remedies beside biodiversity conservation and their characterization.

We are grateful to the members of the Scientific Advisory Board for their encouragement, helpful suggestions, and their constructive criticism.

Editors

LIST OF CONTRIBUTORS

Amar Deep, Experimental and Public Health Laboratory, Department of Zoology, University of Lucknow, Lucknow-226007, U.P., India.

Amit Kumar, Experimental and Public Health Laboratory, Department of Zoology, University of Lucknow, Lucknow-226007, U.P., India.

Ankita, Section of Fishery Science & Aquaculture, Department of Zoology, Aligarh Muslim University, Aligarh-202002, U.P., India.

Arun Ratn, Environmental Toxicology & Bioremediation Laboratory, Department of Zoology, University of Lucknow, Lucknow-226007, U.P., India.

Dr. A. K. Singh, Directorate of Coldwater Fisheries, Bhimtal -263136, Nainital, India.

Dr. Ajay Kumar, Department of Medical Gastroenterology, King George's Medical University, Lucknow-226003, U.P., India.

Dr. Anubha Shukla, Environmental Toxicology & Bioremediation Laboratory, Department of Zoology, University of Lucknow, Lucknow-226007, U.P., India.

Dr. Aysha Aziz Faridi, Department of Biosciences, Jamia Millia Islamia, New Delhi-110025, India.

Dr. Baby Tabassum, Toxicology Laboratory, Department of Zoology, Govt. Raza P.G. College, Rampur, 244901, U.P., India.

Dr. C.K. Mandal, Zoological Survey of India, FPS Building, Kolkata-700016, India.

Dr. Kalpana Singh, Department of Zoology, University of Lucknow, Lucknow-226007, U.P., India.

Dr. M. Afzal Khan, Section of Fishery Science and Aquaculture, Department of Zoology, Aligarh Muslim University, Aligarh-202002, U.P., India.

Dr. Manoj Kumar, Environmental Toxicology & Bioremediation Laboratory, Department of Zoology, University of Lucknow, Lucknow-226007, U.P., India.

Dr. Md. Arshad, Molecular Endocrinology Lab, Department of Zoology, University of Lucknow, Lucknow-226007, U.P., India.

Dr. Mohammad Israil Ansari, Department of Botany, University of Lucknow, Lucknow-226007, U.P., India.

Dr. Neera Yadav, Molecular and Human Genetics Lab, Department of Zoology, University of Lucknow, Lucknow-22600, U.P., India.

Dr. Orus Ilyas, Department of Wildlife Sciences, Aligarh Mulsim University Aligarh-202002, U.P., India.

Dr. Prince Tarique Anwar, Department of Zoology, Aligarh Muslim University, Aligarh-202002, U.P., India.

Dr. Priya Bajaj, Toxicology Laboratory, Department of Zoology, Govt. Raza P.G. College, Rampur -244091, U.P., India.

Dr. Rustam Ali, Department of Zoology, S.P. Mukherjee Degree College, Bhadohi-221401, U.P., India.

Dr. Shahid Bin Zeya, Department of Zoology, Aligarh Muslim University, Aligarh-202002, U.P., India.

Dr. Somenath Ghosh, Biochemistry Laboratory, Department of Zoology, University of Allahabad, Allahabad-211002, U.P., India.

Dr. Suchit Swaroop, Department of Zoology, University of Lucknow, Lucknow-226007, U.P., India.

Dr. V. P. Sharma, Senior Principal Scientist, Head Quality Assurance and Prof. AcSIR, CSIR-Indian Institute of Toxicology Research, Lucknow- 226001, U.P., India.

Farah Akram, Department of Wildlife Sciences, Aligarh Muslim University, Aligarh -202002, U.P., India.

Farah Bano, Department of Zoology, University of Lucknow, Lucknow-226007, U.P., India.

Farheen Jameel, Department of Zoology, University of Lucknow, Lucknow-226007, U.P., India.

Graish Kumar, Department of Zoology, University of Lucknow, Lucknow-226007, U.P., India.

Humaira Rani, Amity Institute of Biotechnology, Amity University, Lucknow-226 028, U.P., India.

Irfan Paswal, Section of Fishery Science & Aquaculture, Department of Zoology, Aligarh Muslim University, Aligarh-202002, U.P., India.

Juhi Rais, Molecular Endocrinology Lab, Department of Zoology, University of Lucknow, Lucknow-226007, U.P., India.

Masood Nadeem, Section of Fishery Science & Aquaculture, Department of Zoology, Aligarh Muslim University, Aligarh-202002, U.P., India.

Md. Nurul Hasan, Zoological Survey of India, FPS Building, Kolkata-700016., India.

Mohammad Hashim, Toxicology Laboratory, Department of Zoology, Govt. Raza P.G. College, Rampur-244091, U.P., India.

Mohd Gufran Khan, Section of Fishery Science and Aquaculture, Department of Zoology, Aligarh Muslim University, Aligarh-202002, U.P., India.

Nagma, Section of Fishery Science and Aquaculture, Department of Zoology, Aligarh Muslim University, Aligarh-202002, U.P., India.

Neelam Singh, Aquatic Toxicology Research Laboratory, Department of Zoology, University of Lucknow, Lucknow-226007, U.P., India.

Nida Idrees, Toxicology Laboratory, Department of Zoology, Govt. Raza P.G. College Rampur, 244901, U.P., India.

Priyanka Agarwal, Department of Zoology, University of Lucknow, Lucknow-226007, U.P., India.

Prof. A.K. Sharma, Department of Zoology, University of Lucknow, Lucknow-226007, U.P., India.

Prof. Chandana Haldar, Pineal Research Laboratory, Department of Zoology, Banaras Hindu University, Varanasi-221005, U.P., India.

Prof. Kamal Jaiswal, Department of Applied Animal Sciences, Babasaheb Bhimrao Ambedkar University Vidya Vihar, Raebareily Road Lucknow- 226025, U.P., India.

Prof. M. Serajuddin, Department of Zoology, University of Lucknow, Lucknow-226007, U.P., India.

Prof. M. Moshahid A. Rizvi, Department of Biosciences, Jamia Millia Islamia, New Delhi-110025, India.

Prof. Madhu Tripathi, Department of Zoology, University of Lucknow, Lucknow-226007, U. P., India.

Prof. Monisha Banerjee, Molecular and Human Genetics Lab, Department of Zoology, University of Lucknow, Lucknow-226007, U.P., India.

Prof. Ravi S. Pandey, Biochemistry Laboratory, Department of Zoology, University of Allahabad, Allahabad-21102, U.P., India.

Prof. Sunil P. Trivedi, Environmental Toxicology & Bioremediation Laboratory, Department of Zoology, University of Lucknow, Lucknow-226007, U.P., India.

Reema Sonker, Department of Zoology, University of Lucknow, Lucknow-226007, U.P., India.

Rifhat Aara, Department of Zoology, Aligarh Muslim University, Aligarh-202002, U. P., India.

Rita Verma, Department of Applied Animal Sciences, Babasaheb Bhimrao Ambedkar University, Vidya Vihar, Raebareily Road, Lucknow- 226025, U.P., India.

Robeena Sarah, Toxicology Laboratory, Department of Zoology, Govt. Raza P.G. College, Rampur-244901, U.P., India.

Saima Amjad, Department of Zoology, University of Lucknow, Lucknow-226007, U.P., India.

Salman Khan, Section of Fishery Science & Aquaculture, Department of Zoology, Aligarh Muslim University, Aligarh-202002, U.P., India.

Saurabh Kumar, Department of Zoology, University of Lucknow, Lucknow-226007, U.P., India.

Shama Parveen, Molecular and Human Genetics Lab, Department of Zoology, University of Lucknow, Lucknow-226007, U.P., India.

Sunil Kumar Rai, Post-Doctoral Fellow, Department of Biochemistry and Molecular Medicine, School of Medicine, University of California, Davis, CA, USA.

Syeda Uzma Usman, Department of Zoology, Aligarh Muslim University, Aligarh-202002, U.P., India.

Tarique Mahmood, Department of Zoology, University of Lucknow, Lucknow-226007, U.P., India.

Virendra Atam, Department of Medicine, King George's Medical University, Lucknow-226003, U.P., India.

Yashika Awasthi, Environmental Toxicology & Bioremediation Laboratory, Department of Zoology, University of Lucknow, Lucknow-226007, U.P., India.

Zarreen Syed, Department of Wildlife Sciences, Aligarh Muslim University, Aligarh-202002, U.P., India.

❒

CONTENTS

SECTION-II
FUNCTIONAL BIOLOGY

Section-I
XENOBIOTICS AND ENVIRONMENT

Chapter-1

QUALITY MANAGEMENT SYSTEM: RELEVANCE TO FRONTIER AREAS OF XENOBIOTICS

V. P. Sharma*

Sr Principal Scientist, Head Quality Assurance & Prof AcSIR

CSIR-*Indian Institute of Toxicology Research, Lucknow*

**Email of Corresponding Author:* vpsitrc1@rediffmail.com*

ABSTRACT

Chemicals of different origin may act to mimic estrogens and be detrimental to reproduction and development in humans and animals. Xenobiotics are foreign substances or exogenous chemicals which are available in the nature at high concentrations polluting the environment. Few microbes are capable of breaking down of xenobiotics to some extent but generally they are non-degradable in nature. They may affect biodiversity, survival of eukaryotes and poses health hazards to humans and environment. These compounds are persistent and may remain in the environment for many years leading to bioaccumulation or biomagnifications. Xenobiotics associated with production of plastics, pesticides, paints, textile, cosmetics, agrochemicals, pharmaceuticals etc can't be eliminated from our lives. There is ubiquitous human exposure to EDCs, heavy metals, organotin owing to usage of plasticizers, stabilizers and colourants in plastic-based consumer products. Microorganisms have always been the easiest and most efficient tool used for elimination of abundant or harmful chemicals. Genetically modified species may be used in the process of bioremediation. Risk is likelihood of an adverse effect resulting from a given exposure and response to the chemical depends upon duration, route of exposure, toxicokinetics, dose response relationship and susceptibility of the individual.

QMS may be expressed as the organizational structure, policies, procedures, processes and resources needed to implement management. Quality assurance refers to administrative and procedural activities implemented in a system so that requirements and goals for a product, service or activity are fulfilled. It is never an accident; it is always the result of high intention, sincere effort, intelligent direction and skillful execution and represents the wise choice of many alternatives. Moreover, it is not merely an act but a habit. We have been challenging for the global economy due to complicated economic developments and unconventional monetary policies for boosting activities in all priority sectors. One of the permanent quality objectives of an organization in tune to process or system-based monitoring should be focused towards the continual improvement of its overall performance, leveraging clear and concise process, performance measures. Assessments of quality in scientific fields are vital for different reasons viz reducing adverse events, optimizing efficiency and enhancing satisfaction.

The International Standards [ISO/IS/IEC 17025:2017] adopts a number of management and technical principles. The OECD principles of GLP are a set of principles that may define a quality system using SOPs for conduction of non-clinical health and environmental safety testing for submission to appropriate regulatory authorities in support of the registration, licensing or regulation of chemical and related products. We need to have holistic planning with selection of courses, experienced academicians, competent manpower and latest infrastructure facilities as per compliance to International standards viz ISO/OECD/GLP/ASTM/FAO/EU/WHO etc.

Organizations need to understand futuristic customer needs or requirements and attempt to exceed the expectations of customers for long term relationship and attaining excellence in services. We need to enhance abilities of team with concerted efforts, help in personality development by training them to the highest level so that they are leaders of tomorrow with confidence and creativity. Efficient QMS helps to generate confidence in R & D, render best services to customers along with International recognition and systematic monitoring. We should make an endeavour for excellence in our field of expertise and include holistic approaches with sustainability.

Keywords: Biodiversity, Endocrine Disrupting Chemicals (EDCs), Improvement, Management, Standard Operating Procedures (SOPs), Quality.

1. INTRODUCTION

Quality management system is vital for improvement and basically it pertains to collection of business processes focused on consistently meeting customer requirements and enhancing their satisfaction. It is aligned with an organization's purpose and strategic direction (ISO/IEC/BIS 9001:2015). This may help in business to attain greater consistency in the activities involved in providing products or services and/or reduce expenditures. Moreover, that may lead to efficiency improvements by use of time and resources in appropriate manner.

2. WHAT ARE XENOBIOTICS?

Xenobiotics are chemical substances which are found within an organism that is not naturally produced or expected to be present within the organism. It may also include substances which are available or detected in much higher concentrations than are usual. Natural compounds may also become xenobiotics if they are taken up by another organism, such as the uptake of natural human hormones by fish found downstream of sewage treatment plant outfalls, or the chemical defenses produced by some organisms as protection against predators. They are not limited to toxic industrial chemicals and the study to explore the understanding of effects of xenobiotics on the human body continues.

3. PHTHALATES IN VARIED MATRICES: WATER, SOIL, AIR AND BLOOD

The phthalate levels have been measured in six different brands of bottled water exposed to temperatures ranging from -18° and 40° C and sunlight for 45 days (Jeddi et al 2016). Di-2-ethyl hexyl Phthalate [DEHP] and Di-butyl Phthalates [DBP] contaminants were detected at different storage conditions in all the bottled water samples during the storage time. Phthalates may lead to infertility and higher incidences of congenital malformations of the genital tract. The EDCs may be detected in variable concentration in different matrices throughout the globe depending on exposure pattern and duration. The sophisticated hyphenated techniques like HPLC, GC-MS-MS, and LC-MS etc are available to monitor and take initial remedial actions towards preventive approaches. Green chemistry concepts are also being adopted worldwide for efficient monitoring.

4. REPRESENTATIVE XENOBIOTICS AND DISEASES

As the pathogen grows, it reaches a critical density in which the amount of available nutrients is no longer enough to sustain the population. It may form a toxin in the hopes of poaching nutrients from the harmed and dead bacterial cells. Another growing branch of xenobiotics research focuses on foods with concern on artificial sweeteners and their effect on our microbes. These sugar substitutes are considered to be an important addition to some weight-reduction programs. The use of chemicals such as saccharin and aspartame may lead to an increase in body fat and make a person more prone to the development of

Type 2 diabetes. The body and its microbes are unable to use such chemicals for food. Moreover, bacteria suffer the most as they have a relatively short lifespan as without real sugar, they simply die off or have to look to other options for nutrients viz fats, proteins and starches. The change in food supply eventually causes a shift in the diversity of the microbes in our guts. Sometimes, the bacteria which are normally useful in keeping our metabolism and immunity balanced are the ones we lose. In their place, other bacteria may take over and start to produce chemicals known to trigger imbalance in our bodies. The most prominent is called propionate. According to selected researchers' point of view, an increase in this particular chemical has been associated not only with weight gain and a higher risk of diabetes but also irritable bowel syndrome, mood dysfunction and even possibly autism spectrum disorder.

Foreign molecules may have effects on our microbial population elsewhere in the body. In the pulmonary region the smoke of cigarettes may tends to turn bacteria more pathogenic and lead to an increase in the risk of infection and pneumonia. The use of chemical antiperspirants may change the microbial population on the skin and allow even worse-smelling bacteria to find a comfortable environment. The field of xenobiotics, microbes and our health are growing, and we may expect to find more revelations on impact of chemicals in our lives and well-being. We need to collect and interpret short- and long-term studies to explore the implications with changing lifestyles.

5. IMPLICATIONS OF ENVIRONMENTAL FACTORS

Few studies implicate the environmental factors and phthalates exposure in the pathogenesis of autism spectrum disorders. Bisphenol A or phthalates which belong to the category of endocrine disrupting chemicals [EDCs] and advanced glycation end products may affect human health in everyday industrialized life. The acute or prolonged prenatal exposure to EDCs that mimic endogenous hormones may contribute to the altered fetal programming and in consequence may cause to polycystic ovary syndrome and other adverse health effects. These are also reported to interfere with metabolic alterations viz insulin resistance, obesity, compensatory hyperinsulinemia etc. which may exacerbate the polycystic ovary syndrome phenotype and contribute to consequences such as Type 2 diabetes and cardiovascular diseases. Even cosmetics contain a vast number of chemicals (viz parabens, phthalates, polycyclic aromatic hydrocarbons, siloxanes etc) and several of them are not under the regulatory purview of the Food and Drug Administration [FDA]. Many of the EDCs are mainly used in packaging industries, pesticides, food constituents are release into the environment from different sources. The clinical evidences, experimental models, epidemiological studies suggest that they have major risks for humans by targeting different organs and systems in the body besides affecting other flora and faunas. Prenatal DEHP exposure may induce neurodevelopment toxicity associated with the neural stem cell niche and corticogenesis (Komada et al 2016). Biobased composites are being developed for environmental sustainability and reducing the impact of toxicants (Sharma 2015).

6. NEED & STRATEGIES FOR EFFICIENT QUALITY MANAGEMENT SYSTEM FOR STUDIES AND MONITORING

Efficient Quality Management System is indeed the need of the hour to understand the water crisis in India (ADB 2009). Water needs to be managed as a common pool community to achieve food security, support livelihood and ensure equitable and sustainable developments which are closely interconnected. We need to develop appropriate precision technologies and policies to help farmers in water stressed areas to conserve water. It is perceived that solely relying on the engineered focused approach does not adequately incorporate the benefits and opportunities of integrated water management (V. Dutta et al 2016).

7. CONCLUSION

Thus, a more holistic range of preventive measure and options are necessary. As we anticipate seeing India among the top five global scientific powers by 2020 we need to assist Science and Technology based high risk innovations through new mechanisms, facilitate the stakeholder's partnership for scaling success of Research & Developments to trigger changes in the mindset and value systems for rewarding the performances, respect and recognize for creating wealth from derived knowledges.

Thus, it is important to have capacity building of trained personnel with concerted efforts. We must have skill development and personality development programmes of highest level so that leaders of tomorrow are created with confidence. Efficient QMS is anticipated to generate confidence in R & D, render best services to stakeholders and regulators with International recognition and systematic monitoring. We need to have better understanding so as to meet needs of next century and strive hard to exceed the expectations of customers for long term relationship and attaining excellence.

ACKNOWLEDGEMENT

Author is thankful to the management CSIR-Indian Institute of Toxicology Research, Lucknow for encouragement and support in the academic pursuit. The R & D interactions with Dr. M. Serajuddin, faculty and students of University of Lucknow helped in shaping the manuscript with holistic and future perspective needs to highlight the impact of EDCs.

REFERENCES

Asian Development Bank Water Resources Development in India: Critical Issues and Strategic Options; 2009.

Central Pollution Control Board Status of water supply, Waste Water Generation and Treatment in Class I and Class II cities of India; Ministry of Environments and Forests and Climate Change; 2009.

M.Komada, Y. Gendai, N. Kagawa, T. Nagao Prenatal exposure to DEHP impairs Development of the Mouse Neocortex Toxicol Letters Jul 27; 259: 69-79 doi 101016/j. toxlet; 2016.

M.Z. Jeddi, N. Rastkari, R. Ahmadkhaniha, M. Yunesian Endocrine Disruptor Phthalates in Bottled Water: Daily Exposure and health risk assessment in pregnant and lactating women, Environ Monitor Assessment Sep 188(9):534 doi 10.1007/s 1066-016-5502-1; 2016.

M.Z. Jeddi, L. Janani, A.H. Memari, S.A. khondzadeh, M. Yunesian. The role of Phthalate esters in Autism Development: A systematic Review Environ Res Aug 24:151-493-504 doi 10.1016/j envres 2016.

A.Z. Rutkowska, E. Diamanti-kandarakis, Polycystic Ovary Syndrome and environmental toxins Fertil Steril, pii S0015-0282(16)62728-0 doi 10.1016/fertnstert; 2016.

V.P. Sharma, Biobased Composites : Future Trends and Environmental Aspects (Chapter 17) in the book entitled Biobased Composites for High Performance Materials : From Strategy to Industrial Application Editor : Wirasak Smitthipong, Rungsima

Chollakup and Michel Nardin published by CRC Press ; www.crcpress.com Taylor and Francis Group; pages 303-314, 2015.

V.P. Sharma, Plastics of the Future- Innovations for improvement and sustainability with special relevance to Biomedical Applications published in Biodegradable Polymeric Nanocomposites, Advances in Biomedical Applications Editor Dilip Depan CRC Press, Taylor and Francis Group; pages 221-230; Submitted earlier and downloaded e copy by 130.70.02.22; 2015.

V. Dutta, K.K. Shukla, S. Chander and R.C. Sobti, Water Policy in India: Building Blocks from Synergy with Science, Technology Innovation policy for Inclusive Growth Working paper series no. 06, ISBN 978-93-86110-01-5; Narendra Publishing House, Delhi, DST Centre for Policy Research, BBAU, 2016.

❐

Chapter-2

HUMAN HEALTH AND AGRICULTURAL CROPS CONTAMINATION WITH XENOBIOTICS

Humaira Rani[1] and Mohammad Israil Ansari[2*]

[1] *Amity Institute of Biotechnology, Amity University, Lucknow-226 028, India*

[2] *Department of Botany, University of Lucknow, Lucknow-226 007, India*

**Email of Corresponding Author: ansari_mi@lkouniv.ac.in*; ansari_mi@hotmail.com*

ABSTRACT

Human health is an important issue with advancement of life style. Xenobiotics are very toxic to human health. Chemicals and waste water are very serious problem as environmental pollutants. Industrial polluted water irrigation to the agricultural crops resulted heavy metals contamination. These heavy metals accumulation has caused deposition of heavy metals amount in agricultural crops in majority of urban and industrial areas. Industrial polluted and sewage water for irrigation of agricultural crops have more potential to accumulate heavy metals from contaminated soil. Xenobiotic toxic compounds have a negative impact on human health. Regular monitoring of these heavy metals in the industrial polluted, sewage water as well as in the agricultural crops chain is an important concern for all of us. A holistic approach is necessary to overcome this problem by various biological and non-biological techniques as available in the form of remediation as well as proper disposal of xenobiotic contaminated materials.

Keywords: Human Health; Xenobiotic; Agriculture; Environment; Heavy metals; Industrial.

1. INTRODUCTION

Human health is very important. With increase in industry there is alarming situation for all of us. Environmental pollution and pollutants, originating by human activities continuously antagonize plant population with an extensive assortment of xenobiotic compounds or its interaction with other compounds, relating to chemical structures that are at least moderately unusual for plant. The reaction of plant to these chemical encounters and its response to stimuli are being categorized with the perspective to its stress physiology, toxicology and eco-toxicology. The extensive genomic study through proteomics and transcriptomics has revealed that gene expression alterations play a significant role in plant provocations to xenobiotics (Ramel et al 2009; Weisman et al 2010). These gene expression alterations follow unique arrays of coordination and regulation within the organism (Jennings et al 2009). Thus, it appeared that effects of xenobiotic could not only comprise physiological and biochemical interference (Fufezan et al 2002; Anetar et al 2008; Liu et al 2009; Ramel et al 2009), but also have an impact at genetic level (Unver et al 2010; Ramel et al 2007, 2009).

Xenobiotics, in plants might causes injury, thus herbicide are used to shield cereal crops which do not cause plant injury, as a result productive yield could be generated (Skipsey et al 2011; Riechers et al 2010). Several studies related to transcriptomics elucidated that safener also affect at genetic level, resulting in large-scale alteration of gene expression (Riechers et al 2010; Skipsey et al 2011). Plants are capable to prompt responses to innumerable natural products, like reactive oxygen species, hormones, toxic compounds, regulators, pathogen association and allelochemicals.

Toxicity is generally attributed to act upon macromolecules or cellular targets (Ge et al 2008; Peng et al 2011; Jin et al 2011) and herbicides are usually intended on the basis of definite protein targets, such as components of photosystem I and II. Signaling effects and transcriptomic of xenobiotics in plants incorporate not only corresponding gene system and defenses mechanism (Ramel et al 2007) but also partial responses (DeRidder et al 2002) and even toxic and death of cell responses (Wang et al 2012). Further, investigation related to transcriptomic and signaling effects are particularly useful for understanding xenobiotics that do not possess targets the assortments of vigorously toxic and inert compounds (Nobels et al 2011; Wang et al 2012).

2. ENVIRONMENTAL CONTAMINATION WITH XENOBIOTICS

Xenobiotics are principal source of environmental toxicity. Due to heavy metals in the surroundings causes harmful effects to plants and animals along with humans. This is a call of consideration for administering authorities as well as for common public. Food crops could be exposed to heavy metal by polluted top soil or atmospheric diaspora of such metals from industrialized areas, though human may be affected by heavy metals through ingestion of adulterated food stuffs (Machiwa 2010). These heavy metals (Cu, Mn, Fe, Zn) are considered to be valuable for plants when consumed in quantities which assist the normal growth and development of plants (Machiwa 2010; Aziz et al 2015). Heavy metals are beneficial within the limits, but when exceeded beyond limits, it becomes toxic to the plants (Satpathy et al 2014). High levels of heavy metal in the soil, inhibits the normal growth of plant and leads to fall in crop yield (Chibuike and Obiora 2014; Nnaji and Igwe 2014). The heavy metals like arsenic, chromium and cadmium have been found to be more toxic to plants.

Heavy metals are non-degradability for a long time in the soil (Zhao et al 2015). Food crops like rice which are cultivated in waterlogged surroundings are even more wide-open to heavy metal toxicity due to the source from the soil and water. Among the most consumed crops in world rice is one of them, which may expose population widely to heavy metal toxicity. Microelements like Cu, Zn, Fe and Mn are beneficial to humans also, but when consumed in minute quantities but when they surpass the adequate confines in the body, they may cause health related issues (Fernández-Luqueño et al 2013). Heavy metal toxicity may lead to fever, hypertension and may even cause kidney disorder, cancer and DNA damage (Vanita et al 2014). Alterations in DNA methylations results to epigenetic silencing of gene expression due to heavy metal toxicity (Vanita et al 2014).

The threat to human revelation to heavy metals due to food stuffs is more prone when the food crop is cultivated on soil contaminated with heavy metals. Contamination of agricultural soil by heavy metal toxicity is due to meager practices of irrigation and use of agrochemicals such as fertilizers, pesticides and herbicides rich in heavy metals (Vanita et al 2014). It was reported that mining areas are more prone to heavy metal toxicity contamination in soil. Due to leaching process from mining sites heavy metals reach to agricultural soils causing contamination. On the other hand, during rainy season due to runoff water to the agricultural fields near the mining sites deposits tailing and waste, containing heavy metal contamination leading to high deposition and soil contamination (Matthews and Kakulu 2012).

3. HUMAN HEALTH AND XENOBIOTICS

It is very important issue in our country. The xenobiotics are biologically active artificial chemicals which may have an impact on health of humans (Jaffe 2005). Investigational, epidemiological studies and other confirmations propose that such toxic biologically active chemicals may cause severe and antagonistic health effects. Xenobiotics adversely affect the immune system which is highly susceptible. This immunotoxicity might cause deprived confrontation for infection, tumors formation and even cancer which could not be detected by our immune system (Banerjee et al 2008; Eugine and Vincent 2016). The elevated lipophilicity and tenacity of several xenobiotics significantly improves their biomagnification, thus impending probable and serious health hazards on predators at different trophic levels of food chain (Wu 1999). In our body liver plays a significant role in detoxification of body cells and as a result it is chronically wide-open to exposure of xenobiotics and their derivatives which are toxic (El-Saad and Elgerbed 2010).

Number of xenobiotics easily available in environment target our nervous system (Anetor et al 2008). The xenobiotics usually damage neurons or cells which transmits signals to the brain resulting in failure of nervous system (Descamps and Deschamps 2012). Neurotoxicity is usually expressed as multiple syndromes and its effects reliant upon the nature, level and period of the acquaintance of xenobiotics to nerve cells (Julvez and Grandjean 2009). Generally heavy metals, biphenyls, polychlorinated organic compounds cause environmental pollution. The xenobiotics are also proficient in mitochondrial disrupting, as a result free ROS are generated. Generally mitochondrial dysfunction is associated with diseases like, heart disease, diabetes, Parkinson's, metabolic syndrome, Alzheimer's, depression and even cancer (Ozaydin 2017). Genotoxicity is another health factor related to xenobiotics (Wu 1999). Xenobiotic compounds which directly or indirectly interact with DNA results in alteration of gene and imparts long-term effects and may even cause reproductive dysfunction (Garcia et al 2012). Kidney is also prone to xenobiotics as all toxic heavy metals like Hg, Pb, Cd, Pd and As are potentially nephrotoxic (Soderland et al 2010; Sarah and Bridges 2017). Thus, these nephrotoxic elements as associated with chronic disorders of endocrine, excretory, and metabolic function of the kidney (Voss et al 2003). Lipids are also susceptible to oxidative alterations by ROS. Lipid peroxidation is one of the major causes of disorders and diseases like cardiovascular disease, cancer and neurodegenerative diseases that is been done through modification of the fluidity, and membrane permeability (Singh et al 2017).

4. REMEDY OF XENOBIOTICS CONTAMINATION

Government must have strong and strict regulation for xenobiotic. Policies at national level must be made and reinforced in order to protect agricultural soil from heavy metals contamination. Industrial effluents should not surpass the acceptable parameters. Policies for mining should be reinforced for the reason to encourage appropriate utilization of areas of mining. Apart from this use of agrochemicals such as fertilizers, pesticides and herbicides rich in heavy metals should be avoided. Policies from government would be made in order to protect soil pollution. Regulatory bodies like food and drug administration should make proper acceptable limits for toxic metals in food stuff and strict action should be made against defaulters. Further role of social worker may also be important by educating the surrounding people.

5. CONCLUSION

Xenobiotics are very severe problem for human health. These are the compounds which are superfluous to the biosphere. As the xenobiotics are easily available in water and soil through industrial and sewage waste, so these xenobiotic compounds can reach to the agricultural field crop easily. These are easily available to microorganisms and plants and reach us and other animals very easily. Once they enter in human body or animal that give adverse effect of health as a result of interaction with the biomolecules and imparting its toxic effects as a result several disorders. Therefore, a holistic approach is necessary to overcome this problem. To overcome this problem various biological and non-biological techniques as available in the form of remediation as well as proper disposal of xenobiotic contaminated materials.

REFERENCES

Anetor JI, Anetor GO, Iyanda AA, Adeniyi FAA. 2008. Environmental Chemicals and Human Neurotoxicity: Magnitude, Prognosis and Markers. African Journal of Biomedical Research, 11: 1-12.

Aziz RA, Rahim SA, Sahid I, Idris WMR and Bhuiyan AR. 2015. Determination of heavy metals uptake in soil and paddy plants. American-Eurasian J. Agric. Environ. Sci. 15: 161-164.

Banerjee B D, Chakraboti A, Suke SG, Ahmed R S and Tripathi AK. 2008. Xenobiotic-induced Immune Alterations: Implications in Health and Disease.Indian Journal of Biochemistry and Biophysics 45:7-15.

Chibuike GU and Obiora SC. 2014. Heavy metal polluted soils: Effect on plants and bioremediation methods. Appl. Environ. Soil Sci. vol. 2014: 1–12.

DeRidder BP, Dixon DP, Beussman DJ, Edwards R, Goldsbrough PB. 2002. Induction of glutathione-S-transferases in Arabidopsis by herbicide safeners. Plant Physiology 130, 1497–1505.

Descamps CM and Deschamps F. 2012. Nervous system disorders induced by occupational and environmental toxic exposure. Open Journal of Preventive Medicine 2(3):272-278.

El-Saad AA and Elgerbed AS. 2010. Dimethoate-Induced Hepatotoxicity in Rats and The Protective Roles of Vitamin E and N-Acetylcysteine. Egyp. J. Exp. Biol. (Zool.) 6(2):219-230.

Eugine M and Vincent T. 2016. Prevalence of Persistent Organic Pollutants in Blantyre – Malawi. American Journal of Environmental Protection 4(3): 61-66.

Fernández-Luqueño F, López-Valdez F, Gamero-Melo P, Luna- Suárez S, Aguilera-González EN, Martínez AI, García- Guillermo M, Hernández-Martínez G, Herrera-Mendoza R, Álvarez- Garza MA, Pérez-Velázquez IR. 2013. Heavy metal pollution in drinking water – a global risk for human health: A review. Afr. J. Environ. Sci. Technol 7:567-584.

Fufezan C, Rutherford AW, Krieger-Liszkay A. 2002. Singlet oxygen production in herbicide-treated photosystem II. FEBS Letters 532, 407–410.

Ge C, Wan D, Wang Z, Ding Y, Wang Y, Shang Q, Ma F, Luo S. 2008. A proteomic analysis of rice seedlings responding to 1,2,4-trichlorobenzene stress. Journal of Environmental Sciences 20, 309–319.

Garcia FP, Ascencio SC, Oyarzun JG, Hernandez AC and Alavarado PV. 2012. Pesticides: classification, uses and toxicity. Measures of exposure and genotoxic risks.

Journal of Research in Environmental Science and Toxicology 1(11): 279-293.

Jennings LK, Chartrand MM, Lacrampe-Couloume G, Lollar BS, Spain JC, Gossett JM. 2009. Proteomic and transcriptomic analyses reveal genes upregulated by cis-dichloroethene in Polaromonas JS666. Applied and Environmental Microbiology 11, 3733–3744.

Julvez J and Grandjean P. 2009. Neurodevelopmental Toxicity Risks Due to Occupational Exposure to Industrial Chemicals during Pregnancy. Industrial Health 47:459-468.

Jaffe R. 2005. Xenobiotics: Managing Toxic Metals, Biocides, Hormone Mimics, Solvents and Chemical Disruptors, DK6006- C029.

Jin XF, Shuai JJ, Peng RH, Zhu B, Fu XY, Tian YS, Zhao W, Han HJ, Chen C, Xu J, Yao QH, Qu SC, Xiong AS. 2011. Identification of candidate genes involved in responses of Arabidopsis to polychlorinated biphenyls based on microarray analysis. Plant Growth Regulation 65, 127–135.

Liu H, Weisman D, Ye YB, Cui B, Huang YH, Colo´n-Carmona A, Wang ZH. 2009. An oxidative stress response to polycyclic aromatic hydrocarbon exposure is rapid and complex in Arabidopsis thaliana. Plant Science 176, 375–382.

Matthews COA and Kakulu S. 2012. Impact of mining and agriculture on heavy metal levels in environmental samples in Okehi local government area of Kogi State. Int. J. Pure Appl. Sci. Technol 12:66-77.

Machiwa JF. 2010. Heavy metal levels in paddy soils and rice (*Oryza sativa* (L)) from wetlands of Lake Victoria Basin, Tanzania. Tanz. J. Sci. 36:59–72.

Nobels I, Spanoghe P, Haesaert G, Robbens J, Blust R. 2011. Toxicity ranking and toxic mode of action evaluation of commonly used agricultural adjuvants on the basis of bacterial gene expression profiles. PLoS ONE 6, e24139.

Nnaji JC and Igwe OU. 2014. Fractionation of heavy metals in soil samples from rice fields in New Bussa, Nigeria. Int. J. Chem. Tech. Res. 6(14): 5544-5553.

Ozaydin F N. 2017. Xenobiotics and Chronic disease. Journal Toxicology 1(4): 1-2.

Peng RH, Xu RR, Fu XY, Xiong AS, Zhao W, Tian YS, Zhu B, Jin XF, Chen C, Han HJ, Yao QH. 2011. Microarray analysis of the phytoremediation and phytosensing of occupational toxicant napththalene. Journal of Hazardous Materials 189, 19–26.

Riechers DE, Kreuz K, Zhang Q. 2010. Detoxification without Intoxication: herbicide safeners activate plant defense gene expression. Plant Physiology 153, 3–13.

Ramel F, Sulmon C, Cabello-Hurtado F, Taconnat L, Martin-Magniette ML, Renou JP, El Amrani A, Coue´e I, Gouesbet G. 2007. Genome-wide interacting effects of sucrose and herbicide-mediated stress in *Arabidopsis thaliana:* novel insights into atrazine toxicity and sucrose-induced tolerance. BMC Genomics 8, 450.

Ramel F, Sulmon C, Bogard M, Couе´e I, Gouesbet G. 2009. Differential dynamics of reactive oxygen species and antioxidative mechanisms during atrazine injury and sucrose-induced tolerance in *Arabidopsis thaliana* plantlets. BMC Plant Biology 9, 28.

Soderland P, Lovekar S, Weiner DE, Brooks DR and Kaufma JS. 2010. Chronic Kidney Disease Associated with Environmental Toxins and Exposures. Advances in Chronic Kidney Disease 17(3):254-264.

Sarah E. Orr SE. and Bridges CC. 2017. Chronic Kidney Disease and Exposure to Nephrotoxic Metals-review. International Journal of Molecular Sciences 18: 1-35.

Skipsey M, Knight KM, Brazier-Hicks M, Dixon DP, Steel PG, Edwards R. 2011. Xenobiotic responsiveness of *Arabidopsis thaliana* to a chemical series derived from a herbicide safener. Journal of Biological Chemistry 286, 32268–32276.

Satpathy D, Reddy MV and Dhal S P. 2014. Risk Assessment of Heavy Metals Contamination in Paddy Soil, Plants, and Grains (*Oryza sativa* L.) at the East Coast of India. BioMed Res. Int. vol. 2014: 1–11.

Unver T, Bakar M, Shearman RC, Budak H. 2010. Genome-wide profiling and analysis of Festuca arundinacea miRNAs and transcriptomes in response to foliar glyphosate application. Molecular Genetics and Genomics 283, 397–413.

Ulrich RG. 2003. The toxicogenomics of nuclear receptor agonists. Current Opinion in Chemical Biology 7, 505–510.

Vanita C, Piar C, Avinash N, Kaur KJ, Pakade BY. 2014. Evaluation of heavy metals and its genotoxicity in Agricultural soil of Amritsar, Punjab, India. Int. J. Res. of Chem. Environ 4: 20-28.

Voss JU, Roller M and Mangelsdorf I. 2003. Nephrotoxicity of Organic Solvents - Evaluation of the Literature, Federal Institute for Occupational Safety and Health, Germany, Berlin.

Wang Y, Peng X, Xu W, Luo YB, Zhao W, Hao J, Liang Z, Zhang Y, Huang KL. 2012. Transcript and protein profiling analysis of OTA-induced cell death reveals the regulation of the toxicity response process in *Arabidopsis thaliana*. Journal of Experimental Botany 63, 2171–2187.

Wu RS. 1999. Eutrophication, Water Borne Pathogens and Xenobiotic Compounds: Environmental Risks and Challenges. Marine Pollution Bulletin 39, 11-22.

Weisman D, Alkio M, Colo´n-Carmona A. 2010. Transcriptional responses to polycyclic aromatic hydrocarbon-induced stress in *Arabidopsis thaliana* reveal the involvement of hormone and defense signalling pathways. BMC Plant Biology 10, 59.

Zhao K, Fu W, Ye Z, Zhang C. 2015.Contamination and spatial variation of heavy metals in the soil-rice system in Nanxun County, Southeastern China. Int. J. Environ. Res. Public Health 12: 1577-1594.

❑

Chapter-3

STUDY ON THE BEHAVIOURAL PATTERN OF SNAKE-HEADED FISH *CHANNA PUNCTATUS* DUE TO EXPOSURE OF CADMIUM

Nida Idrees, Baby Tabassum* and Robeena Sarah

Toxicology Laboratory, Department of Zoology, Govt. Raza P.G. College Rampur, 244901, U.P., India.

**Email of Corresponding Author: dr.btabassum@gmail.com*; nidaidrees999@gmail.com*

ABSTRACT

The adverse impact of heavy metals on human health as well as environment is the major worldwide concern. Heavy metals enter in environment through various natural and anthropogenic activities of human which polluted the aquatic environment. The aquatic flora and fauna are highly affected by these toxic metal concentrations in water and bioaccumulation in organisms. Fish considered as a good indicator of heavy metal pollution in water reservoirs. Channa punctatus is an edible fish which is commonly found in India. It is cheap and highly proteinaceous but due the high concentration of these metals, the fish may not safe as a food source. Cadmium Chloride is a heavy metal which is highly toxic and mostly accumulated in kidney causes renal tubular damage in fishes including humans. Other effects of the metal exposure in humans are disturbances of calcium metabolism, hypercalciuria bone fracture, formation of stones in the kidney, psychological disorders, damage to central nervous system, diarrhea, stomach pain, severe vomiting, reproductive failure and even infertility and DNA damage. High exposure can lead to lung cancer and prostate cancer. The fish showed remarkable changes in behaviour i.e. increased swimming, restlessness, surfacing and hyperactivity, when exposed to lower concentrations of $CdCl_2$. On the higher concentrations showed uncoordinated behaviour such as erratic and jerky swimming, attempt to jump out of water, frequent surfacing and gulping of air, decrease opercula movement and secretion of mucus on the body and gills followed by exhaustion and death.

Keywords: Environmental pollution, health effect, cadmium, Channa punctatus, behavioural changes.

1. INTRODUCTION

The number of factories and inhabitants has increased rapidly due to industrialization and urbanization. Freshwater contamination with a wide range of pollutants has become a matter of serious concern over the last few decades (Canli et al., 1998; Dirilgen, 2001; Vutukuru, 2005). The natural aquatic systems extensively are polluted with heavy metals released from household, industrial and other man-made activities (Velez and Montaro, 1998). Heavy metal contaminations have adverse effects on the ecological balance of the recipient environment and a diversity of aquatic organisms (Vosyliene and Jankaite, 2006). The equilibrium of the aquatic environment and organisms is disturbed by these actions (Farombi et al., 2007). It is important to know the biological status of the system to manage the aquatic ecosystem, especially when evaluating the impact of a chemical that create a stressor on the aquatic biota. Bioavailability is a dynamic process with two different phases: a physico-chemically driven desorption and a physiologically driven uptake

(Embrahimpour and Mushrifah, 2008). The atmospheric deposition, geological matrix erosion and anthropogenic sources, such as manufacturing discharges, and mining wastes can introduce heavy metals into the aquatic ecosystem. Pollution by cadmium and other heavy metals in freshwater reservoirs (water, sediment and fish) have been paid attention for considerable time (USATSDR, 1999). In the water system, fish are the inhabitants that cannot escape from the harmful effects of these pollutants they usually absorb metal directly from contaminated water or indirectly from feeding. Fish may act as sentinel organism for indicating the potential for exposure of human population to pollutants in water reservoir and recognized as major vectors for contaminant transfer to humans (Kasherwani et al., 2009). Fishes are relatively sensitive to changes in their surrounding environment. Fish health may therefore reflect and give a good indication of the status of specific aquatic ecosystem (Mokhtar et al., 2009). The toxic pollutant may affect the quality of water, feeding, swimming, delayed hatching and maturation period of fish (Atif et al., 2005). Behaviour is considered a very important individual's level response that is the result of molecular, physiological and ecological processes (Weis, 2005). The change in behaviour due to these heavy metals usually occurs on the acute and sub acute level of exposure.

2. CADMIUM BACKGROUND AND ITS TOXICITY

Cadmium is the heavy metal with no known beneficial and biological properties that supports life; there is no evidence that it is either biologically necessary or any beneficial characteristics (Eisler, 1985). Even at low concentrations it is highly toxic to all life, including plants, fish, birds, mammals (including humans), and microorganisms. It creates toxicity and causes cancer, birth defects and genetic diseases. In the study done by Borgmann et al., 2005 revealed that used comparative acute toxicity testing of 63 heavy metals, cadmium was the most toxic metal. In its elemental form Cadmium does not break down but it can change form into different compounds and species. Some of them can bind strongly to soil or sediment particles, depending primarily on the acidity of the surrounding water. In sulfide ore bodies, cadmium is usually associated with sphalerite (a zinc sulfide mineral). Other sources of cadmium are coming from mining of coal and other fossil fuels and shale. It released in the atmosphere through volcanic eruptions. It can enter the atmosphere during the burning of coal and household wastes and can enter water around a sulfide mine as a byproduct of acid mine drainage.

In unpolluted waters, natural cadmium concentrations are generally less than 1 μg/L or 1 part per billion (ppb). Most of the cadmium is released from anthropogenic sources such as mining and smelting of sulfide ores, fuel combustion, and application of phosphate fertilizers or sewage sludge. As it is a non-degradable, cumulative pollutant, continued releases are of global concern (ATSDR 2008). World health organization, the threshold body set the cadmium limit in drinking water is 0.003 mg/l.

3. BEHAVIOURAL CHANGES IN CHANNA PUNCTATUS DUE TO CADMIUM EXPOSURE

The behavioural condition studied by Kaushal and Mishra, 2011 on *C. punctatus* in the control and exposed media were noted for 24-hr. The fish showed a marked change in their behaviour when exposed to various concentration of the toxicant as shown in table 1. Many remarkable changes were seen during the exposure. Fish has shown the blinking of eyes and continuously jumping. They showed un-coordinate Swimming, restlessness and loss of equilibrium throughout the exposure.

Table 1. Comparative behavioral responses given by freshwater fish *Channa punctatus* after exposure of different cadmium compound

S.No.	Behavioural response		Various cadmium compounds		
		Control	$CdCl_2$	$CdSO_4$	$Cd(NO_3)_2$
1.	Blinking of eye	-	++++	++	+
2.	Jumping	-	++++	+	+
3.	Un-coordinate Swimming	-	++++	++++	++++
4.	Schooling	-	++	+++	++++
5.	Restlessness	-	++++	++++	++++
6.	Loss of balance	-	++++	++++	++++

Table 2. Morphological changes of freshwater fish *Channa punctatus* of different cadmium compounds.

S.No.	Morphological changes		Various cadmium compound		
		Control	$CdCl_2$	$CdSO_4$	$Cd(NO_3)_2$
1.	Discoloration of skin	-	++++	++++	++++
2.	Lesions on skin	-	++++	+++	+++
3.	Shedding of scale	-	++++	++++	++
4.	Mucus secretion	-	++++	+++	++
5.	Sedimentation of chemical on body	-	++++	++++	++++
6.	Muscular bleeding	-	++++	++	++
7.	Clumping of gills	-	++++	+++	++
8.	All fins damage	-	++++	++	++

Table 2 shows the morphological changes of freshwater teleost fish *Channa punctatus* of different cadmium compound.1s i.e. $CdCl_2$, $CdSO_4$ and $Cd(NO_3)_2$. Discoloration of skin and lesions on skin was seen. Shedding of scales was also seen. Mucus secretion was revealed throughout the exposure. The sedimentation of chemical was existed on the fish body. Fins were slowly damaged as the increase of exposure time. Gills were clumped, and muscular bleeding was seen in the exposure of cadmium.

The other experiment was done by Batool et al. 2014 results that the fish exposed to 96 hr of cadmium exposure. Fish showed remarkable changes in behaviour. The introduction of chemical in the test media showed that fish try to jump out of aquarium to avoid the chemical followed by increased swimming, surfacing and hyper activity. The exposed cadmium concentrations showed strong and direct relationship with somersaulting activity followed by convulsions, equilibrium status, hyperactivity, fin movement and swimming rate for *Channa marulius* with R2 value of 0.838; 0.796; 0.784; 0.749; 0.701 and 0.644, respectively. The fish, *W. attu* showed this relationship in the following order hyperactivity> somersaulting activity> Equilibrium status> swimming rate> fin movement> convulsions. The relationship computed between copper concentrations and fish behaviour was presented in figure 3. The computed R2 values showed that the relationship trend between fish *(C. marulius)* behaviour and metal (Copper) concentrations was somersaulting activity>

Equilibrium status> hyperactivity> swimming rate> convulsions> fin movement. The same for *W. attu* remained as hyperactivity (0.882)> Equilibrium status (0.840)> somersaulting activity (0.827)> fin movement (0.771)> swimming rate (0.696)> convulsions (0.493). The *C. marulius* in both Cadmium and Copper exposed test mediums showed maximum relationship with somersaulting activity while the same for *W. attu* was maximum for hyperactivity. The high values of R2 (Coefficient of determination) computed for each regression equation reveals high reliability of these regression models. No behavioural changes were observed in the control group of fish that showed normal behaviour throughout the experiment. The reaction and survival of aquatic animal not only depended on the biological state of animals and physiochemical characteristics of water but also on the kind, toxicity, type of exposure to the toxicants.

4. CONCLUSION

Cadmium is a toxic heavy metal and its all compounds poses toxicological effects on the snakehead fish *Channa punctatus*. The higher concentration of the toxicant may because of adoptive capacity of fish to heavy metal contaminated environment. On the acute and sub acute level of exposure, *C. punctatus* showed several behavioural changes such as mucous secretion, damage of fins, jumping, misbalance of the body and many other changes has been seen. At higher concentration, the toxic metal may prove lethal to fish as well as human and other organisms. The fish mortality may have resulted by absorption, bio-accumulation of cadmium compounds or greater activity of chemical in body of fish.

ACKNOWLEDGEMENT

Authors are acknowledged the UGC's Maulana Azad National Fellowship for financial assistance to conduct this study.

REFERENCES

Agency for Toxic Substances and Disease Registry (ATSDR). 2008. ATSDR Medical Fact Sheet; Division of Toxicology and Environmental Medicine ToxFAQs; Accessed November 19, 2010. http://www.atsdr.cdc.gov/tfacts5.pdf.

Atif, F., Parvez, S., Pandey, S., Ali, M., Kaur, M, Rehman, H., Khan, H.A., Raisuddin, S., 2005. Modulatory effect of cadmium exposure on deltamethrin induced oxidative stress in *Channa punctatus* Bloch. Archives of Environmental Contamination and Toxicology, 49: 371-377.

Batool, M., Abdullah S., Javed, M. 2014. Behavioral Intoxication of *Channa marulius* and *Wallago attu* during Acute Exposure of Cadmium and Copper. Pak. j. life soc. Sci. 12(1): 7-15.

Borgmann, U., Couillard, Y., Doyle, P., Dixon, D.G. 2005. Toxicity of sixty-three metals and metalloids to *Hyalella azteca* at two levels of water hardness: Environmental Toxicologyand Chemistry. 24(3): 641–652.

Canli, M., Ay, O., Kalay, M. 1998. Levels of heavy metals (Cd, Pb, Cu and Ni) in tissues of *Cyprinus carpio, Barbus capito* and *Chondrostoma regium* from the Seyhan River. Turkish Journal of Zoology. 3: 149-157.

Dirilgen, N. 2001. Accumulation of heavy metals in fresh water organisms: Assessment of toxic interactions. FAO. Fischer Technology. 212: 1-13.

Ebrahimpoure, M., Alipour, H., Rakhshah, S. 2010. Influence of water hardness on acute toxicity of copper and zinc on fish. Toxicology and Industrial Health, 26: 361-365.

Eisler, R. 1985. Cadmium Hazards to Fish, Wildlife and Invertebrates: A Synoptic Review. U.S. Fish and Wildlife Service Biological Report 85 (1.2), Contaminant Hazard Reviews Report No. 2. 46 pp.

Farombi, E.O., Adelowo, O.A. Ajimoko, Y.R. 2007. Biomarkers of oxidative stress and heavy metal levels as indicators of environmental pollution in African Cat fish *(Clarias gariepinus)* from Nigeria Ogun River. International Journal of Environmental Research and Public Health. 4: 158-165.

Kasherwani, D., Lodhi, H.S., Tiwari, J.K., Shukla, S., Sharma, U.D., 2009. Cadmium toxicity to catfish, *Heteropneustes fossilis* (Bloch). Asian Journal of Experimental Sciences, 23: 149-156.

Kaushal, B. T., Mishra, A. 2011. A comparative toxicity analysis of cadmium compounds on morphological and behavioral aspects in air breathing freshwater fish *Channa punctatus*. I.J.S.N. 2(2): 266-269.

Mokhtar, M.B., Ahmad, Z.A., Vikneswaran, M., Sarva, M.P. 2009. Assessment level of heavy metals in *Paenaeus mondon* and *Oreochromis mossambicus* spp. In selected Aquaculture ponds of high densities development area. European Journal of Scientific Research. 30:348-360.

USATSDR, (United States Agency for Toxic Substances Disease Registry). 1999. Toxicological Profile for Cadmium. Published by US. Department of Health and the Human Services. Public Health Services, 229-240.

Velez, D., Montoro, R. 1998. Arsenic speciation in manufactured seafood products: a review. Journal of food Protection. 9: 1240-1245.

Vosyliene, M.Z., Jankaite, A. 2006. Effect of heavy metal model mixture on rainbow trout biological parameters. Ekologija, 4: 12-17.

Vutukuru, S.S. 2005. Acute effects of Hexavalent chromium on survival, oxygen consumption hematological parameters and some biochemical profiles of the India major carp, *Labeo rohita*. International Journal of Environmental Research and Public Health. 2:456-462.

Weis, J.S. 2005. Does pollution affect fisheries? Book critique. Environmental Biology and Fishes. 72: 357-359.

❒

Chapter-4

ASSESSMENT OF SOME ENDOCRINE DISRUPTING COMPOUNDS (BISPHENOL-A AND Γ-HCH) ON GONADAL DEVELOPMENT AND SEX DIFFERENTIATION IN INDIAN MAJOR CARP, *LABEO ROHITA* (HAM.1822)

Rita Verma[1*], A.K. Singh[2] and Kamal Jaiswal[1]

[1] Department of Applied Animal Sciences Babasaheb Bhimrao Ambedkar University, Vidya Vihar, Raebareily Road, Lucknow-226025 (UP), India

[2]Directorate of Coldwater Fisheries Bhimtal -263136 (Nainital), India

**Email of Corresponding Author: ritaverma.nbfgr@gmail.com*

ABSTRACT

This study describes laboratory-based exposure experiments conducted with Bisphenol-A, γ-HCH on an Indian major carp Labeo rohita. 200 fish specimens of Labeo rohita weighting 150±10.2 g was exposed to different doses (0.2 mg/L, 0.3 mg/L, 0.5 mg/L) of Bisphenol-A (BPA) and γ-HCH (0.2 mg/L, 0.3 mg/L, 0.5 mg/L) for 15, 30, 60 days. The effect of these compounds on RNA/DNA ratio, estradiol, testosterone, aromatase and vitellogenin level was estimated and histological changes in gonads was recorded. Our result showed that RNA/DNA ratio was significantly low and less than 1 which indicating poor development of fishes with in 3 months. Bisphenol-A induced the estradiol level significantly ($p<0.5$) and γ-HCH significantly ($p<0.01$) decreased the 17-β estradiol level in serum. The level of Vitellogenin significantly increased in Bisphenol-A groups. The results of this study suggested how Bisphenol-A and γ-HCH caused reproductive disturbance creating hormonal imbalance i.e., endocrine disruption relating the consequences of the release of estrogenic or androgenic substances in wastewater effluent in natural waters.

Keywords: EDCs, Bisphenol-A, γ-HCH, RNA/DNA Ratio, Aromatase

1. INTRODUCTION

Natural aquatic bodies like rivers/streams are now reported to be contaminated with many pesticides and organic chemicals such as polycyclic aromatic hydrocarbons (PAHs), organochlorine pesticides (OCPs) (Singh et al. 2005; Singh et al. 2007; Verma et al. 2013) and plasticizers like phthalic acid ester (Srivastava et al. 2010). These pollutants are reported to reach aquatic environment through domestic effluents (Kinnberg et al. 2000; Shioda and Wakabayashi 2000; Knorr and Braunbeck 2002; Schwaiger et al. 2002; Tanaka and Grizzle 2002; Gagne et al. 2005; Harris et al. 2005; Tijani et al. 2013), livestock waste (Ying et al. 2002; Jobling et al. 2002; Ankley et al. 2003; Singh and Srivastava 2013; Tijani et al. 2013), agricultural runoff (Cespedes et al. 2004), pharmaceuticals wastes (Gehring et al. 2002a; Ying et al. 2002) and personal care products or personal hygiene products, food and beverage containers (James et al. 2013; Singh et al. 2013). Such substances have been reported as a persuasive environment contaminator (Ying et al. 2002; Barnhoorn et al. 2004; Jurajda et al. 2010a) continually entering aquatic animal such as fish through respiration, osmoregulation, and intake of contaminated food. Effect of such plasticizers, pesticides and pollutants has been reported to result in physiological and behavioral abnormalities in aquatic organisms via impairment of endogenous hormones levels and consequently disturbing the next progeny (WHO/PICS 2002). These compounds can act at the cellular level, involve steroid receptor binding or block steroid receptor binding or disrupting the biosynthesis and or metabolism of steroids (Keith et al. 2000; Makynen et al. 2000; Jensen et al. 2001; Metzler 2001; Maclatchy et al. 2003; Sharpe and Irvine 2004), altering steroid synthesis inhibition of cytochrome P450 isoforms or steroid metabolism. Cytochrome P450 aromatase is member of the cytochrome P450 superfamily. The CYP 1, CYP 19 gene affected by the exposure of these. These genes encode an enzyme controlling the synthesis of estrogen and expressed in brain and gonad of vertebrates (Chiang et al. 2001).

Bisphenol-A (BPA) is an industrial chemical, used to manufacture polycarbonate and numerous plastic articles and leach to salt and fresh water. Low levels of BPA have also been found to cause biological effects and its mode of action appears to mimic that of the female hormone, estrogen. BPA therefore belongs to a group of chemicals termed "hormone disruptors" or "endocrine disruptors" that are able to disrupt the chemical messenger system in the body. BPA is classified as endocrine disruptor compound (EDC) comprising of a vast group of additives, or plasticizers, used in the manufacture of synthetic materials. BPA is an industrially important compound used in many plastic applications, for example, plastic pipes, epoxy resins, and coating, and monomer in the manufacture of polycarbonate (infant feeding bottles, tableware, microwave, returnable water and milk bottles etc.) and epoxy resins used in internal protective lining for food and beverage cans, coating on metal lids for glass jars and bottles and surface coating on drinking water storage tanks and is discharged into the environment at manufacturing plants throughout the world (Drastichova et al. 2005; Kitamura et al. 2005; Ramirez 2007; Vandenberg et al. 2009). Kwak et al. (2001) described that BPA has an estrogenic potency. Similarly, Hexachlorocyclohexane (HCH) is a commercial insecticide and is available as a group of manufactured chemicals that do not occur naturally in the environment. Various HCH isomers; alpha (α), beta (β), delta (δ) and gamma (γ) are available. Although technical-grade HCH is no longer used as an insecticide, α-, β-, γ-, and δ-HCH have been found in the soil and surface water at hazardous waste sites because they persist in the environment (Zhan et al. 2000; Druga et al. 2009).

Endocrine disrupting compounds (EDCs) renowned as synthetic or natural compounds that could mimic or block the natural hormones in the endocrine systems of animals. Endocrine disruption in aquatic animals is well documented and studied. It is also documented that how these chemicals disturb the endocrine regulation in fish. (Jensen et al. 2001; Belt et al. 2002; Kinnberg and Toft 2003; Kazeto et al. 2004; Filby et al. 2007; Schung et al. 2011). These days due to human intervention our Indian river contaminated with kinds of Endocrine disruptor compound. (Srivastava et al. 2010; Roy et al. 2011; Singh et al. 2007). Histopathological examination is decisive for the examine the disruption in fish and reproductive hormone level is the biomarker for the determination of endocrine disruption activity (Navas et al. 2005). Widespread use of plasticizers and pesticides has been regarded as an important factor related to the decline of fish abundance in freshwater systems (Burkhardt-Holm et al. 2005; Kidd et al. 2007). The aim of this study was to determine the impact of different doses of Bisphenol-A and HCH on the gonadal development, sex differentiation and reproductive hormone levels in an Indian major carp *Labeo rohita* so as to make out the scale of the endocrine disruption activities of these chemicals as such compounds are reported from rivers/streams.

2. MATERIALS AND METHODS

Test Species and Experimental Design:

Actively moving immature fishes of *Labeo rohita* (total length, 20±1.6cm, weight, 150±10.2g, n=100) with no visible sign of disease were collected from the grow-out pond, Lucknow and were acclimatised to the laboratory conditions in large tanks for two weeks. They were treated with potassium permanganate for the elimination of any fungal and bacterial infection. Adequate aeration was provided by using air bowlers and optimum water quality parameters were maintained during the acclimatisation period. Fish were fed adult zooplankton twice daily and also commercially available fish food pellets (Tyio Pvt. Ltd. India).

Two treatment groups were set up to study the effect of Bisphenol A (99% pure 133027 lot # MKBG4149V; Sigma Chemicals, USA) and HCH (Catalogue no. 33856 lot no. SZBA329XV; Sigma Chemicals, USA) on the gonadal development and reproductive physiological changes in the *Labeo rohita*. After the acclimation period, 100 immature *Labeo rohita* of 150± 10.2 g weight and 20±1.6 cm length was divided equally and kept into four glass aquaria containing 150 L water in triplicate. Fish of each aquarium were given 0.2 mg/L, 0.3 mg/L and 0.5 mg/L of Bisphenol A. Similarly, in other group of experiment fish of similar length and weight were maintained in triplicate and were exposed to 0.2 mg/L, 0.3 mg/L, 0.5 mg/L HCH. Every 10th day, the water of each aquarium was replaced with fresh water containing similar dose of Bisphenol A or HCH. The test chemical Bisphenol was diluted in carrier solvent DSMO (1%) whereas HCH carrier solvent used was acetone (Naciff et al. 2005). These diluents were also mixed in control group as vehicle solution. After three months of exposure, length and weight of each group of fishes was recorded with the help of callipers digital scale and digital electronic balance respectively. The gonado somatic index (GSI) was calculated with the help of the formula (Singh et al. 2015)

$$\mathrm{GSI} = \frac{W1}{W2} \times 100$$

Where, W1 = total weight of gonad
W2 = total weight of fish

For hormonal analysis, both groups of fishes were anesthetized with 0.01% tricane methane sulfonate (MS-22, Sigma Chemical Ltd. USA) and blood was collected with the help of 1 ml syringe from the caudal fin region and it was centrifuged at 5000 rpm for 5 minute to separate the serum for hormonal assay; serum Aromatase, 17-β estradiol, testosterone and vitellogenin level in the fish was estimated by with the help of sandwich ELISA kit (by Enzo -Life Sciences, India). The OD of each processed samples was taken at 405 nm. Serum Vitellogenin level was assayed using sandwich ELISA kit (Blue gene cat no E10V005C) and the readings were taken at 450 nm wavelength. Testosterone and 17-β Estradiol levels were calculated as follow Calculate the average net optical density for each standard and sample by subtracting the average NSB OD from the average OD for each standard and sample. Average Net OD = Average OD – Average NSB OD. Calculated the binding of each pair of standards well as a percentage of the maximum binding wells (Bo), using the formula: %age Bound = Net OD / Net Bo OD x 100 using legit- log paper, plot the percentage Bound (B/Bo) versus concentration of testosterone and 17-β estradiol standards. The concentration of these hormones of the unknowns was determined by interpolation.

Histology:

The fishes of control and treated groups were gross dissected after blood sampling, gonads and liver of all the experimental and control fishes were excised, fixed in 10% buffered formalin and allowed to fix over night at room temperature. The fixed tissues were processed after approximately 24 h. Fish samples were removed from the fixative and were washed with 70% methanol and dehydrated through a graded methanol series (50%, 70%, and 100%). Fixed and cleared samples were infiltrated with melted Paraplast Plus paraffin (Paraffin wax Merk Bangalore, India) at 60°C. The paraffin was allowed to harden overnight at room temperature, and paraffin blocks were prepared with the help of automated block making machine (Leica, Japan) and the blocks were stored at 4 °C until sectioning. Sections were cut using a rotary microtome (Leica, Japan). Serial sections were cut at 5 μm. The sections were floated out onto a water bath at 40°C to properly stretch the ribbons/tissue, and placed on slides followed by drying at 55°C. Slides were then stained with hematoxylin (EMS, Hatfield, PA, USA) for 3 min, processed through acid alcohol and ethanol washes, and then stained in 1% Eosin Y (EMS, Hatfield, PA, USA) in 80% ethanol for 1 min. Slides were then dehydrated through an ethanol series (70, 95, and 100%) and cleared in xylene. Slides were preserved under glass cover slips to dry. Images of the gonad on each slide were recorded using a Nikon P80 ZOOM digital camera attached to an Olympus BX41 microscope (Optical Analysis Corporation, Nashua, NH, USA).

RNA/DNA Ratio- Spectrophotometric method:

First, the muscle samples (100 mg ml^{-1}, w/v) were homogenized for 5 min in 5% trichloroacetic acid (TCA) at 90°C and then centrifuged at 5000 rpm for 20 min. For the determination of RNA, 2.0 ml of distilled water and 3.0 ml of orcinol reagent was added in 1.0 ml of supernatant. The reaction mixture was kept in boiling water bath for 20 min. The greenish-blue colour thus developed was read at 660 nm in a spectrophotometer. For DNA determination, 1.0 ml of distilled water and 4.0 ml of freshly prepared diphenylamine reagent were added to 1.0 ml of the supernatant. The reaction mixture was kept on a boiling water bath for 10 min. The blue colour developed was measured at 600 nm. Standard curves for RNA and DNA were drawn using different concentrations of yeast RNA and calf thymus DNA, respectively. The values were expressed as μg/100 mg fish muscle tissue on dry basis. For the calculation of results, average the duplicate readings for each standard, control and samples and subtract the average zero standard optical density. Create a standard curve on log-log graph paper, with concentration on the y-axis and absorbance on the x-axis. Draw the best fit straight line through the standard points and it can be determined by regression analysis.

Statistical analysis:

The data of this study were calculated for statistics and expressed as mean ± Standard deviation (SD). Different treatment values were analyzed using ANOVA for comparison and statistical values. The statistical significance test was calculated using SPSS 16.0 Software.

3. RESULTS

The experimental fishes were found to reveal survival % of 70% to 80% after BPA and γ-HCH treatments to Labeo rohita. The length and weight of γ-HCH treated fishes were found depressed as compare to control. GSI, the main sign of gonadal development was decreased during the experimental period by the exposure of BPA and γ-HCH (Fig

1; Fig 2). GSI value in control group of fish ranged from 0.2 to 0.35 and in Bisphenol-A treated group it decreased it was 0.2 to 0.25. Further, it was observed that fish exposed to Bisphenol A and γ-HCH showed significant decreased RNA/ DNA ratio (Fig 3; Fig 4). DNA concentrations ranged from 0.356 microgram to 0.299 microgram at increasing level of Bisphenol A. The average DNA concentration was 0.317 micro grams. The RNA concentration significantly decreased with increasing BPA and γ-HCH concentration, which was 0.449 to 0.240 microgram at increasing doses of Bisphenol- A and γ-HCH. The average RNA concentration was 0.319 micro grams. The RNA/DNA ratios showed a similar decreasing pattern with increasing the concentration of BPA and γ-HCH.

BPA treated fish was found to have significantly ($P<0.05$) reduced serum testosterone level which were 566.6± 19.8; 537.2± 10.9; 531.25± 19.9 especially in high dose when compared with the control fish 1535.7± 5.38; 1505± 1.83; 1587.8 ± 12.7 during 30, 60 and 90 day respectively (Fig 5). However, opposite result was observed in case of γ-HCH treatment where serum testosterone continuously increased after its exposure and its values in high dose treatment were 1683± 11.8pg/ml, 1844± 10.9pg/ml, 1933.2± 14.5pg/ml after 30, 60 and 90 days (Fig 6). Further, Serum 17-β estradiol was found significantly ($P<0.05$) increased and remained more or less similar i.e., 1005± 20pg/ml; 1088± 18p.5g/ml; 1012± 18.5pg/ml after high dose BPA treatment to fish as compared to the control fish values of 297.5± 6.2pg/ml, 314± 12pg/ml and 355.5± 15.5pg/ml during 30, 60 and 90 (Fig 7) day of treatment but γ-HCH has decreased the value but it is not significant ($P>0.05$) compare to control (Fig 8).

Serum Vitellogenin hormone also increased when compare to the control values of 995.4 ± 16.1 pg/ml; 919.8± 20.1 pg/ml and 996.8± 19.5 pg/ml during 30, 60 and 90 day of experiment. However, the increase was over ten times (10323± 20.0 pg/ml) after high dose of of BPA treatment and highly significant (($P<0.05$) (Fig 9). To the contrary HCH treatment slightly depressed the serum vitellogenin levels which were 912.8± 14.5 5 pg/ml; 940± 4.5 5 pg/ml and 807.5± 8.55 pg/ml at 30, 60 and 90 day (Fig 10). Further, it was observed that fish exposed to Bisphenol A and γ-HCH showed increased ($P<0.05$) serum aromatase levels which were 113.9± 13 pg/ml, 150.5± 14.5 and 213.5± 8.9 as compared to the control values of 79.7± 5.2pg/ml, 79.9± 4.5pg/ml and 79.7± 15pg/ml at 30, 60 and 90 days respectively (fig 11). Similar increased level of aromatase values was also observed in the case of HCH treatment but it was less pronounced (Fig 12).

Effect of Bisphenol-A and γ-HCH on Gonadal Histology of *Labeo rohita:*

In BPA treatment group histological observation indicated clear cut difference from the control group. Histological structure of testis showed different stage of spermatogenesis. In controlled condition cross section of testis showed ducts, Sertoli cells (SE), tunica albuginea (TA) and it revealed the presence of spermatogonia (SG) spermatids (ST) and spermatozoa (SZ) (Fig 13a) the testes of BPA exposed fish showed disorganization of tubular elements semeniferous tubules degenerated when treated with 0.5mg/L BPA (Fig 13b; 13c). BPA treated testes showed developing oocyte (OC) after 60 days treatment of 0.5mg/L and a clear mature oocyte was observed in developing testes having number of spermatocytes surrounding it. In controlled condition, immature ovary showed the perinuclear stage and vitellogenic stage. In perinuclear stage the number of nucleoli increased and arranged along the inner side of the nuclear membrane. In vitellogenic stage, oocyte size was found increased. The nucleus was convoluted and follicular layer were well developed.(Fig 14 a) In BPA treated fish, ovular cell were not clear and no perinuclear condition was observed and yolk vesicles were surrounded by the nucleus of the ovule and degenerate corion layer (Fig 14 b). Histological section of ovary of *Labeo rohita* HCH Treated showing developing ST (Spermatocytes) and developing oocyte Cortical Alveolar (CA) (Fig 15a) developing oocyte at high of HCH shows showing developing Spermatocytes (ST), Developing oocyte Migrating nucleus (MN), Vitellogenic follicles (V) showing destroying premature oocyte (POC), Oocyte going to atresia (AOC) (Fig 15 b), another ovary section shows developing ovary showing mature oocyte (MOC), ST (Spermatocytes) (Fig 15c).

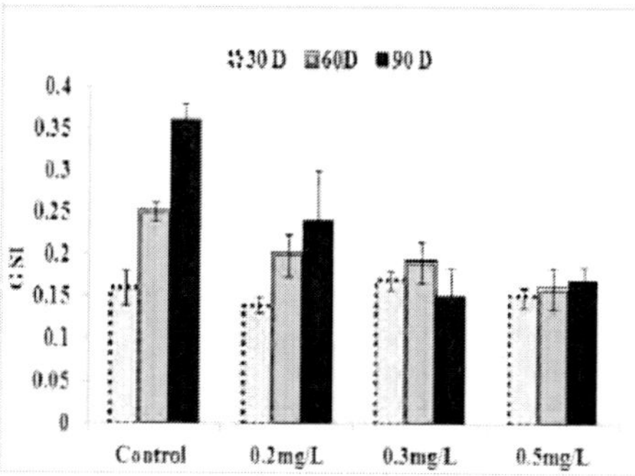

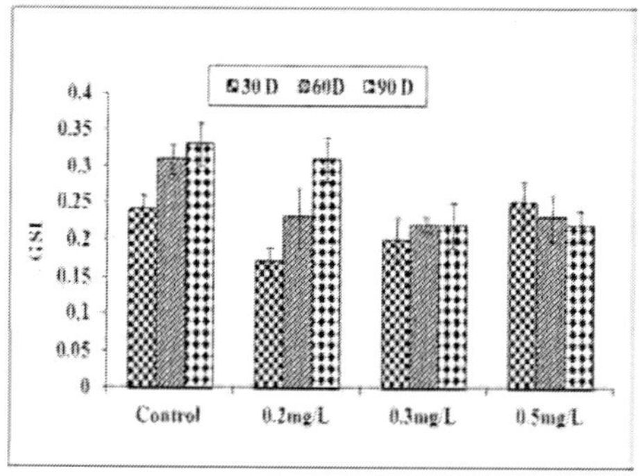

Fig.1. GSI of *Labeo rohita* exposed to 0.2mg/L, 0.3mg/L and 0.5mg/L Bisphenol A for 90 days. (Y axis) represents the Value of GSI. The data are presented as the mean ± standard deviation (n = 10). Statistically significant differences according to a one–way ANOVA (P < 0.05)

Fig.2. GSI of *Labeo rohita* exposed to 0.2mg/L, 0.3mg/L and 0.5mg/L γ-HCH for 90 days. (Y axis) represents the Value of GSI. The data are presented as the mean ± standard deviation (n = 10). Statistically significant differences according to a one–way ANOVA (P < 0.05)

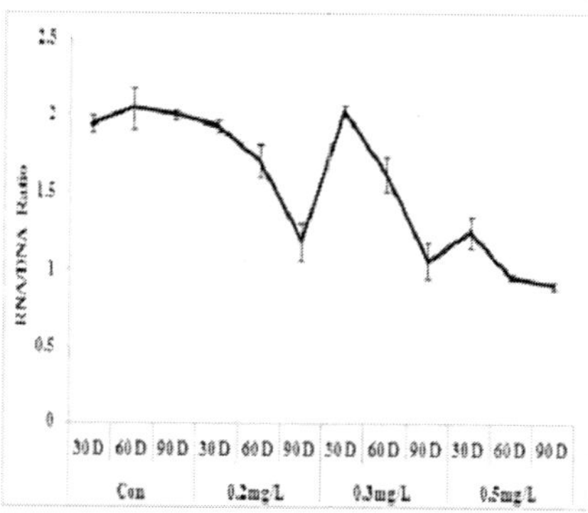

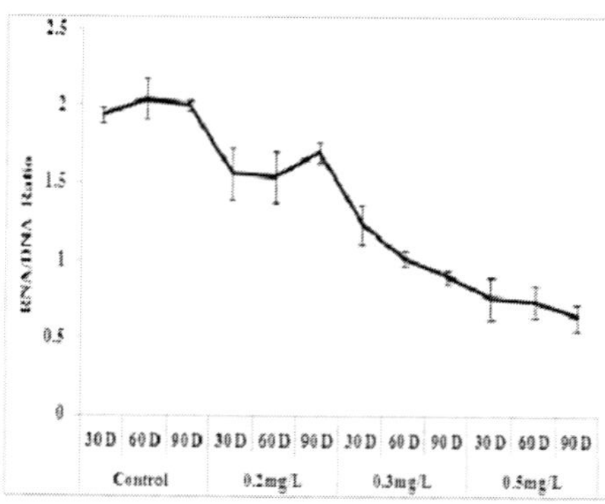

Fig.3. Value of RNA/DNA ratio in *Labeo rohita* exposed to 0.2mg/L, 0.3mg/L and 0.5mg/L Bisphenol-A for 90 days. (Y axis) represents the ratio. The data are presented as the mean ± standard deviation (n = 10) statistically significant differences according to a one–way ANOVA (P < 0.05)

Fig.4. Value of RNA/DNA ratio in *Labeo rohita* exposed to 0.2mg/L, 0.3mg/L, and 0.5mg/L γ-HCH for 90 days. (Y axis) represents the ratio the data are presented as the mean ± standard deviation (n =10). Statistically significant differences according to a one–way ANOVA (P < 0.05)

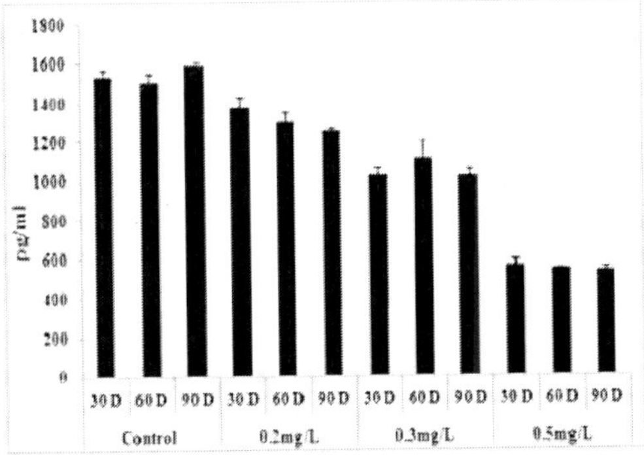

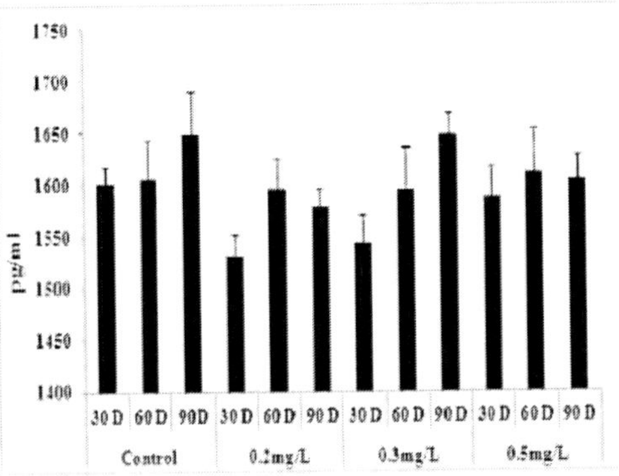

Fig.5. Value of Testosterone *Labeo rohita* exposed to 0.2mg/L, 0.3mg/L and 0.5mg/L Bisphenol A for 90 days. (Y axis) represents the Value of testosterone in pg/ml. The data are presented as the mean ± standard deviation (n = 10). Statistically significant differences according to a one–way ANOVA ($P < 0.05$)

Fig.6. Value of Testosterone *Labeo rohita* exposed to 0.2mg/L, 0.3mg/L and 0.5mg/L γ-HCH for 90 days. (Y axis) represents the Value of testosterone in pg/ml. The data are presented as the mean ± standard deviation (n = 10). Statistically significant differences according to a one–way ANOVA ($P < 0.05$)

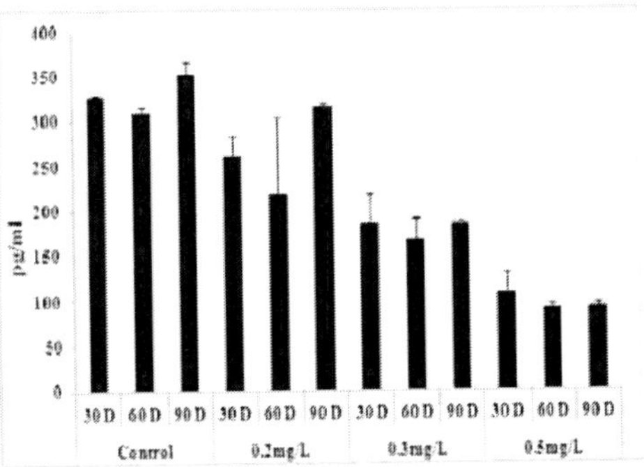

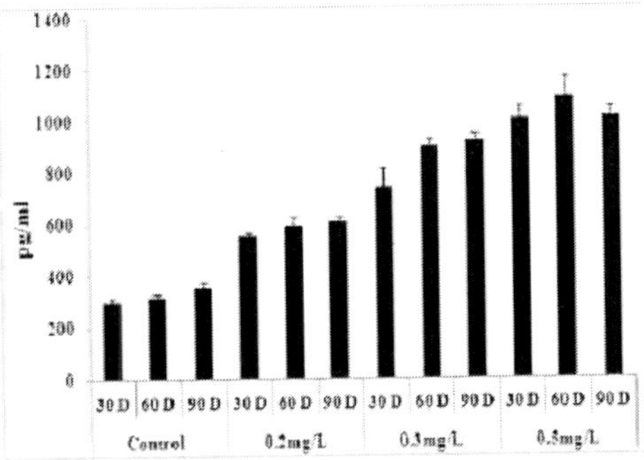

Fig.7. Value of Estradiol *Labeo rohita* exposed to 0.2mg/L, 0.3mg/L and 0.5mg/L Bisphenol A for 90 days. (Y axis) represents the Value of Estradiol value in pg/ml. The data are presented as the mean ± standard deviation (n = 10). Statistically significant differences according to a one–way ANOVA ($P < 0.05$)

Fig.8. Value of Estradiol *Labeo rohita* exposed to 0.2mg/L, 0.3mg/L and 0.5mg/L γ-HCH for 90 days. (Y axis) represents the Value of Estradiol value in pg/ml. The data are presented as the mean ± standard deviation (n = 10). Statistically significant differences according to a one–way ANOVA ($P < 0.05$)

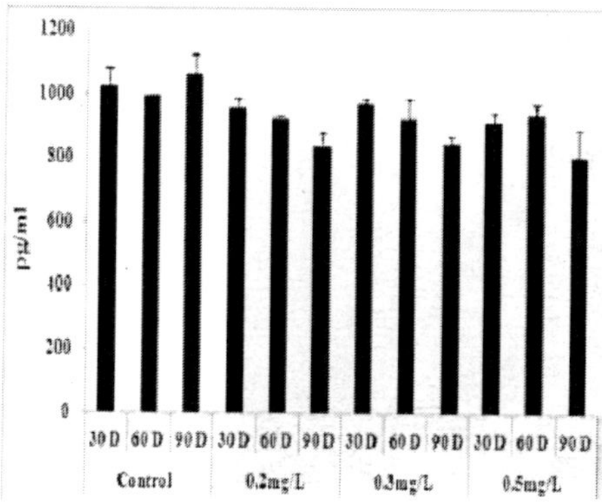

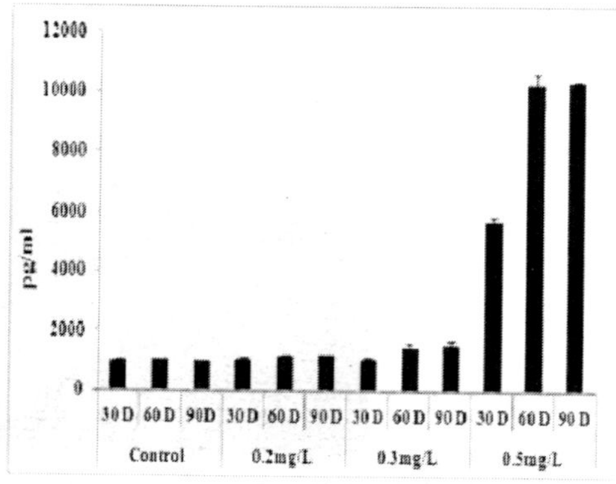

Fig.9. Value of Vitellogenin *Labeo rohita* exposed to 0.2mg/L, 0.3mg/L and 0.5mg/L Bisphenol-A for 90 days. (Y axis) represents the Value of vitellogenin in pg/ml. The data are presented as the mean ± standard deviation (n = 10). Statistically significant differences according to a one–way ANOVA (P < 0.05)

Fig.10. Value of Vitellogenin *Labeo rohita* exposed to 0.2mg/L, 0.3mg/L and 0.5mg/L γ-HCH for 90 days. (Y axis) represents the Value of vitellogenin in pg/ml. The data are presented as the mean ± standard deviation (n = 10). Statistically significant differences according to a one–way ANOVA (P < 0.05)

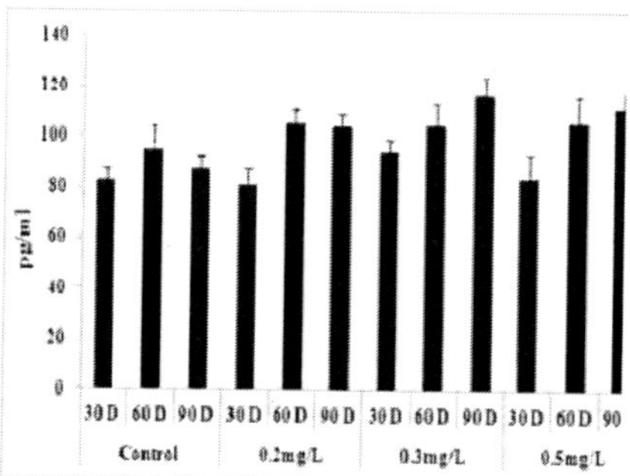

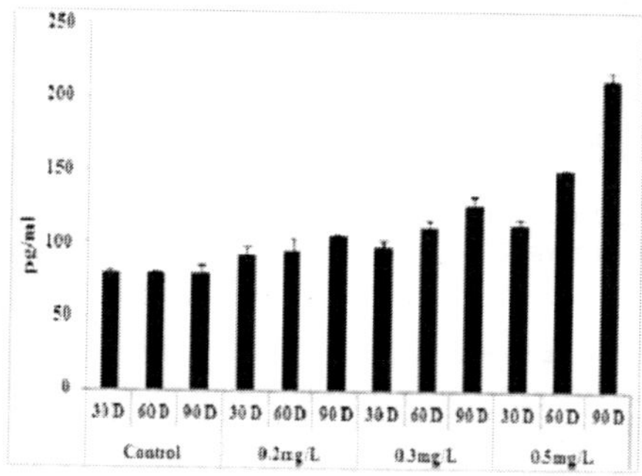

Fig.11. Value of Aromatase *Labeo rohita* exposed to 0.2mg/L, 0.3mg/L and 0.5mg/L Bisphenol A for 90 days. (Y axis) represents the Value of aromatase in pg/ml. The data are presented as the mean ± standard deviation (n = 10). Statistically significant differences according to a one–way ANOVA (P < 0.05)

Fig. 12. Value of Aromatase *Labeo rohita* exposed to 0.2mg/L, 0.3mg/L and 0.5mg/L γ-HCH for 90 days. (Y axis) represents the Value of aromatase in pg/ml. The data are presented as the mean ± standard deviation (n = 10). Statistically significant differences according to a one–way ANOVA (P < 0.05)

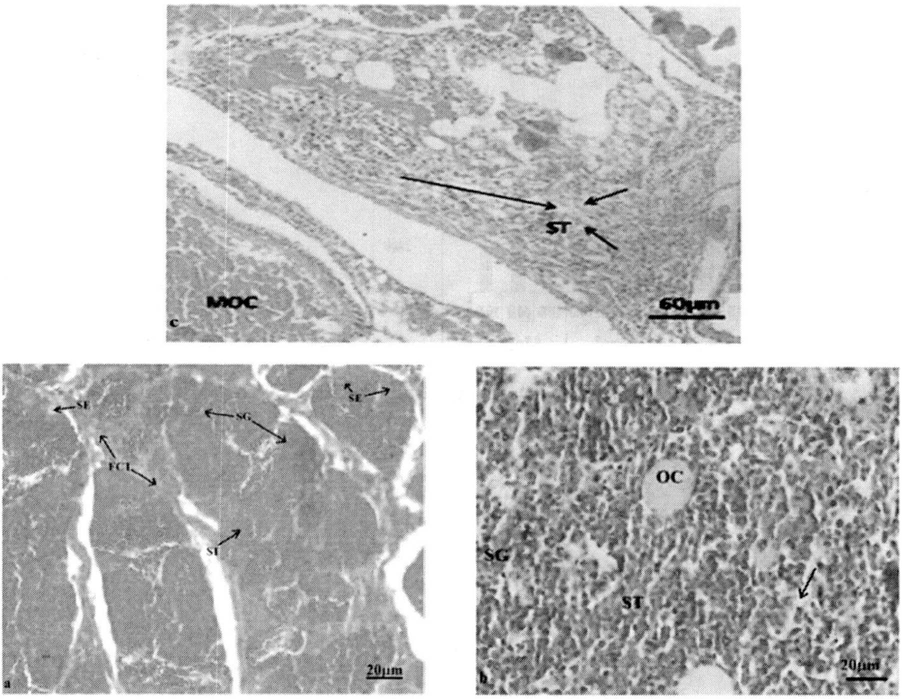

Fig.13a. Histological control testes of *Labeo rohita* testes showed cysts form of spermatocytes (SC), spermatogonia (SG), Sertoli cell (SE) fibrous connective tissue (FCT), all stage of spermatogonia developed in a lobule (semeniferous tubule) Bar scale = 20µm.

Fig.13b. Histological section of testes of *Labeo rohita* Bisphenol A Treated testis showing developing oocyte (OC), ST (Spermatocytes), SG (Spermatogonia). Arrow shows degenerated semeniferous tubule. H&E scale bar 20 µm. c. Developing testes of *Labeo rohita* showing mature oocyte (MOC), ST (Spermatocytes). Arrow (→) shows degenerated semeniferous tubule H&E X- bar 60 µm.

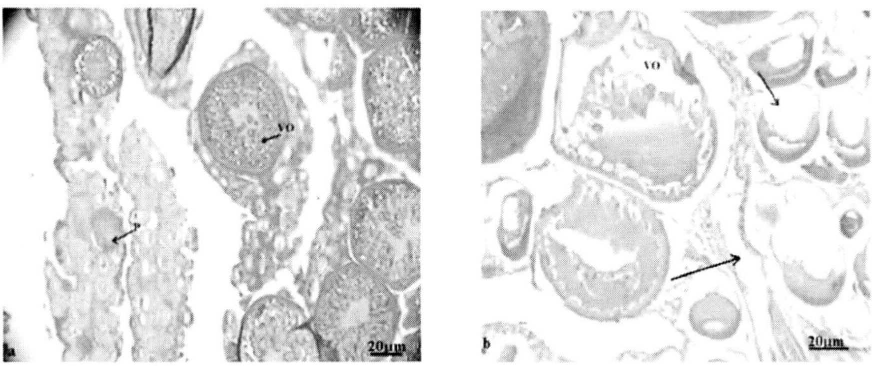

Fig.14a. Histological section of control ovary of *Labeo rohita*. Control ovary showing perinuclear oocytes and Vitellogenic ovule X Bar 20 µm.

Fig.14b. Bisphenol A treated ovary showing no perinuclear oocyte and arrow showing degenerated corion layer H&E scale bar 20 µm

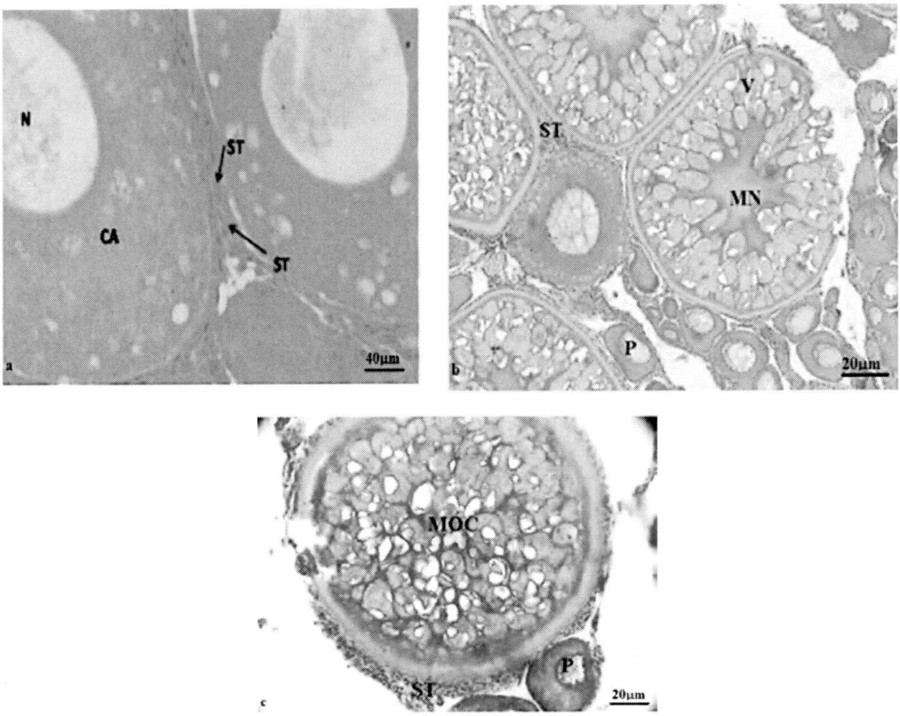

Fig.15a. Histological section of ovary of *Labeo rohita* HCH Treated showing developing ST (Spermatocytes) and developing oocyte Cortical Alveolar (CA) showing developing Spermatocytes (ST), Developing oocyte Migrating nucleus (MN), Vitellogenic follicles (V).

Fig.15b. showing destroying Premature Oocyte (POC), Oocyte going to atresia (AOC).

Fig 15c. Developing Ovary showing Mature Oocyte (MOC), ST (Spermatocytes) H&E X- bar 20 µm.

4. DISCUSSION

The results of this study delineated that exposure of BPA and γ-HCH to *Labeo rohita* depressed the gonado-somatic index. EDCs exposed fish feather minnow has been reported to have noticeably lower GSI than control (Rey, 2006). GSI decreases have been reported in adult rainbow trout and carp exposed to estrogens either in water or via their food (Komen et al. 1989; Jobling et al. 1996; Gimeno et al. 1998). Results of this study along with previous literature show that GSI can be suitably considered for detection of estrogen mimics (Jobling et al. 1996; Bjorkblom et al. 2009). Similar results have been documented by many authors showing that exposure of estrogenic and anti-androgenic contaminants in the laboratory and in fish inhabiting contaminated sites reduced GSI (Panter et al. 1998; Ankley et al. 2001; Sepulveda et al. 2002; Patino et al. 2003; Zha et al. 2008; Sun et al. 2009).

Further, the results of this study revealed that fish exposed to Bisphenol A and γ-HCH significantly decreased RNA/ DNA ratio. Similar observation has been reported by where inhibitory effects of fenvalarate on DNA, RNA and protein content has been shown to have tremendous toxicity of pyrethoidal compound on the main biochemical machinery of the fresh water catfish, *Clarias batrachus* (Tripathi et al. 2001). Our results showed that effect of high dose of HCH affected the DNA, RNA and RNA/DNA ratio so very lower concentration may affect fishes suggesting that the poor growth of experimental fish was due to imbalance in RNA/DNA ratio. The percentage reduction of nucleic acid level was statistically significant in tissues of the fish (Rathod and Kshirsagar 2010; Raksheskar 2012). Further, it is reported that fish exposed to Dimethoate (organophosphate) exhibited a decrease in nucleic acid (DNA and RNA) content (Anderson et al. 1994; Tripathi 2003). Thus, the RNA/DNA ratio can be considered as a biomarker showing early warning signals for detection of effects any EDCs at the species level.

In this study we found increased level of estradiol after BPA treatment and also with HCH. Similar results have been shown by Howdeshell and Saal (2006) that BPA increases serum estradiol level in Bisphenol-A treated fetus. BPA treatment altered the normal sex ratio towards females depending on the concentrations there is a clear evidence that BPA induces feminization in Xenopus laevis tadpoles, revealing an estrogenic potency of bisphenol-A that influences sexual development in amphibians (Levy et al. 2004). There are many other similar reports suggesting decreased testosterone level in mature male common carp exposed to 1000mg/L BPA (Mandich et al. 2007), in juvenile turbot *Psetta maxima* exposed to 59mg/ L BPA (Labadie and Budzinski 2006) and in larvae of brown trout *(Salmo trutta)* exposed to 50mg/L BPA (Bjerregaard et al. 2008). Iwamatsu et al. 2006a reported that aromatizable testosterone in high concentrations may induce a significant increase in E2 content in embryos of medaka *Oryzias latipes* and can cause paradoxical sex reversal. Similar observations were made in zebrafish (Orn et al. 2003). In the African catfish *(Clarias gariepinus)*, in vivo treatments with androgens inhibited C17–20 lyase activity in the testis (Cavaco et al. 2001; Schulz and Miura 2002).

Hormonal regulation between estrogen and androgen is necessary in sexual differentiation of teleost fish and it depend upon activity of some steroid synthesis including enzyme particularly on cytochrome P450, aromatase complex. This enzyme directly participates in the regulation of sexual differentiation because aromatase inhibitor treated female fish have various degree of masculinization *Oncorhynchus mykiss* (Guiguen et al. 1999), zebra fish *Danio ranio* (Kwon et al. 2000a; Uchida et al. 2005) *Orinochronis niloticus* (Singh et al. 2011) common carp (Singh et al. 2013). Further, Vitellogenin hormones is the key hormone for the synthesis of egg cells and zrp Protein and it is the biomarker for EDCs estimation. Serum aromatase value did not significantly ($P>0.05$) change in HCH treated fish while it was significantly high after BPA treatment.

5. CONCLUSION

It is concluded that exogenous Estrogen and androgen can change the biometrics, physiology and gonadal development of fish. The results obtained on the biometric parameters, biochemical data, GSI and gonadal structures demonstrated that Bisphenol A and HCH disrupt the endocrinology of fish and disturb the hormonal balance in body and decreased the fertility of fish however, the role of xenoestrogen very little is known about in fishes the mechanism by which Bisphenol A and α-HCH contributes to this process. In the present investigation, it can be concluded that under exposure to different types of EDCs the DNA and RNA content decreased in the tissues of the fish, *Labeo rohita* leading to a decrease in protein synthesis and cellular degradation. So, the level of vitellogenin hormone, estradiol is a biomarker in toxicological test, which represents intersex or testicular dysfunction such as alternations in steroidogenesis, sperm production and quality, or presence of ovatestis in fishes.

ACKNOWLEDGEMENT

Authors are thankful to Parasitology Lab, Department of Applied Animal Sciences, Babasaheb Bhimrao University (A Central University) Vidhya Vihar Lucknow, 226025 for giving us an opportunity to complete the research work.

REFERENCES

Ankley GT, Jensen KJ, Kahl MD, Korte JJ, Makynen EA. (2001) Description and evaluation of a short-term reproduction test with the fathead minnow *(Pimephales promelas)*. Environ Toxicol Chem 20: 1276-1290.

Ankley GT, Jensen KM, Makynen EA, Kahl MD, Korte JJ, Hornung MW, Henry TR, Denny JS, Leino RL, Wilson VS, Cardon MC, Hartig PC, Gray LE (2003) Effects of the androgenic growth promoter 17-trenbolone on fecundity and reproductive endocrinology of the fathead minnow. Environ Toxicol Chem 22: 1350–1360.

Anderson SL, Wild GC (1994) Linking genotoxic responses and reproductive success in ecotoxicology. Environ Health Perspect 102:9-12.

Barnhoorn IEJ, Bornman MS, Pieterssse GM, van Vuren JHJ (2004) Histological evidence of intersex in feral sharptooth cat-fish *(Clarias gariepinus)* from an estrogen-polluted watersource in Gauteng, South Africa. Environ Toxicol 19:603–608

Belt VDK, Wester PW, VanDerVen LT, Verheyen R, Witters H, (2002) Effects of ethynylestradiol on the reproductive physiology in zebrafish *(Danio rerio)*: time dependency and reversibility. Environ Toxicol Chem 21: 767-775.

Bjerregaard, L.B., Lindholst, C., Korsgaard, B., Bjerregaard, P. (2008) Sex hormone concentrations and gonad histology in brown trout *(Salmo trutta)* exposed to 17beta-estradiol and bisphenol A, Ecotoxicology. 17: 252-263.

Burkhardt-Holm P, Giger W, Güttinger H, Peter A, Scheurer K, Suter MJF, Ochsenbein U, Segner H, Staub E (2005) Where have all the fish gone? The reasons why fish catches in Swiss rivers are declining. Environmental Science and Technology 39: 441 A - 447 A.

Cavaco BM, Barros L Pannett AA (2001) The hyperparathyroidism- jaw tumour syndrome in a Portugues Kindred. QJM 94:213-22.

Cespedes R, Petrovic M, Raldua D, Saura U, Pina B, Lacorte S (2004) Integrated procedure for determination of endocrine-disrupting activity in surface waters and sediments by use of the biological technique recombinant yeast assay and chemical analysis by LC-ESI-MS. Anal Bioanal Chem 378: 697–708.

Chiang LFE, Yan YL, Guiguen Y, Postlethwait J, Chung BC (2001) Two Cyp 19(P450 Aromatase) Gene on Duplication zebrafish chromosome is expressend in ovary or Brain Mol Biol Evol 18(4):542-550.

Drastichova J, Svobodova Z, Groenland M, Dobikova R Ilabek V, Weissova D, Szotkowsk M (2005) An Effect of Exposure to Bisphenol A and 17ß-estradiol on the Sex Differentiation in Zebrafish *(Danio rerio)* Acta Vet Brno 74: 287–291.

Druga M, Druga M, Stef D, Moldovan C (2009) Researches concerning HCH residues in feed and beef meat in a west Romanian county area, Journal of Agroalimentary Processes and Technologies. 15 (4): 536-539.

Filby AL, Neuparth T, 2 Thorpe KL, Owen R, Galloway TS, Tyler CR (2007) Health Impacts of Estrogens in the Environment, Considering Complex Mixture Effects. Environ

Health Perspect 115(12): 1704–1710. Published online 2007 Aug 30. doi: 10.1289/ehp.10443.

Gagne F, Andre C, Blaise C (2005) Increased vitellogenin gene expression in the mussel *Elliptio complanata* exposed to estradiol-17β. Fresenius. Environ Bull 14: 861–866.

Gehring M, Weltin D, Tennhardt L, Vogel D, Bilitewski B (2002a) Elimination of Bisphenol A During Wastewater Treatment: Detailed Mass Balances Derived from Laboratory Experiments. Proceedings of the 2002 AWWA Endocrine Disruptors & the Water Industry Symposium, April 18 – 20, 2002, Cincinnati, Ohio, USA.

Gimeno S, Komen H, Gerritsen AGM, Bowmer T, (1998) Feminisation of young males of the common carp, *Cyprinus carpio*, exposed to 4-tert-pentylphenol during sexual differentiation. Aquat Toxicol 43: 77–92.

Guiguen Y, Baroiller JF, Ricordel MJ, Iseki K, Mcmeel OM, Martin SA, Fostier A (1999) Involvement of estrogens in the process of sex differentiation in two fish species: the rainbow trout *(Oncorhynchus mykiss)* and a tilapia *(Oreochromis niloticus)*. Mol Reprod Dev 54: 154–162.

Harris RM, Kirk CJ, Waring RH (2005) Non-genomic effects of endocrine disruptors: inhibition of estrogen sulfotransferase by phgenols and chlorinated phenols. Mol Cell Endocrinol 244: 72–74.

Howdeshell KL, Saal VFS (2006) Developmental exposure to bisphenol A: Interaction with endogenous estradiol during pregnancy in mice diallylphthalate, oil, and oil spiked with alkylphenols. Environmental Contamination & Toxicology 50(4):552-561. ISSN: 0090-4341.

Iwamatsu T, Kobayashi H, Sagegami R, Shuo T (2006a) Testosterone content of developing eggs -and sex reversal in the medaka *(Coryzias latipes)* General of Comperative Endocrinology 145: 67-74.

Jensen KM, Korte JJ, Kahl MD, Pasha MS, Ankley GT (2001) Aspects of basic reproductive biology and endocrinology in the fathead minnow *(Pimephales promelas)* Comp Biochem Physiol C 128: 127–141.

Jobling S, Sheahan D, Osborne JA, Matthissen P, Sumpter JP (1996) Inhibition of testicular growth in rainbow trout *(Oncorhynchus mykiss)* exposed to estrogenic alkylphenolic chemicals. Environ Toxicol Chem 15: 194–202.

Jobling S, Coey S, Whitmore JG, Kime DE, Van Look KJW, McAllister BG et al (2002). Wild intersex roach *(Rutilus rutilus)* have reduced fertility. Biol Reprod 67:515–524.

Jurajda P, Janac M, Valova Z, Streck G (2010a) Fish community in the chronically polluted middle Elbe River. Folia Zoologica 59: 57–168.

Kazeto Y, Place AR, Trant JM (2004) Effects of endocrine disrupting chemicals on the expression of CYP19 genes in zebrafish *(Danio rerio)* juveniles. Aquat Toxicol 69:25–34.

Keith LH, Jones TL, Needham LL Eds, (2000) Analysis of Environmental Endocrine disruptor, American Chemical Society.

Kinnberg K, Korsgaard B, Bjerregaard P, Jespersen A (2000) Effects of nonylphenol and 17b-estradiol on vitellogenin synthesis and testis morphology in male platyfish *Xiphophorus maculatus*. J Exp Biol 203: 171–181.

Kinnberg K, Toft G (2003) Effects of estrogenic and antiandrogenic compounds on the testis structure of the adult guppy *(Poecilia reticulata)*. Ecotoxicol Environ Saf 54: 16–24.

Kitamura S, Suzuki T, Sanoh S, Kohta R, Jinno N, Sugihara K, Yoshihara S, Fujimoto N, Watanabe H, Ohta S (2005) Comparative study of the endocrine-disrupting activity of bisphenol A and 19 related compounds. Toxicol Sci 84:249–259.

Knorr S, Braunbeck T (2002) Decline in reproduction success, sex reversal and developmental alterations in Japanese medaka *(Oryzias latipes)* after continuous exposure to octylphenol. Ecotox Environ Safety 51:187-196.

Komen J, Lodder PAJ Huskens F, Richter CJJ, Huisman EA (1989) Effect of oral administration of 17alfa-methyltestosterone and 17βestradiol on gonadal development in common carp (*Cyprinus carpio* L). Aquaculture 78:349-363.

Kwon JY, Haghpanah V, Kogson-Hurtado LM, McAndrew BJ, Penman DJ (2000a) Masculinization of genetic female Nile tilapia *(Oreochromis niloticus)* by dietary administration of an aromatase inhibitor during sexual differentiation. J Exp Zool 287:46-53.

Kwak HI, Bae MO, Lee MH, Lee YS, Lee BJ, Kang KS, Chac CL, Sung HJ, Shin JS, Kim JH, Mar WC (2001) Effects of nonylphenol, bisphenol A, and their mixture on the viviparous swortail fish *(Xiphophorus helleri)*. Environ Health Perspect 20: 787-795.

Levy G, Lutz I, Kruger A, Kloas W (2004) Bisphenol A induces feminization in Xenopus laevis tadpoles. Environ Res 94: 102–111.

Labadie P, Budzinski H (2000) Alteration of steroid hormone balance in juvenile turbot *(Psetta maxima)* exposed to nonylphenol, bisphenol-A, tetrabromodiphenyl ether 47, American Zoologist 40(3): 429-437.

MacLatchy DL, Courtenay SC, Rice CD, Van der Kraak GJ (2003) Development of a short-term reproductive endocrine bioassay using steroid hormone and vitellogenin end points in the estuarine mummichog *(Fundulus heteroclitus)*. Environ Toxicol Chem 22: 996–1008.

Makynen EA, Khal MD, Jensen KM, Tietge JE, Wells KL, Van DKG, Ankley GT (2000) Effect of mammalian antiandrogen vinclozolin on the development and reproduction of the Fathead Minnow *(Pimephales Promelas)* Aquat Toxicol 48: 461-475.

Mandich A, Bottero S, Benfenati E, Cevasco A, Erratico C, Maggioni S, Massari A, Pedemonte F, Vigano L (2007) In vivo exposure of carp to graded concentrations of bisphenol A. Gen Comp Endocrinol 153: 15-24.

Metzler M, Pfeiffer E (2001) Chemistry of natural and anthropogenic endocrine active compounds in 'Endocrine Disruptors Part I' (Ed.M. Metzler.) 63–80 (Springer: Berlin).

Naciff JM, Hess JA, Overmann GJ, Torontali SM, Carr GJ, Tiesman JP, Foertsch LM, Richardson BD, Martinez JE, Daston GP (2005) Gene Expression Changes Induced in the Testis by Transplacental Exposure to High and Low Doses of 17a-Ethynyl Estradiol, Genistein, or Bisphenol A. Toxicological Sciences 86(2): 396–416 doi:10.1093/toxsci/kfi198.

Navas JM, Merino R, Jimenez B, Riverac J, Abad E, Zanuy S, Carrillo M (2005) Organochlorine compounds in liver and concentrations of vitellogenin and 17-estradiol in plasma of sea bass fed with a commercial or with a natural diet. Aquatic Toxicology 75: 306–315.

Orn S, Holbech H, Madsen TH, Norrgren L, Petersen GI (2003) Gonad development and vitellogenin production in zebrafish *(Danio rerio)* exposed to ethinylestradiol and methyltestosterone. Aquatic Toxicology, 65: 397–411

Panter GH, Thompson RS, Sumpter JP (1998) Adverse reproductive effects in male fathead minnows *(Pimephales promelas)* exposed to environmentally relevant concentrations of the natural oestrogens, oestradiol and oestrone. Aquat Toxicol 42:243–253.

Patino, R., Goodbred, S.L., Draugelis-Dale, R., Barry, C.E., Foott, J.S., Wainscott, M.R., Gross, T.S. and Covay, K.J. (2003). Morphometric and histopathological parameters of gonadal development in adult common carp from contaminated and reference sites in

Lake Mead, Nevada, Journal of Aquatic Animal Health. 15:55–68.

Raksheskar GA (2012) Influenced of cypermethrin on DNA, RNA and RNA/DNA ratio in gills of the freshwater fish *Channa striata*. Bioscience Discovery 3(1):17-19.

Ramirez AJ (2007) Determination of Pharmaceuticals and Personal Care Products in Fish Using High Performance Liquid Chromatography-Tandem Mass Spectrometry and Gas Chromatography-Mass Spectrometry. A PhD. Thesis

Ray EJ (2006) Effect of p,p' DDE on reproduction and biomarkers of endocrine disruption in fatherad minnows *(pimephales promelas)*. MSc. Thesis University of florida

Rathod ND, Kshirsagar RV 2010, quantification of nucleic acid from fresh water fish *punctius arenatus* (day) exposed to pesticides International Journal of Advanced Biotechnology and Research. Vol1, Issue 1, 2010, pp 43-51.

Schulz RW, Miura T (2002) Spermatogenesis and its endocrine regulation. Fish Physiol Biochem 26: 43–56.

Schwaiger J, Mallow U, Ferling H, Knoerr S, Braunbeck T, Kalbfus W, Negele RD (2002) How estrogenic is nonylphenol? A transgenerational study using rainbow trout *(Oncorhynchus mykiss)* as test organism. Aquat Toxicol 59: 177-189.

Schug TT, Janesick A, Blumberg B, Heindel JJ (2011) Endocrine disrupting chemicals and disease susceptibility. The Journal of Steroid Biochemistry and Molecular Biology 127: 204–215.

Sepulveda, M.S., Johnson, W.E., Higman, J.C., Denslow, N.D., Schoeb, T.R. Gross, T.S. (2002). An evaluation of biomarkers of reproductive function and potential contaminant effects in Florida largemouth bass *(Micropterus salmoides floridanus)* sampled from the St. Johns River, Science of the Total Environment. 289: 133144.

Sharpe RM, Irvine DS (2004) How strong is the evidence of a link between environmental chemicals and adverse effects on human reproductive health? BMJ 328: 447-451.

Shioda T, Wakabayashi M (2000) Effect of certain chemicals on the reproduction of medaka (Oryzias latipes). Chemosphere 40(3): 239-43.

Singh AK (2013) Introduction of modern endocrine techniques for the production of monosex population of fishes. General & Comparative Endocrinology 181: 146–155.

Singh AK, Ansari A, Singh R (2011) Possible impact of pharmaceutical wastes on aquatic environment. In: Water Pollution and Management (Eds) D.S. Malik, Sunil Kumar and Umesh Bharti. Biotech Books, New Delhi. Pp 371-386.

Singh AK, Singh R (2013) In vivo response of melatonin, gonadal activity and biochemical changes during aromatase inhibited sex reversal in common carp Cyprinus carpio (L). Animal Reproduction Science 136: 317-325.

Singh AK, Srivastava PP, Verma R, Ansari A, Srivastava S, Kumar D (2015) Effect of dietary administration of letrozole and tamoxifen on gonadal development, sex differentiation and biochemical changes in common carp (Cyprinus carpio L). Reproduction, Fertility and Development. http://dx.doi.org/10.1071/RD13234 (Australia).

Singh AK, Srivastava R (2013) An overview of gonadal disruption in fishes inhabiting natural aquatic environments and future research need. In Advances in Fish Research (Ed.) Narendra Publishing House, Delhi 103-115.

Singh KP, Malik A, Mohan D, Takroo R (2005a) Distribution of persistent organochlorine pesticide residues in Gomti River, India. Bull Environ Contam Toxicol 74:146–154.

Singh, K.P., Malik, A., Sinha, S. 2007. Persistent organochlorine pesticide residues in soil and surface water of northern Indo- Gangetic alluvial plains, Environmental Monitoring and Assessment,125:147–155.

Singh, K.P., Singh, Malik, A., Mohan, D. and Takroo, R. 2005. Distribution of persistant organochlorine pesticide residues in Gomti River, India, Bulletin of Environmental Contamination and Toxicology, 74: 146–154.

Singh, S., Singh, P.B. and Singh, V. 2007. Exposure and recovery response of isomers of HCH, metabolites of DDT and estradiol-17β in the female catfish, *Heteropneustes fossilis*, Environmental Toxicology and Pharmacology, 24: 245–251.

Srivastava, A., Sharma, P.V. Tripathi, R. Kumar, R., Patel, D.K., Mathur P.K. 2010. Occurrence of phthalic acid esters in Gomti River Sediment, India. Environ Monit Assess, 169:397–406. DOI 10.1007/s10661-009-1182-4.

Sun C, Y Chang C., Tszeng, S. Yeh and N. Su, 2009. Reproductive biology of blue marlin *(Makaira nigricans)* in the western Pacific Ocean. Fish. Bull. 107: 420-432.

Sun H, Shen OX, Wang XR, Zhou L, Zhen SQ, Chen XD. 2009. Anti-thyroid Hormone Activity of Bisphenol A, Tetrabromobisphenol A and Tetrachlorobisphenol A in an improved Reporter Gene Assay. Toxicol In Vitro

Tanaka JN, Grizzle JM (2002) Effects of nonylphenol on the gonadal differentiation of the hermaphroditic fish, *Rivulus marmoratus*. Aquatic Toxicology 57:117-125.

Tijani JO, Fatoba OO, Petrik LF (2013) A Review of Pharmaceuticals and Endocrine-Disrupting Compounds: Sources, Effects, Removal, and Detections. Water, Air, and Soil Pollution 224: 1770-1798.

Tripathi G, Harsh S, Verma P, 2001. Fenvalerate induced macromolecular changes in the catfish, *clarius batrachus*. Environ. Biol., 23: 143-146.

Tripathi PK, Srivastava VK, Singh A (2003) Toxic effect of dimethoate (Organophosphate) on metabolism and enzyme system of fresh water teleost fish *Channa punctatus*. Asian Fisheries Sci 16: 349-359.

Uchida K, Yonezawa M, Nakamura S, Kobayashi T, Machida T. (2005) Impaired neurogenesis in the growth retarded mouse is reversed by T3 treatment. Neuroreport 16:103–106. [PubMed: 15671855].

Verma R, Singh AK (2013) Endocrine disruptor in aquatic bodies: Issues and challenges. Journal of Ecology Environment and Biology 1:17-21.

Vandenberg LN, Maffini MV, Sonnenschein C, Rubin BS, Soto AM (2009) Bisphenol-A and the Great Divide: A Review of Controversies in the Field of Endocrine Disruption Endocrine Reviews 30(1):75–95.

WHO [World Health Organization]. 2002. Global Assessment of the State-of-the-Science of Endocrine Disruptors. Geneva: International Programme on Chemical Safety.

Ying GG, Kookana RS, Ru YJ (2002) Occurrence and fate of Hormone steroids in the environment. Environment International 28:545-551.

Zha J, Sun L, Zhou Y, Spear P, Ma M, Wang Z (2008). Assessment of 17a-ethinylestradiol effects and underlying mechanisms in a continuous, multigeneration exposure of the Chinese rare minnow *(Gobiocypris rarus)*. Toxicology and Applied Pharmacology 226: 298–308.

Zhan W, Xu Y, Li AH, Zhang J, Schramm KW, Kettrup A (2000) Endocrine Disruption by Hexachlorobenzene in Crucian Carp *(Carassius auratus gibelio)* Bull Environ Contam Toxicol 65:560–566.

❒

Chapter-5

BIOACCUMULATION OF ARSENIC IN LIVER & KIDNEY OF *CHANNA PUNCTATUS*

Robeena Sarah, Baby Tabassum* and Nida Idrees

Toxicology Laboratory, Department of Zoology, Govt. Raza P.G. College, Rampur (U.P.), India.

**Email of Corresponding Author: dr.btabassum@gmail.com*; sarah.sarah102@gmail.com*

ABSTRACT

Arsenic contamination is reported to be a noticeable hazard in terms of the affected population of human beings and also aquatic life. Recent studies indicated that groundwater is severely contaminated with arsenic concentrations above the maximum permissible limit of W.H.O. in drinking water. Moreover, increased levels found suggested that arsenic is present in aquatic organisms which are used for human consumption and thus, accumulating in the food chain. The cause of such accumulation is mainly due to illegal dumping of wastes from various small-scale industries, scrapping and illegal burning of e-waste near the banks of the river. Concentration of metals becomes toxic to the fish when its level exceeds the permissible level. Gills, liver and kidneys accumulate heavy metals in higher concentration in comparison to muscles, which exhibit lowest levels of metals accumulation. Among different organs, liver accumulates higher concentrations of metals comparatively and has been used widely to investigate the process of bioaccumulation. Therefore, some precautionary steps should be taken by the responsible citizens to create awareness amongst the local people so as to conserve the environment from the drastic effects of arsenic.

Keywords: Bioaccumulation, Arsenic, C. punctatus, liver and kidney.

1. INTRODUCTION

Pollution of the aquatic environment by inorganic chemicals has been considered a major threat to the aquatic organisms including fishes. The agricultural drainage water containing pesticides and fertilizers and effluents of industrial activities and runoffs in addition to sewage effluents supply the water bodies and sediment with huge quantities of inorganic anions and heavy metals (ECDG 2002). Heavy meals concentrations in the aquatic organism depict the past as well as the current pollution load in the environment in which the organism lives (Ravera et al. 2003). The concentrations of heavy metals in organs of fish show that the aquatic environment is polluted (Farkas et al. 2000). Level of trace metals in different organs of fish is used as an index of metal pollution in an ecosystem, which is considered as an important tool to highlight the role of elevated level of metals in aquatic organisms (Tarrio et al. 1991). Concentration of metals becomes toxic to the fish when its level exceeds the permissible level (Mansour and Sidky, 2002). Thus, fish not only indicates the pollution status of aquatic ecosystem but have significant impact on the food web (Chi et al. 2007). Gills, liver and kidneys accumulate heavy metals in higher concentration in comparison to muscles, which exhibit lowest levels of metals accumulation. Among different organs, liver accumulates higher concentrations of metals comparatively and has been used widely to investigate the process of bioaccumulation. Kidneys also play a vital role in excretion of trace metal ions (Wepener et al. 2001). As the liver is one of the most important detoxifying organs in fish, accumulation of arsenic in this organ would surely have some negative effects on survival and therefore, to survive, there would be an activation of detoxication pathways (Das et al., 1998). Bioaccumulation of heavy metals in a river may be of great consequences to the present and future generation of the nation as a whole if the extent of the exposure is high.

2. ARSENIC

Arsenic is toxic to both plants and animals and inorganic arsenicals are proven carcinogens in humans (Ng, 2005; Murcott, 2012). The toxicity of arsenic to human health ranges from skin lesions to cancer of the brain, liver, kidney, and stomach (Smith et al., 1992). Short term exposure can cause nausea and vomiting, decreased production of red and white blood cells, abnormal heart rhythm, damage to blood vessels, and a sensation of "pins and needle" in hands and feet. Long term exposure to arsenic may lead to darkening of the skin and the appearance of small "corns" or "warts" on the palms (EPA 1998). Generally inorganic arsenic species are more toxic than organic form to living organism, including humans and other animals (Goessler and Kuehnett, 2002; Ng, 2005). Ingestion of arsenic trioxide of 70-80 mg has been reported to be fatal for humans (Vallee et al., 1960).

3. BIOACCUMULATION OF ARSENIC IN *C. PUNCTATUS*

Liver and Kidney are vital organs in vertebrates which perform detoxification mechanism, protein synthesis, and excretion of nitrogenous waste and homeostatic functions, respectively. Among aquatic organisms, fish are appropriate bioindicators of bioaccumulation. The accumulation and distribution patterns of toxic metals in fish tissues depend on their rates of uptake and elimination and may cause various physiological defects and mortality when accumulation reaches a substantially high level (Kalay and Canli 2000). Arsenic does not readily bioconcentrate in aquatic species. It is typically water soluble and does not combine with proteins. Planktivorous fish are more likely to concentrate as than omnivorous or piscivorous fish (USEPA 1999). Williams et al. (2006) studied on arsenic bioaccumulation in freshwater fish and reported the ratio of bioconcentration factor and bioaccumulation factors ranging from 0.1 to 3097 L/kg. Many investigators studied on accumulation of arsenic in fish's organs and they found high accumulation of arsenic occurs in liver and gills than the other parts of the fish (Allen et al. 2004; Bears et al. 2006; Hamdi et al. 2009). Fish exposed to as for a longer period in the laboratory or in natural As-contaminated habitats exhibited histopathological lesions in the liver, gallbladder, and kidney, which in turn affect the hepatic and renal functions (Pedlar et al. 2002).

Liver:

Liver is the major organ involved in the regulation of metabolic functions. Most of the biotransformation of inorganic As takes place in the liver (Rossman 2003). Arsenic is actively metabolized in the liver and has the tendency to accumulate here as well as reported in different teleosts such as rainbow trout (Cockell et al. 1991). Lam et al. (2006) and Li et al. (2016) found metabolic and histopathological liver when the zebra fish exposed with arsenic. Extreme degenerative changes in the hepatopancreas of *Channa punctatus* were found to occur within 48 hr of exposure to the nonlethal levels of As (Roy and Bhattacharya 2006). The degenerative changes were characterized by vacuolation and apoptosis of the hepatocyte, pyknosis in many of the necrotic cells, peliosis hepatitis in the tissue (Ahmed et al. 2013).

Kidney:

Kidney is the major route of As excretion, as well as a major site of conversion of pentavalent Arsenic. Roy and Bhattacharya (2006) found during their investigation that 1/20 and 1/10 LC50 doses of As_2O_3 resulted in discrete pathology in the trunk kidney, the effects being highly significant at both treatments. In both 1/20 (3.8 mg/L) and 1/10 LC50 (7.6 mg/L) of As treatment it was evident that on the very first day there was shrinkage in the glomerulus with a resultant increase in Bowman's space followed by enlargement of the glomerulus and normalization of Bowman's space during the last phase. Increase in Bowman's space suggests an increase in the filtration rate and consequently in urine volume.

4. CONCLUSION

Fishes are known for their ability to concentrate heavy metals in their tissues and since they play important role in human nutrition, they need to be carefully screened to ensure unnecessary high level of these toxic trace metals which were transferred to man through consumption (Adeniyi and Yusuf, 2007). Regular monitoring of arsenic levels and their associated health effects in aquatic organisms may not only provide insight into overall aquatic health but may also lookout for potential impacts on food chain.

REFERENCES

Adeniyi, A. A and Yusuf, K. A. (2007). Determination of heavy metals in fish tissues, water and bottom Sediments from Epe and Badagry Lagoons Lagos Nigeria. Environmental Monitoring Assessment, 37: 451- 458.

Ahmed MK, Habibullah-Al-Mamun M, Parvin E, Akter MS, Khan MS (2013) Arsenic induced toxicity and histopathological changes in gill and liver tissue of freshwater fish, tilapia *(Oreochromis mossambicus)*. Exp Toxicol Pathol 65(6):903–909.

Allen HE (ed) (2002) Bioavailability of metals in terrestrial ecosystems: importance of partitioning for bioavailability to invertebrates, microbes and plants. Society for Environmental Toxicology and Chemistry, SETAC Press, New York, p 158.

Bears H, Richards JG, Schulte PM (2006) Arsenic exposure alters hepatic arsenic species composition and stress-mediated gene expression in the common killifish *(Fundulus heteroclitus)*. Aquat Toxicol 77:257–266.

Bhattacharya T, Roy AK, Bhattacharya S (2006) Blood glucose and hepatic glycogen interrelationship in *Channa punctatus*: a parameter of non lethal toxicity bioassay with intestinal pollutants. Indian J Exp Biol 25:539–541.

Cockell KA, Hilton JW, Bettger WJ (1991) Chronic toxicity of dietary disodium arsenate heptahydrate to juvenile rainbow trout *(Oncorhynchus mykiss)*. Arch Environ Contam Toxicol 21(4):518–527.

Das, D., Sarkar, D., Bhattacharya, S., 1998. Biomed. Environ. Sci. 11, 187–195.

ECDG, "European Commission DG ENV. E3 Project ENV. E.3/ETU/0058. Heavy Metals in Waste," Final Report, 2002.

EPA. (1998). Toxicological Review of Baruins and Compounds. U. S. Environmental Agency.

Farkas A, Salanki J and Varanka I (2000) "Heavy Metal Con- centrations in Fish of Lake Balaton, Lakes and Reser- voirs," Research and Management, Vol. 5, No. 4, pp. 271-279.

Goessler W., Kuehnelt D. (2002). Analytical methods for the determination of arsenic and arsenic compounds in the environment. Environmental chemistry of Arsenic, 27-50.

Hamdi M, Sanchez MA, Beene LC, Liu Q, Landfear SM, Rosen BP, Liu Z (2009) Arsenic transport by zebrafish aquaglyceroporins. BMC Mol Biol 10(1):104.

Kalay M, Canli M (2000) Elimination of essential (Cu, Zn) and nonessential (Cd, Pb) metals from tissues of a freshwater fish *Tilapia zilli*. Turk J Zool 24:429–436.

Lam SH, Wu YL, Vega VB, Miller LD, Spitsbergen J, Tong Y, Zhan H, Govindarajan KR, Lee Mathavan S, Krishna Murthy KR, Buhler DR, Liu ET, Gong Z (2006) Conservation of gene expression signatures between zebrafish and human liver tumors and tumor progression. Nat Biotechnol 24:73–75.

Li C, Li P, Tan YM, Lam SH, Chan EC, Gong Z (2016) Metabolomic characterizations of liver injury caused by acute arsenic toxicity in zebrafish. PLoS ONE 11(3):e0151225.

Mansour SA and Sidky MM (2002) "Ecotoxicological Stud- ies: Heavy Metals Contaminating Water and Fish from Fayum Governorate, Egypt," Food Chemistry, Vol. 78, No. 1, pp. 15-22.

Murcott S. (2012). Arsenic contamination in the world an international sourcebook IWA monographs. ISBN: 9781780400389, pages: 344.

Ng JC (2005). Environment contamination of arsenic and its toxicological impact on humans. Environmental chemistry, 2(3): 146-60.

Pedlar RM, Ptashynski MD, Evans R, Klaverkamp JF (2002) Toxicological effects of dietary arsenic exposure in lake whitefish *(Coregonus clupeaformis)*. Aquat Toxicol 57:167–189.

Ravera RC, Beone GM, Dantas M and Lodigiani P (2003) "Trace Element Concentrations in Freshwater Mussels and Macrophytes as Related to Those in Their Environ- ment," Journal of Limnology, Vol. 62, No. 1, pp. 61-70.

Rossman TG (2003) Mechanism of arsenic carcinogenesis: an integrated approach. Mutat Res 533:37–65.

Smith A.H., Hopenhayn-Rich C., Bates M.N., Goeden H.M., Hertz-Picciotto I., Duggan H.M., Wood R., Kosnett M.J., Smith M.T. (1992). Cancer risks from arsenic in drinking water. Environmental Health Perspectives, 97: 259-67.

Tarrio J, Jaffor M, Ashraf M (1991) "Levels of Selected Heavy Metals in Commercial Fish from Five Fresh Water Lake in Pakistan," Toxicological & Environmental Chemistry, Vol. 33, No. 1-2, pp. 133-140.

Q.-Q. Chi, G.-W. Zhu, A. Langdon, "Bioaccumula- tion of Heavy Metals in fishes from Taihu Lake, China," Journal of Environmental Sciences, Vol. 19, No. 12, 2007, pp. 1500-1504.

U. S. Environmental Protection Agency (USEPA) (1999) Biological assessment of the Idaho water quality standards for numeric water quality criteria for toxic pollutants for the U.S. Fish and Wildlife Service and the National Marine Fisheries Service. USEPA Region 10 Office, Seattle, WA.

Vallee B.L., Ulmer D.D., Wacker W.E.C. (1960). Arsenic toxicology and biochemistry. Arch. Ind. Health, 21:132-51.

Wepener W, Vurenvan JHJ, Preezdu HH (2001) "Up- take and Distribution of a Copper, Iron and Zinc Mixture in Gill, Live Rand Plasma of a Freshwater Teleost, *Tilapia sparrmanii*," Water SA, Vol. 27, No. 1, pp. 99- 108.

Williams L, Schoof RA, Goodrich-Mahoney JW (2006) Arsenic bioaccumulation in freshwater fishes. Hum Ecol Risk Assess 12(5):904–923.

❒

Chapter-6

CATHARANTHUS ROSEUS (L.): A REMARKABLE PHYTOREMEDIATOR FOR HEAVY METALS *VIZ.*, Cd, Pb AND Ni

Baby Tabassum*, Mohammad Hashim and Priya Bajaj
Toxicology Laboratory, Department of Zoology
Govt. Raza P.G. College, Rampur (UP) India - 244091
**Email of Corresponding Author: dr.btabassum@gmail.com*;*
mohammadhashim043@gmail.com

ABSTRACT

Environmental contamination by heavy metals from anthropogenic and industrial activities has caused considerable irreparable damage to aquatic ecosystems. Several different physicochemical and biological processes are commonly employed to remove heavy metals from industrial wastewater before their discharge into the environment. Contaminated soil and waterposes a major environmental and human health problem. In the present study Catharanthus roseus (locally called Sadabahar a common weed) were grown in pots filled with garden soil. The seedlings were collected from the sterilized soils. Pot experiments were conducted using the metal solutions prepared by dissolving in distilled water (aqueous solutions) to prepare stock solution of 850 ppm for each metal of Cadmium, lead and nickel. The three metal solutions were added to the pots in vitro condition for a period of 75 days alternately days. In addition, a set of control blank experimental pots were also maintained. Accumulations of the heavy metals were analyzed after 25, 50 and 75 days respectively in leaves, stem and roots of Catharanthus roseus by AAS. The results showed that Cadmium content was found high in roots and low in leaves by 25th day. There was no change in Pb accumulation in leaf after 50th day. Stem concentration increased to 70.04 mg/kg and root concentration was increased to 92.21 mg/kg biomass; many changes observed in the roots. In the 75th day only minimum change were noticed in leaf, stem and root. Whereas, after 50th day root, stem and leaf accumulations was increased. After 75th day high accumulation was noticed in the leaf, stem and root almost doubled than the 50th day accumulations, but a high accumulation difference noticed in the root. Finally, after the total experimental period it was concluded that root accumulation was higher compared to stem and leaves. Hence, It was concluded that the Catharanthus roseus was a good phytoremediator of heavy metal like Cadmium lead and nickel.

Keywords: Heavy metal, Catharanthus roseus, Water, Soil, Environment

1. INTRODUCTION

The biochemical impact of metal ions on the cells is as diverse as their chemical nature. There is present earth`s crust, approximately 90 elements, among which 80% are metals and 60% are heavy metals with specific weights higher than 5 g cm^{-3}.Metals are natural component or complex compound in soil and Water. Some of these metals are micronutrients necessary for plant growth, like as Zn, Cu, Mn, Co, while others have unknown biological function, like as Cd, Pb, Ni, Cr, As, and Hg (Subhashini V. et al.,2013). Some metals, for example, Zn and Cu, are essential for normal plant growth and development as they serve as structural and functional components of specific proteins. Other metals, for example, Cd and Pb, have no known function in plants although Cd requirement for carbonic anhydrase from marine diatoms has been reported (Lasat, M. M, 2000).

Environmental contamination by heavy metals from anthropogenic and industrial activities has caused considerable irreparable damage to ecosystem and human health. Thus Several different physicochemical and biological processes are commonly employed to remove heavy metals from environment. Contaminated soil and water pose a major environmental and human health problem. But all biological and physical processes, those plants carry out helps in the uptake, degradation or breakdown of organic contaminants by internal and external metabolic processes taking place in the plants(Shirin imam, 2017). It involves the use of plants to uptake, accumulate and degrade contaminated compound inside its tissue. Phytoremediation has the benefit of the unique and selective uptake heavy metals potential *Catharanthus roseus* plant by root system and shoot system, together with the translocation, accumulation of heavy metals, and pollutant contaminants storage and degradation abilities of the entire Plants.

According to Sinha et al., (2004) the plants act both as "accumulators" and "excluders". Accumulators survive despite concentrating contaminants in their aerial tissues. They biodegrade or biotransform the contaminants into inert forms in their tissues, the excluders restrict contamination uptake into their biomass. Plants to be used for phyto-extraction should have: (a) tolerance to high concentrations metals (b) high metals-accumulation capability, (c) heavy biomass (d) ability to grow fast and a (e) profuse root system. The success of phytoremediation depends especially on the plants ability to accumulate biomass rapidly and store large quantities of the uptakemetals in the shoot tissue (Blaylock et al., 1997; Mcgrath, 1998; Blaylock and Huang, 2000). Heavy metals uptake, by plants using phytoremediation technology, seems to be a prosperous way to remediate heavy metals contaminated environment. It's applicable at sites containing organic, nutrient or metal pollution that can be accessed by the root and shoot of plants and degraded, accumulate or metabolized in place.

2. MATERIALS AND METHODS

Sample Collection:

Plants were grown in pots filled with garden soil. The seedlings were collected from the uncontaminated soils. All the selected seedlings of Catharanthus roseus were of uniform size and free of any disease symptoms. Nickel and Lead were selected for the study, the uptake was estimated in root, stem and leaves for every 25 days for a total periodof 75 days. In addition, a set of control blank experimental pots was also maintained.

Solution Preparation:

The metal solutions prepared by dissolving in distilled water to prepare stock solution of 850 ppm for each metal.

Metal Analysis:

The calibration curves for each metal were also prepared. A blank reading was taken to incorporate necessary correction factor. The heavy metal solutions of 5mg/L was prepared from the stock and administered to the plants and care was taken to avoid leaching of water from the pots. The metal uptake was estimated once in every 25 days. The sample plants were removed from the pots and washed under a stream of water and then with distilled water. The collected plants were air dried, then placed in a dehydrator for 2-3 days and then oven dried for four hours at 100° C. The dried samples of the plant were powdered and stored in polyethylene bags. The powdered samples were subjected to acid digestion.

1gm of the powdered plant material were weighed in separate digestion flasks and digested with HNO_3 and HCl in the ratio of 3:1. The digestion on hot plate at 110ºC for 3-4 hours or continued till a clean solution was obtained.After filtering with Whatman No. 42 filter paper the filtrate was analyzed for the metal contents in AAS (Simarzdu 6800).

3. RESULTS AND DISCUSSION

In the 25th day Nickel accumulation was observed that it was increased than the control plants. In this period of experiments high accumulation was observed in the stem part than roots and leaves. After 50th day root, stem and leaf accumulations was increased and high accumulation difference was observed in the leaf part. After 75th day high accumulation was observed in the leaf, stem and root almost doubled than the 50th day accumulations, a high accumulation difference observed in the root part. While no change in Pd accumulation in leaf after 50th day but root and stem accumulations was increased. After 75th day increase accumulation was observed in the leaf, stem and root almost doubled than the 50th day accumulations.

Plant Part	Control	25th day	50th day	75th day
Leaf	*22.03*	*22.05*	*22.65*	*23*
Stem	*45.63*	*46.16*	*70.04*	*71.89*
Root	*23.05*	*65.04*	*92.21*	*92.89*
Total Accumulation	*90.71*	*133.25*	*184.9*	*187.78*

Table 1. Total Accumulation of Cadmium (mg/kg) in *Catharanthus roseus* during the Experimental period.

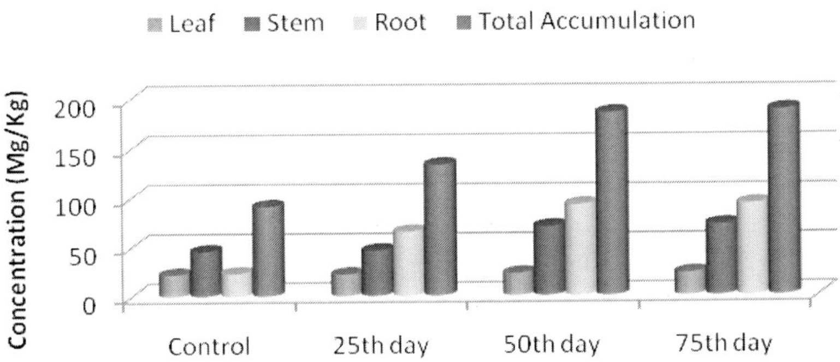

Accumulation of Cadmium in *Catharanthus roseus in the experimental period*

Plant Part	Control	25th day	50th day	75th day
Leaf	*28.03*	*28.5*	*28.49*	*28.98*
Stem	*62.05*	*67.86*	*68.57*	*69.96*
Root	*24.65*	*78.32*	*82.43*	*87.94*
Total Accumulation	*114.73*	*174.68*	*179.49*	*186.61*

Table 2. Total Accumulation of Lead (mg/kg) in *Catharanthus roseus* during the Experimental period.

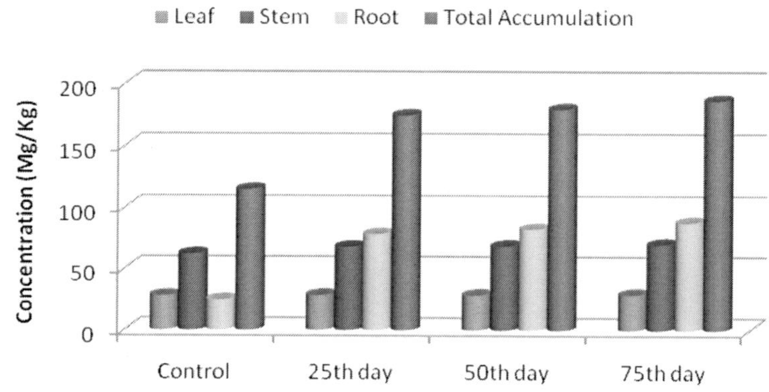

Accumulation of Lead in *Catharanthus roseus in the experimental period*

Plant Part	Control	25th day	50th day	75th day
Leaf	*3.09*	*4.56*	*7.12*	*14.59*
Stem	*6.21*	*10.01*	*10.87*	*24.53*
Root	*6.57*	*7.98*	*9.19*	*26.48*
Total Accumulation	*15.87*	*22.55*	*27.18*	*65.6*

Table 3. Total Accumulation of Nickel (mg/kg) in *Catharanthus roseus* during the Experimental period.

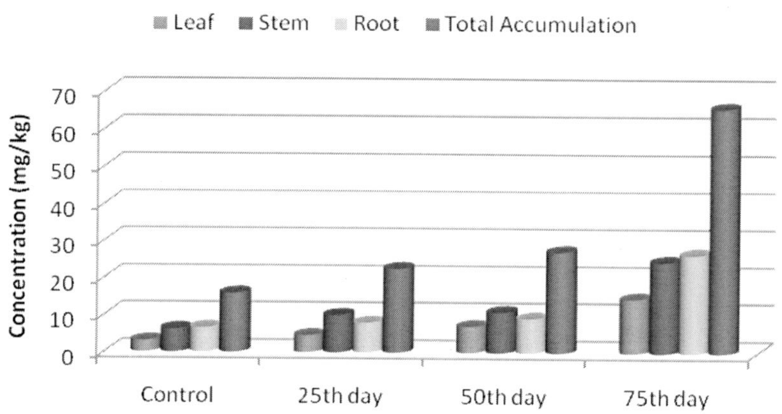

Accumulation of Nickel (mg/kg) in *Catharanthus roseus* during the experimental period

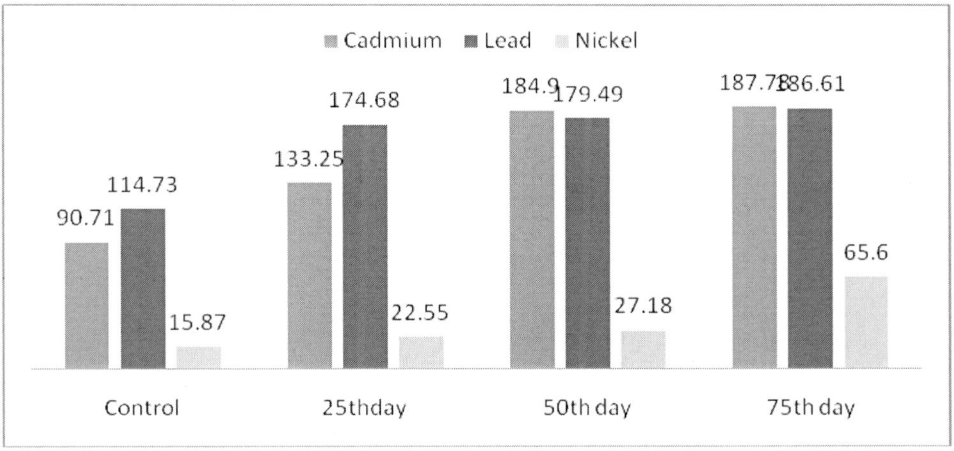

Total Accumulation of heavy metals (mg/kg) in *Catharanthus roseus* during the Experimental period

Metals that are accumulated by plants and largely stored in the roots of plants are indicated by TFvalues < 1 with values > 1 indicating that the metalsare stored in the stems and leaves.A plant is classified as a hyper-accumulator for heavy metals when it meetssome criteria;

- Shoot and root quotient (level of heavy metal in the shoot divide by level of heavy metal in the root) > 1.
- Extraction coefficient gives the proportion of total heavy metal in the soil which is taken up by the plant shoot/aerial part of the plant (Harrison and Chirgawi 1989, Rotkittikhun et al. 2006).
- Higher levels of heavy metals of 10 –500 times the levels in normal plants(uncontaminated plants) according to Allen (1989), more than 1000g/g of copper, lead, nickel,chromium; or more than 100g/g of cadmium or more than 10000g/g of zinc (Shen and Liu 1998,Ginocchio and Baker 2004, Rotkittikhun et al.2006).
- Furthermore, a plant which has high levels of heavy metals in the roots but with shoot/root quotients less than 1 is classified as a heavy metal excluder (Boularbah et al. 2006).
- Metals that are accumulated by plants and translocation from the root to the shoot may depend on the concentration in the root (hardiman et al., 1984). The movement of the heavy metal from the polluted sediments into the roots of the plant and ability to translocation the metals from roots to aerial parts were assessed correspondingly by means of BCF and TF. The ability of plant to tolerate and accumulate heavy metals is useful for phytoextraction and phytostabilization purpose (Yoon et al., 2006) This ratio is an indication of the ability of the plant to translocate metals from roots to the aerial parts of the plants (Marchiolet al., 2004). The plant species was highly accumulated these metals in the roots part, finally it was very useful for phytostabilization.

4. CONCLUSION

Heavy metals uptake by plants using phytoremediator technology seems to be a prosperous way to remediate heavy metals contaminate environment. It was concluded that the *Catharanthus roseus* is a good phytoremediator of heavy metal like Cadmium, Lead and nickel.

Based on the bio-concentration factor and translocation factor, Phytoremediator is a fast-developing field, sustainable and inexpensive Process.

ACKNOWLEDGEMENT

The Authors are thankful to Council for Science and Technology (CST) Uttar Pradesh, India for providing financial assistance to conduct this study

REFERENCES

Allen. S. E. (1989). Chemical Analysis of Ecological Materials. 2nd edition, Blackwell Scientific Publications, Oxford.

Blaylock, M. J. and Huang, J. W. (2000). Phytoextraction of metals. In: I. Raskin and B. D. Ensley (eds). Phytoremediation of toxic metals: using plants to clean up the environment. Wiley, New York. 53–69.

Blaylock, M. J., Salt, D. E., Dushenkov, S., Zakharova, O., Gussman, C., Kapulnik, Y., Ensley, B. D. and Raskin, I. (1997). Enhanced accumulation of Pb in Indian mustard by soil applied chelating agents. Environ. Sci. Tech., 31;860–865.

Boularbah, A., Schwartz, C., Bitton, G., Aboudrar, W., Ouhammou, A. and Morel, J. L. (2006). Heavy metal contamination from miningsites in South Morocco: Assessment of metal accumulation and toxicity in plants. Chemospher.,6.

Hardiman, R. T., Jacoby, B. and Banin, A. (1984). Factors affecting the distribution of cadmium,copper and lead and their effects upon yield and zinc content in bush bean (*Phaseolus vulgaris* L.). Plant Soil., 81: 17–27.

Harrison, R. M. and Chirgawi, M. B. (1989). The Assessment of air and soil as contributors of some trace metals to vegetable plants I. Use of a filtered air growth cabinet. Sci. Total Environ.,83: 13-34.

Lasat, M. M. 2000. Phytoextraction of metals from contaminated soil: a review of plant/soil/metal interaction and assessment of pertinent agronomic issues. Journal of Hazardous Substance Research. 2(5): 1-5.

Marchiol, L., Sacco, P., Assolari, S., and Zerbi, G. (2004). Reclamation of polluted soil: phytoremediation potential of crop-related Brassica species. Water Air Soil Pollut., 158:345–356.

McGrath, S. P. (1998). Phytoextraction for soil remediation. In: R. R. Brooks (ed). Plants that hyper accumulate heavy metals. CAB International, New York. 109–128.

Rotkittikhun, R., Kruatrachue, M., Chaiyarat, R., Ngernsansaruay, C., Pokethitiyook, P., Paijitprapaporn, A. and Baker, A. J. M. (2006). Uptake and accumulation of lead by plants from the Bo Ngam lead mine area in Thailand.Environ. Pollut.,144: 681- 688.

Shen, Z. G. and Liu, Y. L. (1998). Progress in the study on the plants that hyper accumulate heavy metal. Plant Physiol. Comm., 34: 133-139.

Shirin Imam. 2017 Phytoremediation: A Green method to combat environmental pollution. International Journal of Engineering Sciences & Research Technology, 6(8): 2277-9655.

Sinha, R. K., Heart, S. and Tandon, P. K. (2004). Phytoremediation: role of plants incontaminated site management, in Book of Environmental Bioremediation Technologies,pp. 315–330, Springer, Berlin, Germany.

Subhashini, V. and Swamy, A.V.V.S. (2013) Phytoremediation of Pb and Ni Contaminated Soils Using *Catharanthus roseus* (L.). Universal Journal of Environmental Research and Technology, 3, 465-472.

Yoon, J. M., Van Aken, B. and Schnoor, J. L. (2006). Leaching of contaminated leaves following uptake and phytoremediation of RDX, HMX, and TNT by poplar. International Journal of Phytoremediation., 8(1): 81-94.

❐

Chapter-7

XENOBIOTICS AND CANCER

Ajay Kumar[1*], Amar Deep[1,2], Virendra Atam[1] and Suchit Swaroop[2]

[1]*Department of Medicine, King George's Medical University, Lucknow*

[2]*Experimental and Public Health Laboratory, Department of Zoology, University of Lucknow*

**Email of Corresponding Author: drajaymd12345@gmail.com*; jsa.amardeep@gmail.com*

ABSTRACT

The "xeno" in "xenobiotics" comes from the Greek word xenos meaning guest, friend, or foreigner. Xenobiotics are the chemicals found but not produced in organisms or the environment. Some naturally occurring chemicals (endobiotics) become xenobiotics when present in the environment at excessive concentrations. Xenobiotics are mostly produced by human activities. Some organisms may also form them as a part of their defense system, e.g., mycotoxins, bacterial and herbal toxins, etc., and xenobiotics become harmful when entering the food chain. Certain xenobiotics entering the human body deregulate important cellular- and organ-signaling pathways because they mimic physiological substrates and also promote carcinogenesis or cause genotoxicity by interaction with genetic information. Contemporary human exposure to xenobiotics is unavoidable, as they are omnipresent. Common concern about xenobiotics is due to their toxicity to living organisms, potentially causing cancer and potential to cause environmental effects on a global level.

Keywords: Bile duct epithelium (BDE).

1. INTRODUCTION

Xenobiotics are a wide range of external chemicals exposed to human body; both purposely (e.g., medicines) and accidentally (e.g., environmental contaminants) and these foreign molecules are known as xenobiotics. Xenobiotics are mostly produced by human activities. Exposure to some xenobiotics is voluntary because of their anticipated beneficiary effects on human health (e.g., drugs, antibiotics, dietary supplements like antioxidants, etc.). Daily use of xenobiotics such as food ingredients (dyes, stabilizers, emulsifiers, salt compounds, preservatives, etc.), cosmetics and personal care products (makeup, hair dyes, soaps, perfumes), and household products (chlorine bleach, bug sprays, cleaners, etc.) further increases exposure. Nevertheless, the positive role of xenobiotics as components of modern technologies enabling further progress of human civilization is to be underlined. At the present time, it is not possible to divide xenobiotics into straight forward categories, e.g., into good versus bad ones, and to simply eliminate the latter from the living environment.

2. XENOBIOTICS AND CANCER

Various xenobiotics promote carcinogenesis or cause genotoxicity by interaction with genetic information. The majority of procarcinogens undergo metabolic conversion to become ultimate carcinogens. The extent of carcinogen production is dependent on the ratio between the activation and detoxification capacity. Thus, understanding the inter-individual variability in XMEs may shed more light on the molecular basis of diseases apparently caused by the interaction between the genetic background and environment. Experts in molecular epidemiology now evaluate the prognostic value of the reported associations between the genetic variants in the genes encoding XMEs and various diseases, especially cancer. Although the concept of the association between the metabolic capacity modified by the genetic variation in XMEs and disease caused by the metabolically activated procarcinogens seemed reasonable, recent meta-analyses and genome wide scans (GWAS) have shown that the task will be extremely difficult [Souce k et al. 1995]. The association studies usually used small numbers of patients due to problems with recruiting. Moreover, each cancer type is a distinct disease with distinct etiology so pooling patients with different diagnoses is questionable. Identification of individual chemicals and levels of exposure in the studied individuals is greatly complicated by the long latency period from exposure to onset of cancer (usually decades). Finally, the interplay between different groups of low-penetrance genes (such as XMEs) most probably multiplies the effect of associations [Gillam et al. 2005]. The final goals of molecular epidemiology are to predict cancer susceptibility based on the inter individual variability and to prevent interacting exposures. In the individuals already suffering from cancer, individualized therapy tailored to the XMEs over-expressed in tumor cells shall be available in near future [Soucek et al. 1995]. There is increasing evidence identifying the crucial role of numerous dietary components in modifying the process of carcinogenesis. The varied effects exerted by nutrient and non-nutrient dietary compounds on human health and cancer risk are one of the new challenges for nutritional sciences. In the present paper, an attempt is made to review the most recent epidemiological data on interactions between dietary factors and metabolic gene variants in terms of cancer risk. The majority of case-control studies indicate the significant relationship between cancer risk and polymorphic xenobiotic metabolising enzymes in relation to dietary components. The risk of colorectal cancer is associated not only with CYP2E1 high-activity alleles, but also GSTA1 low-activity alleles, among consumers of red or processed meat. Genetic polymorphisms of NAT1 and NAT2 may be also a breast-cancer susceptibility factor among postmenopausal women with a high intake of well-done meat. [Zheng et al. 1999]. On the other hand, phytochemicals, especially isothiocyanates, have a protective effect against colorectal and lung cancers in individuals lacking GST genes. Moreover, polymorphism of GSTM1 seems to be involved in the dietary regulation of DNA damage. The European Prospective Investigation into Cancer and Nutrition study shows a significant inverse association between the polycyclic aromatic hydrocarbon-DNA adduct level and dietary antioxidants only among GSTM1-null individuals. [Palli et al. 2004] However, the absence of a modulatory effect of polymorphic xenobiotic metabolising enzymes and diet on the development of cancer has been indicated by some epidemiological investigations. Studies of interactions between nutrients and genes may have great potential for exploring mechanisms, identifying susceptible populations/individuals and making practical use of study results to develop preventive strategies beneficial to human health.

3. XENOBIOTIC-INDUCED LIVER INJURY

The liver is prone to xenobiotic-induced injury becauseof its central role in xenobiotic metabolism, its portallocation within the circulation, and its anatomic and physiologic structure. The liver plays a key role in the detoxification and elimination of xenobiotics. Excretion into bile followed by passage via bile ducts into the intestine and subsequent fecal elimination comprises a major route for the elimination of xenobiotics and their metabolites. Under certain conditions, xenobiotics (including therapeutic drugs) and their metabolites can directly damage the bile duct epithelium (BDE) or elicit immune responses against BDE cells (cholangiocytes). Drug-induced cholestasis is usually acute and self-limited; after drug withdrawal, symptoms rapidly disappear and full recovery usually occurs within 3 months [Marschall et al. 2007]. Xenobiotic hepatotoxicity is a particularly important area-of-need for therapeutics. Drug-induced liver injury, for example, is the most common cause of acute liver failure (ALF) and ALF-induced deaths in many countries. Most drugs are not intrinsically toxic to the liver but can cause injury secondary to the production of a hepatotoxic drug metabolite, a process known as bioactivation [Vessey et al. 1996; Dahm et al. 1995]. Because gastrointestinal absorption is enhanced by lipid solubility, most xenobiotics are highly lipophilic compounds, which are poorly excreted by the kidneys [Vessey et al. 1996]. The liver plays a critical role in promoting excretion of these compounds by transforming them into metabolites of greater water solubility. Hepatocellular necrosis is a hallmark of these injuries, but the associated symptoms and histologic pattern of injury are nearly identical to those observed with acute viral hepatitis [Bass et al. 1996; Zimmerman et al. 1993; Scheuer et al. 1988; Barnard et al. 1994]. Intrinsic injuries are predictable, in that a threshold dose exists in all individuals, typically leading to zonal liver cell necrosis accompanied by little or no signs of inflammation. Idiosyncratic liver injuries are usually associated with classic signs of hypersensitivity, including fever or rash, and liver biopsy specimens reveal evidence of monocyticor eosinophilic infiltrates [Lee et al., 1995]. These reactions tend to occur only after repeated exposure, suggesting the need for initial sensitization, and drug rechallenge generally elicits prompt reappearance of symptoms. Both humoral and cellular immune mechanisms have been implicated in these types of liver injuries [Bass et al. 1996].

4. XENOBIOTICS AND LUNG CANCER

Lung cancer is the leading cause of cancer mortality for men and women in the United States andis a growing worldwide problem. Several environmental factors associated with an increased risk of developing lung cancer have been identified, including mainstream cigarette smoking; exposure to radon, polycyclicaromatic hydrocarbons (PAHs), nickel, chromate, arsenic, asbestos, chloromethyl ethers andionizing radiation; and chronic obstructive pulmonary disease with airflow obstruction. The risk of lung carcinoma increases with the number of cigarettes smoked, years of smoking, earlier age of commencing smoking, degree of inhalation, tar and nicotine content, and use of unfiltered cigarettes, and decreases proportionally with the number of years after quitting [Garfinkel et al., 1988]. Lung carcinogenesis results from the interaction between endogenous factors and inhaled compounds.

5. XENOBIOTICS METABOLISM

According to Xu et al. xenobiotics that enter the human body undergo four stages: absorption, distribution, metabolism, and elimination [Xu et al.]. These processes are performed by xenobiotics/drug-metabolizing enzymes (XMEs), including Phase I (oxidation, reduction, or hydrolysis reactions) and Phase II (conjugation reaction) enzyme systems. [Williams et al.]. XMEs are important enzyme families that are present in the liver and extra hepatic organs, including the skin, lung, kidney, intestine, and colon/rectum, that interact with endogenous and exogenous chemicals and xenobiotics [Gundert-Remy et al.]. The xenobiotic metabolism involves the continuous biotransformation steps of oxidation, reduction or hydrolysis of parent substances to reactive or polar groups, such as -NH2, -COOH, and-OH groups (Phase I), followed by conjugated with hydrophilic molecules (Phase II), such as glutathione, glucuronic acid, and sulfate, to increase the hydrophilicity of the substances, there by rendering the substances suitable for renal or intestinal excretion [Mandlekaret al.]. However, in some instances, xenobiotics are metabolized by Phase I enzymes, which can form reactive or mutagenic metabolites that may induce DNA mutation and carcinogenesis. For example, aflatoxin B1 is metabolized to mutagenic aflatoxin-8,9-epoxide by CYP3A4 [Kamdemet al.]. The majority of Phase I reactions are performed by cytochrome P450 (CYP) enzymes, particularly those in the CYP1, CYP2, and CYP3 families, which metabolize a vast range of xenobiotics [Ding et al.].

6. CYTOCHROME P450

Xenobiotics enter in our body through metabolism and excretion. The stage of xenobiotic metabolism contributes to the conversion of the parent substances for either metabolic bioactivation or detoxification [Omiecinskiet al.].The key enzymes for xenobiotic bioactivation are CYPs (Phase I), which catalyze the xenobiotics to generate reactive metabolites,such as a molecular ions, quinone, and epoxide. This is a necessary step and is followed by Phase II conjugation reaction [Attia et al.]. Phase II enzymes, including glutathione S-transferases (GSTs), UDP-glucuronosyl-transferase (UGTs), sulfo-transferases, nicotinamide adenine dinucleotide phosphate hydrogen (NADPH):quinone acceptor oxido-reductase 1, and quinone reductase, have been associated with xenobiotic detoxification and, ultimately, excretion [Gundert-Remy et al.]. Phase II enzymes induced by natural dietary compounds act as a detoxification strategy through the detoxification of reactive xenobiotic metabolites and the promotion of xenobiotic excretion.Since CYPs were first discovered by Klingenberg [Klingenberg et al.] in 1958, the importance and functions of xenobiotics/drug-metabolizing enzymes (XMEs) for biotransformation of xenobiotics and drugs have gradually become appreciated. P450s are heme-thiolate enzymes that participate in the biotransformation of xenobiotics and production of many important endogenous compounds including steroid hormones, prostaglandins, and leukotrienes. Their expression is believed to be highly regulated; some P450s are expressed only in specific tissues and at times in specific cells within the tissue. Similarly, the expression pattern of a number of P450s is different at each developmental stage and different in females and males. Paradoxically, these enzymes catalyse the formation of reactive intermediates of thousands of chemicals that can damage DNA, as well as lipids and proteins. P450 expression can also affect the production of proliferation-promoting molecules derived from arachidonic acid and alters various downstream signal transduction pathways. Reactive intermediates, activated carcinogens, and arachidonic acid derivatives can be precursors to malignancy and cause chemical cancer. In a developed cancer, these enzymes can determine the outcome of anticancer therapy and altered expression of P450s can support cancer progression and cause drug resistance. Since P450s are the "bridges" between the environment and our body, their function can be linked in many ways to carcinogenesis: they activate xenobiotics to ultimate carcinogens but at the same time they metabolize (and in some cases activate) drugs used for cancer treatment. In some tumors, the cancer tissue maintains its drug resistance with altered expression of P450s and this explains why they are potential targets for anticancer therapy. Mostly, those P450s could be linked to chemical cancer, which are involved in xenobiotic metabolism (CYP1-4 family).

7. CONCLUSION

We would like to highlight the importance of regulatory mechanisms in cancer, which harmonize the genetically determined P450 activity with the environmental effects and could be beneficially altered by anticancer drugs in the future. Industrial pollution and food contamination have become increasingly serious due to adverse human health effects. Metabolism of xenobiotics andthe regulation of xenobiotic-related signaling pathways arekey factors for the activation of xenobiotic toxicity. Metabolism of xenobiotics and the first-pass effect are not only carried out in the liver but also in the intestines by gut mucosa and microbiota.

SOURCES AND SELECTION CRITERIA

We present examples on applications of these methods to a range of research questions and studies as published in major clinical journals, including the BMJ, Cochrane, Pubmed, and Google scholar.

AUTHORSHIP & CONTRIBUTORSHIP

Dr. Ajay Kumar wrote the manuscript, with the help from Amar Deep. Amar Deep, Suchit Swaroop and Virendra Atam critically edited and reviewed the manuscript.

CONFLICTS OF INTEREST

All authors: No reported conflicts. "The author(s) declare(s) that there is no conflict of interest regarding the publication of this paper."

REFERENCES

Attia SM. Deleterious effects of reactive metabolites. Oxid Med Cell Longev 2010; 3:238e53.

Barnard GF, Scharf MJ, Dagher RK. Sulfone syndrome in a patient receiving steroids for pemphigus. Am J Gastroenterol1994; 89:2057–9.

Bass NM, Ockner BA. Drug-induced liver disease. In: Zakin D, Boyer TD, eds. Hepatology: a textbook of liver disease, 3rd ed. Philadelphia: WB Saunders,1996:962–1017.

Dahm LJ, Jones DP. Mechanisms of chemically induced liver disease. In: Zakin D, Boyer TD, eds. Hepatology: a textbook of liver disease, 3rd ed. Philadelphia: WB Saunders, 1996:875–90.

Dahm LJ, Jones DP. Mechanisms of chemically induced liver disease. In: Zakin D, Boyer TD, eds. Hepatology: a textbook of liver disease, 3rd ed. Philadelphia: WB Saunders, 1996:875–90.

Ding X, Kaminsky LS. Human extra hepatic cytochromes P450: function in xenobiotic metabolism and tissue-selective chemical toxicity in the respiratory and gastrointestinal tracts. Annu Rev Pharmacol Toxicol 2003; 43:149e73.

Garfinkel L, Stellman SD. Smoking and lung cancer in women: findings in a prospective study. Cancer Res. 1988; 48:6951–5.

Gillam EM (2005) Exploring the potential of xenobiotic-metabolising enzymes as biocatalyst: evolving designer catalysts from polyfunctional cytochrome P450 enzymes. Clin Exp Pharmacol physiol 32:147-152.

Gundert-Remy U, Bernauer U, Blomeke B, Doring B, Fabian E, Goebel C, Hessel S, Jackh C, Lampen A, Oesch F, Petzinger E, Volkel W, Roos PH. Extra hepatic metabolism at the body's internal-external interfaces. Drug Metab Rev 2014; 46:291e324.

Kamdem LK, Meineke I, Godtel-Armbrust U, Brockmoller J, Wojnowski L.

Dominant contribution of P450 3A4 to the hepatic carcinogenic activation of aflatoxin B1. Chem Res Toxicol 2006; 19:577e86.

Klingenberg M. Pigments of rat liver microsomes. Arch BiochemBiophys 1958; 75:376e86.

Lee WM. Drug-induced hepatotoxicity. N Engl J Med 1995; 333:1118–27.

Mandlekar S, Hong JL, Kong AN. Modulation of metabolic enzymes by dietary phytochemicals: a review of mechanisms underlying beneficial versus unfavourable effects. Curr Drug Metab 2006; 7:661e75.

Marschall HU, Wagner M, Zollner G, Trauner M. Clinical hepatotoxicity. Regulation and treatment with inducers of transport and cofactors. Mol Pharm2007;4(6):895-910.

Omiecinski CJ, Vanden Heuvel JP, Perdew GH, Peters JM. Xenobiotic metabolism, disposition, and regulation by receptors: from biochemical phenomenon to predictors of major toxicities. Toxicol Sci 2011;120(Suppl 1): S49e75.

Palli D, Masala G, Peluso M, et al. (2004) The effects of diet on DNA bulky adduct levels are strongly modified by GSTM1 genotype: a study on 634 subjects. Carcinogenesis 25, 577–584.

Scheuer PJ. Drugs and toxins. In: Liver biopsy interpretation, 4thed. London: Bailliere Tindall, 1988:99–105.

Soucek P. (2015) Xenobiotics. In: Schwab M. (eds) Encyclopedia of Cancer. Springer, Berlin, Heidelberg.

Vessey DA. Metabolism of xenobiotics by the human liver. In: Zakin D, Boyer TD, eds. Hepatology: a textbook of liver disease, 3rd ed. Philadelphia: WB Saunders, 1996:257–305.

Vessey DA. Metabolism of xenobiotics by the human liver. In: Zakin D, Boyer TD, eds. Hepatology: a textbook of liver disease,3rd ed. Philadelphia: WB Saunders,1996:257–305.

Williams R. The metabolism and detoxication of drugs, toxic substances and other organic compounds. In: Tecwyn Williams R, editor. Detoxication mechanisms. New York: Wiley; 1959.

Xu C, Li CY, Kong AN. Induction of phase I, II and III drug metabolism/transport by xenobiotics. Arch Pharm Res 2005;28:249e68.

Zheng W, Deitz AC, Campbell DR, Wen WQ, Cerhan JR, Sellers TA, Folsom AR & Hein DW (1999) N-acetyltransferase 1 genetic polymorphism, cigarette smoking, well-done meat intake, and breast cancer risk. Cancer Epidemiol Biomarkers Prev 8,233–239.

Zimmerman HJ, Maddrey WC. Toxic and drug-induced hepatitis. In: Schiff L, Schiff ER, eds. Diseases of the liver, 7th ed.Philadelphia: JB Lippincott, 1993:707–83.

❐

Chapter-8

AMELIORATION OF SODIUM FLUORIDE INDUCED GENOTOXICITY BY ASCORBIC ACID AND CALCIUM IN *HETEROPNEUSTES FOSSILIS*

Neelam Singh and Madhu Tripathi *

Aquatic Toxicology Research Laboratory, Department of Zoology,
University of Lucknow, Lucknow, UP-226007.
Email of Corresponding Author: drmtripathi@gmail.com

ABSTRACT

Fluoride is naturally-occurring component of rocks and soil and also found in air, water, plants and animals. It enters the atmosphere through volcanic emissions and the resuspension of soil by wind. The present study was conducted to confirm the genotoxic potential of fluoride (F) in fresh water fish Heteropneustes fossilis, as well as to evaluate the protective role of ascorbic acid and calcium on genotoxicity induced by sodium fluoride. Fish were divided into nine groups consisting control, fluoride treated and fluoride + ameliorative agent group for 30 days. Total DNA damage induced due to fluoride was analysed by comet assay. Results revealed that ascorbic acid and calcium efficiently ameliorate the toxic effect of F by reducing the DNA damage. On the basis of this study it can be concluded that ascorbic acid and calcium mitigates the genotoxic effects of the fluoride effectively.

Keyword: Sodium Fluoride, Heteropneustes fossilis, Ameliorative agent, Genotoxicity

1. INTRODUCTION

Aquatic environment is polluted by a variety of pollutants that originate from natural and anthropogenic sources which are toxic to the aquatic organisms including fishes. Freshwater invertebrates and fishes, mainly net-spinning caddis fly larvae and upstream migrating adult salmons are more sensitive to fluoride toxicity than marine and estuarine animals (Camargo , 2003). Fluoride toxicity to aquatic invertebrates and fishes increases with increasing fluoride concentration, exposure time, water temperature, and decreases with increasing intraspecific body size and water content of calcium and chloride. The intake of fluoride by ingestion, inhalation and absorption through the skin, damage the cellular structure which results in accumulation of other toxic metals in the body which is real danger to the life because metals showed adverse effects on biomolecular structures of tissues (Handy and Eddy, 1990). When fishes are exposed to excessive fluoride, lipid peroxidation generates many free radicals that could enter the nuclei and cause base modifications and DNA breakage, in addition to causing more lipid peroxidation (Anuradha et al., 2001). In the present study we determined the level of basal DNA damage and the ameliorative role of ascorbic (A.A.) and calcium nitrate ($CaNO_3$) in the alkaline comet assay after treatment with fluoride and along with ameliorative agents (Ascorbic acid and Calcium).

2. MATERIALS AND METHODS

Fresh healthy living fish *Heteropneustes fossilis* were procured from local resource of Lucknow. The weight 50+5g and size 15+10 cm of fishes were uniform. Fish were disinfected with 1% $KMnO_4$ solution and acclimatized under standard laboratory conditions for 15 days. Other physico-chemical characteristics of water during acclimatization of fish were determined by standard method (APHA, 2005) which showed the mean values as follows: temperature 28+ 1.50 C, pH 6.8+0.2, DO 7.2+ 1 mg /L, total hardness 110 – 120 mg/L, alkanity 100mg/L, nitrite 0.10-0.11 mg/L, Nitrate 7.0-7.1 mg/L and fluoride 0.1 mg/L. The fish were divided into nine groups with 24 fish per group in aquaria measuring 60 × 40 × 45 cm according to Table 1.

Groups	Treatment	Duration of Exposure
Control (I)	Untreated	30 days
LD (II)	35mg F/L (1/10thLC $_{50}$)	30 days
HD (III)	70mg F/L (1/5thLC $_{50}$)	30 days
LD+Ca^{++} (IV)	35mg F/L + 15 mg $CaNO_3$ /L	30 days
HD+Ca^{++}(V)	70mg F/L + 15 mg $CaNO_3$ /L	30 days
LD +A.A.(VI)	35mg F/L + 50 mg ascorbic acid/L	30 days
HD +A.A. (VII)	70 mg F + 50 mg ascorbic acid/L	30 days
LD+Ca^{++}+A.A.(VIII)	35mg F/L + 15 mg $CaNO_3$/L+ 50 mg ascorbic acid/L	30 days
HD+Ca^{++}+ A.A. (IX)	70 mg F/L + 15 mg $CaNO_3$/L+ 50 mg ascorbic acid/L	30 days

Table 1: Experimental Design of the Study

The comet assay was performed under alkaline conditions (pH > 13) according to the procedure of Singh et al. (1988). Following 24 h incubation at 37°C, the blood cells were resuspended in 0.75% low melting point (LMP) agarose in PBS (pH 7.4), and placed on microscope slides precoated with 0.5% normal melting point (NMP) agarose. To prevent additional DNA damage, all the steps were conducted under reduced light or in the dark. Cells were lysed for 1 h at 4°C in a buffer containing 2.5 M NaCl, 100 mM EDTA, 1% Triton X-100, 10 mM Tris-HCl, 1% N-lauroylsarcosine sodium, pH 10. Slides were placed in an electrophoresis unit for 40 min to allow DNA to relax in the electrophoretic solution consist ing of 300 mM NaOH, 1 mM EDTA, pH > 13, at 4°C. Then the slides were subjected to electrophoresis (4°C) for 23 min at an electric field strength of 0.86 V/cm, 25 V, 300 mA. After electrophoresis the slides were neutralized in 0.4 M Tris-HCl (pH 7.5). DNA was stained with 80µL 1X Ethidium Bromide. Slides were analyzed using a fluorescence microscope. For visualization of DNA damage slides were observed using 20 × objective fluorescent microscope (Nikon) with a Biovis comet software.

3. RESULTS

Results were expressed as Mean ± SEM. Statistical analysis for all the assays performed by Two Way ANOVA followed by Bonferroni Post Test. For this Graph pad prism 5.01 (USA) versions was used. A total of 50 nucleoids were analyzed for each group to obtain DNA damages. Fluoride induced DNA damage in blood cells of *Heteropneustes fossilis* was quantified in terms of Tail Length and % DNA Tail are presented in table 2 (fig 1) and table 3(fig 2). In addition, images of different DNA migration profile of blood cells were also obtained. The comet images provide the clear visualization and categorization of the level of DNA damage in blood cells are presented in fig 3. For the evaluation of ameliorative effect of ascorbic acid and calcium nitrate to fluoride exposure, Tail DNA and Tail Length levels were recorded as mark of DNA damage in the fluoride treated groups (LD and HD) and comparison was made between control and fluoride+ ascorbic acid, fluoride+$CaNO_3$ and fluoride+ ascorbic acid +$CaNO_3$ treated groups by counting 50 comet images (cells) per sample.

Tail Length (um) in Blood Cells:

After one month % Tail DNA was increased significantly (P***>0.0001) in all fluoride treated groups as compared to control. In LD (I) and HD (II) groups Tail length was increased significantly (P***>0.0001) as compared to control (I). In ameliorative agents treated groups there was a decrease significantly (P***>0.0001) in Tail length in all the groups (LD +Ca++ (IV), HD+Ca++ (V), LD+A.A. (VI), HD+A.A. (VII), LD+Ca+++A.A. (VIII), HD+Ca+++A.A. (IX)) as compared to LD and HD groups (Table .2 and Fig.1).

Percentage Tail DNA in Blood Cells:

After one month % Tail DNA was increased significantly (P***>0.0001) in all fluoride treated groups as compared to control. In LD (II) and HD (III) groups % Tail DNA was increased significantly (P***>0.0001) as compared to control (I). In other groups like LD +Ca++ (IV), LD+A.A. (V), LD+Ca+++A.A. (VI), HD+Ca++ (VII), HD+A.A. (VIII), HD+Ca+++A.A. (IX) groups it was decreased significantly (P***>0.0001) as compared to LD and HD groups (Table. 3 and Fig.2).

Group	One Month	%Change
Control(I)	6.3±1.02	
LD (II)	58.8±1.6***	772.05
HD (III)	79.5±4.7***	1161.90
LD + Ca^{++} (IV)	32.3±1.4***	412.69
HD+ Ca^{++} (V)	40.6±1.29***	544.44
LD+A.A. (VI)	30.4±0.5***	382.53
HD+ A.A. (VII)	46.2±1.7***	633.33
LD+Ca^{++}+ A.A. (VIII)	31.9±1.76***	406.34
HD+Ca^{++}+ A.A. (IX)	34.7±3.67***	450.79

Table 2: Tail Length in Blood Cells of Control, Fluoride treated, Fluoride+A.A. treated and Fluoride+ $CaNO_3$ treated Groups of *Heteropneustes fossilis* after different exposure period. Values are mean ± SEM, *n* = 50, P*<0.0001, P** <0.001 and P* <0.05.**

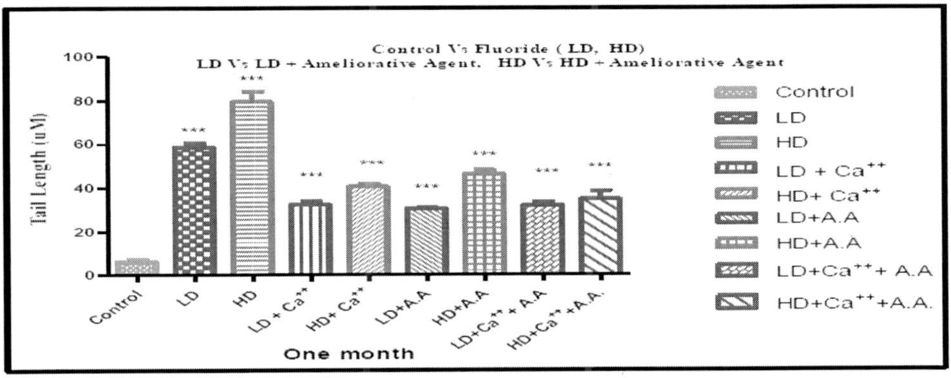

Fig 1: **Effect of Fluoride on Tail length (uM)of DNA damage after One Month Exposure of Fluoride and along with Calcium nitrate ($CaNO_3$) and Ascorbic acid in Blood cells of** *Heteropneustes fossilis.* **Value are mean ± SEM, n = 50, P ***<0.0001, P** <0.001 and P* <0.05.**

Group	One Month	%Change
Control(I)	6.81±1.3	
LD (II)	58.39±5.19***	758.67
HD (III)	68.59±6.9***	908.67
LD + Ca^{++} (IV)	35.154±1.4***	496.32
HD+ Ca^{++} (V)	40.89±3.6***	501.32
LD+A.A. (VI)	29.27±5.9***	330.44
HD+ A.A. (VII)	35.76±4.18***	425.88
LD+Ca^{++}+ A.A. (VIII)	40.14±3.11***	490.29
HD+Ca^{++}+ A.A. (IX)	45.74±2.6***	572.64

Table 3: % Tail in Blood Cells of Control, Fluoride treated, Fluoride+A.A. treated and Fluoride + $CaNO_3$ treated groups of *Heteropneustes fossilis* after different exposure period. Values are mean ± SEM, *n* = 50, P ***<0.0001, P** <0.001 and P* <0.05.

Table 3: % Tail in Blood Cells of Control, Fluoride treated, Fluoride+A.A. treated and Fluoride + $CaNO_3$ treated groups of *Heteropneustes fossilis* **after different exposure period. Values are mean ± SEM,** ***n*** **= 50, P ***<0.0001, P** <0.001 and P* <0.05.**

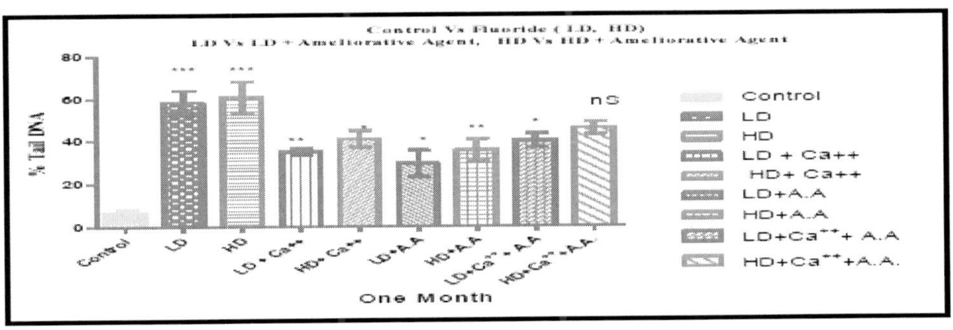

Fig 2: Effect of Fluoride on % Tail DNA of DNA damage after One Month Exposure of Fluoride and along with Calcium nitrate ($CaNO_3$) and Ascorbic acid in Blood cells of *Heteropneustes fossilis*. Value are mean ± SEM, n = 50, P ***<0.0001, P** <0.001 and P* <0.05.

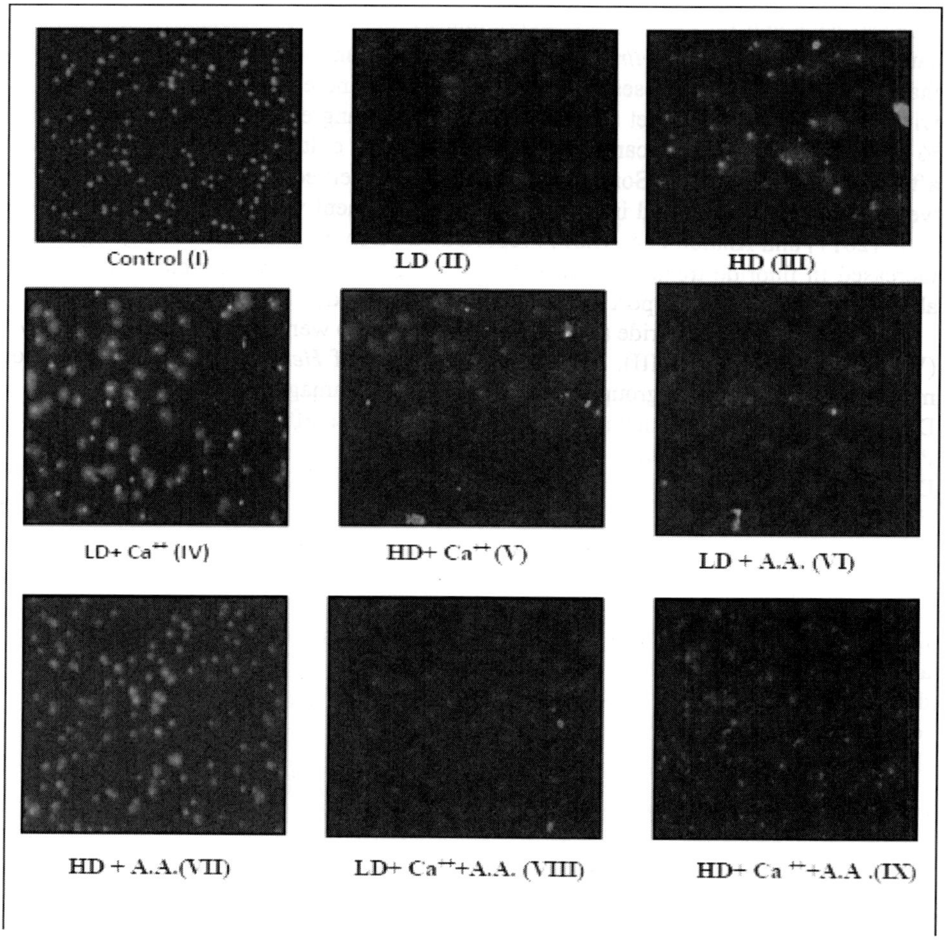

Fig.3. Representative images of Sodium Fluoride induced oxidative DNA damage in Blood cells of *Heteropneustes fossilis* after one month exposure. Typical Comet has been shown in Control and treated groups LD(II), HD(III), LD+Ca++(IV), HD+Ca++(V), LD+A.A.(VI), HD+A.A.(VII), LD+Ca+++A.A.(VIII), HD+Ca+++A.A. (IX).

4. DISCUSSION

DNA damage of blood cells exposed to fluoride markedly increased, especially in the LD (I) and HD (III) groups as compared to control (I) group of *Heteropneustes fossilis*. However, decrease in DNA damage has been observed in the blood cells of fishes exposed to fluoride + ameliorative agent groups (IV, V, VI, VII, VIII and IX) as compared to LD (II) and HD (II) groups. It is clear from the result that the parameters of DNA damage such as Tail Length, % DNA Tail were altered after administration of fluoride. Tail Length, % DNA Tail were increased significantly in LD and HD groups in concentration dependent manner with fluoride dose as compared to control value. Ge et al. (2005) reported significantly higher DNA migration in thyroid cells of rats exposed to fluoride as compared to control.

He and Chen (2006) substantiated the role of fluoride in causing DNA damage in oral mucosal cells and rat hepatocytes. According to them induced oxidative stress, may be cause of DNA damage. *In vitro* study of L-02 cells (Human embryo hepatocyte) revealed that, exposure of higher doses of fluoride, caused an increased percentage of L-02 cells with DNA damage (Wang et al., 2004). Similarly, Zhang et al. (2004) also reported that % of Tail DNA was significantly increased in primary cultured rat hippocampal neurons after treated with fluoride. Song et al. (2015) have observed % Tail DNA and Tail length were significantly increased in liver cells of rat treatment with fluoride. Similarly, Dutta et al. (2017) and Khatun el al. (2017) reported Tail Length and Tail intensity significantly increased in fluoride treated larvae and ovary cells of Drosophila melanogaster. All the above reported studies support the finding of present study.

In present study, fluoride and calcium treated group were LD+Ca++ (IV), HD+CA++ (V), LD+Ca++ +A.A. (VIII), HD+CA+++A.A. (IX) of *Heteropneustes fossilis* for one-month exposure. In these groups, parameters of DNA damage such as Tail Length, % Tail DNA were reduced as compared to LD and HD groups after one month in blood cells. Present finding showed that when fluoride administrated with calcium it was found that DNA damage was less in all groups (IV, V, VII, IX) as compared to LD and HD groups. Fedirko et al. (2010) have reported similarity with present study that calcium and vitamin D3 may decrease oxidative DNA damage in the human colorectal mucosa. Bentle et al. (2006) also observed that calcium play an important role in repairing of DNA by reducing Tail Length in breast cancer cells line. Ogutcu et al. (2014) found that calcium reduces DNA damage by methylation caused in *Vicia Faba* seedlings due to cadmium Stress. All the above-mentioned reports of different workers support the finding of present study regarding the role of Ca++in fluoride exposure groups. Calcium is an ideal antidote for fluoride due to its affinity to form the fluoride an insoluble complex, calcium fluoride, which reduces fluoride absorption (Wagner and Muhler, 1959). Thus, Ca++ administration in present study may be considered to be responsible for lesser damage of DNA in comparison to LD and HD groups. Ascorbic acid (A.A., vitamin C) has a role in maintaining the growth and has antioxidant properties. In the present study, when AA was administrated to the groups along with NaF for one month, two months and three months, a significant protective effect by A.A. was observed by decreasing the genotoxicity. In LD+A.A. (VI), HD+A.A.(VII), LD+Ca+++A.A. (VIII), HD+Ca+++A.A.(IX) groups, all parameters of DNA damage such as Tail Length, % Tail DNA were decreased significantly as compared to LD and HD group in blood cells . Result showed that when fluoride administrated along with A.A., DNA damage was less as compared to control (I). Ameliorative role of ascorbic acid for DNA damage in human lymphocytes against different toxicants such as N-methyl-N'-nitro-N-nitrosoguanidine (MNNG), idarubicin and cisplatin are reported by several workers (Błasiak et al., 2002; Wozniak et al., 2004; Hu et al., 2016). Similarly, Choi et al. (2005) and Kontek et al. (2013) also observed that A.A. decreases DNA damage by reducing % DNA Tail in Type 2 diabetic patient and human cancer cell line (HT29 cells) respectively. The present study reveals therapeutic effects of ascorbic acid, calcium treatment alone or in combination to recover sodium fluoride induced genotoxicity.

5. CONCLUSION

Present study revealed the decrease in the primary DNA damage by calcium and ascorbic acid verified by the parameters like comet tail length and percent DNA in tail of the comet in blood cells of *Heteropneustes fossilis* as compared to fluoride treatment groups.

The present study reveals therapeutic effects of ascorbic acid, calcium treatment alone or in combination to recover sodium fluoride induced genotoxicity. Although, ameliorative role of Ca++ and A.A. in case of fluoride genotoxicity is yet to be explored extensively.

ACKNOWLEDGEMENT

Authors wish sincere thank to the Head of Department of Zoology,University of Lucknow for providing us the laboratory facilities & support. First author express thanks to UGC- BSR fellowship for financial assistance to carry out the research work.

REFERENCES

Anuradha C.D., Kanno S., Hirano S (2001). Oxidative damage to mitochondria is a preliminary step to caspase-3 activation in fluoride-induced apoptosis in HL-60 cells. Free Radic Biol Med. 31:367–373.

Bentle M. S., Reinicke K. E., Bey E. A., Spitz D. R., Boothman D. A. (2006). Calcium dependent modulation of Poly (ADP-Ribose) Polymerase-1 alters cellular metabolism and DNA repair. The Journal of Bio. Chem. 281: 33684 –33696.

Błasiak J., Trzeciak A., Gasiorowska A., Drzewoski J., Małecka P. E. (2002). Vitamin C and quercetin modulate DNA-damaging effect of N-methyl-N '-nitro-N-nitrosoguanidine (MNNG) (2002). Plant Foods Hum Nutr. 57:53-61.

Camargo J.A. (2003). Fluoride toxicity to aquatic organisms: a review. Chemosphere. 50:251-64.

Choi S. W., Benzie I., Lam C.S.Y., et al (2005). Inter-relationships between DNA damage, ascorbic acid and glycaemic control in Type 2 diabetes mellitus. Diabetic Medicine. 22:1347–1353.

Dutta M., Rajak P., Khatun S., Roy S. (2017). Toxicity assessment of sodium fluoride in *Drosophila melanogaster* after chronic sub-lethal exposure. Chemosphere. 166:255–266.

Fedirko V., Bostick R.M., Long Q., Flanders W.D. et al. (2010). Effects of supplemental vitamin D and calcium on oxidative DNA damage marker in normal colorectal mucosa: A randomized clinical trial. Cancer Epidemiol. Biomarkers & Prev. 19:280- 91.

Ge Y., Ning H., Wang S., Wang J. (2005). DNA Damage in thyroid gland cells of rats exposed to long term intake of high fluoride and low iodine. Fluoride 38:318–323.

Handy. R.D, Eddy. F.B., (1990), Functional Ecology New Horizons in Ecotoxicology. 4: 385-392.

He L. F., Chen J.G. (2006). DNA damage, apoptosis and cell cycle changes induced by fluoride in rat oral mucosal cells and hepatocytes. World J Gastroenterol 12:1144-1148.

Hu J., Liebb J. D., Sancara A., Adara S. (2016). Cisplatin DNA damage and repair maps of the human genome at single-nucleotide resolution PNAS. 113: 11507–11512.

Khatun S., Rajak P., Dutta M., Roy S. (2017). Sodium fluoride adversely affects ovarian development and reproduction in *Drosophila melanogaster*. Chemosphere, 186:51-61.

Kontek R., Kontek B., Grzegorczyk K. (2013). Vitamin C modulates DNA damage induced by hydrogen peroxide in human colorectal adenocarcinoma cell lines (HT29) estimated by comet assay *in vitro*. Arch Med Sci. 9:1006-12. doi:10.5114/aoms.2013. 39791.

Ogutcu H., Arslan E., Agar G., Gulluce M., Turan M., and Sahin F. (2014). Protective role of calcium on DNA methylation caused cadmium stress in *Vicia faba* seedlings. International Conference on Agricultural, Ecological and Medical Sciences,

Bali(Indonesia).

Singh N.P., McCoy M.T., Tice R.R., Schneider E.L. (1988). A simple technique for quantitation of low levels of DNA damage in individual cells. Exp. Cell.Res. 175:184–191.

Song G H, Huang F B, Gao J P, Liu M L et al. (2015). Effects of fluoride on DNA damage and caspase-mediated apoptosis in the liver of rats. Biol. Trace. Elem.Res.DOI 10.1007/s12011-015-0265-z.

Wagner M. J., Muhler J. C. (1959). The relationship between fluoride ingestion and urinary calcium. J.Dent.Res. 38:1078-1081.

Wang A.G., Xia T., Chu Q.L., Zhang M. et al. (2004). Effects of fluoride on lipid peroxidation, DNA damage and apoptosis in human embryo hepatocytes. Biomed.Environ. Sci. 17:217-222.

Wozniak K., Czechowska A., Blasiak J. (2004). Cisplatin-evoked DNA fragmentation in normal and cancer cells and its modulation by free radical scavengers and the tyrosine kinase inhibitor STI571. Chem Biol Interact. 147:309-18.

Zhang W.L., Zhao H., Xue L.J., Zhang X.Y., Li G.S. (2004). Intracellular ionized calcium level and oxidative stress in kidneys of Wistar rats after intake of excessive fluoride. Chin. J. End. 23:25-27.

❐

SECTION-II
FUNCTIONAL BIOLOGY

Chapter-9

REPRODUCTION AND ENERGY BUDGET IN GOATS

Somenath Ghosh[1], Sunil Kumar Rai[2], Ravi S. Pandey[a] and Chandana Haldar[3*]

[1]Biochemistry Laboratory, Department of Zoology, University of Allahabad, Allahabad-21102, India.

[2] Post-Doctoral Fellow, Department of Biochemistry and Molecular Medicine, School of Medicine, University of California, Davis, CA, USA

[3]Pineal Research Laboratory, Department of Zoology, Banaras Hindu University, Varanasi-221005, India.

**Email of Corresponding Author: chaldar2011@gmail.com*

ABSTRACT

Reproduction is highly dynamic and most important biological process for perpetuation of species in living beings. Reproduction in both males and females involves gametogenesis and steroidogenesis thus, it is highly energy demanding. In most of the mammals including human the energy allocation for reproduction is well managed due to their opportunistic behavior in reproduction however, it is a matter of concern in seasonal breeders. In the long day breeders (like squirrel, hamsters) the energy allocation pattern is partially documented. Their reproductive phase is divided into two different phases Reproductive Active Phase (RAP) and Reproductive Inactive Phase (RIP). During, RAP the circulatory level of melatonin is low with high reproductive behavior and performances. But during RIP; reproductive performances are low while circulatory level of melatonin is high. The main problem is with the short day breeders (sheep and goats) where the energy allocation for reproduction in sheep is partially documented whereas for goats, the reports are absolutely missing. Further, in some animals (like mice and rats) it is evident that there may be a local hormonal circuit in peripheral endocrine organs (like ovary for GnIH and GnRH), salivary gland GI tract etc. But, no report is available regarding the primary and secondary lymphoid organ can be regarded as endocrine organ and also a site for local micro circuit of hormones. This kind of results or even speculations are totally lacking in any seasonal (including long day or short day) breeders. Thus, in the present chapter we will try to understand the present state-of-art of knowledge available in this field and will also try to summarize some new direction in this area for future research.

Keywords: Mel-PLA-nano-particles, Characterization, Immune effect, Synthesis, Toxicity.

1. INTRODUCTION

Food availability and energetic requirement fluctuate in most of the habitats. Living cells require a continuous supply of fuels for the biosynthesis and metabolism, and hence in odd condition, cessation of eating is a primary response and organisms engage in other behaviors involved in perpetuation of species. In such condition when there is no food available, the organism depends on fuels and nutrients from internal and external source they have (body fat and food storage, respectively). Evolution always favored animals that are able to store a significant quantity of energy inside the body (Nelson et al., 1995). This is necessary to inhibit ingestive behaviors so that animals may engage in activities that improves reproductive success. Some species increase energy intake during a period of intense parental care, whereas others increase both energy intake and storage in anticipation of the birth of offspring. In mammals, the biological processes like lactation in female, parental care and thermoregulation in both sexes are one of the most costly processes. In some mammalian species, females over eat and store the excess fuel during the gestation as lipids that are used in lactation later on. In other mammalian species, fooding is at the basal rate during the gestation but store a lot to be eaten during the energetically demanding period of lactation. Thus, Energy budget management and reproduction are integrated with those which control reproduction (Bronson and Heideman, 1994; Wade and Schneider, 1992). When food is plentiful and energy requirements are low, energy is available for all of the processes necessary for immediate survival. Metabolic signals, hormonal mediators and modulators and neuropeptides give expression to these priorities in at least two interrelated ways. First, they are permissive for the neuroendocrine events that control spermatogenesis, ovulatory cycles and fertility. Second, in many species, the same metabolic signals and chemical messengers that increase the motivation to engage in reproductive behaviors also attenuate the motivation to engage in foraging, hoarding and eating (Buijs et al., 1998; Schneider et al., 2002).

On the other hand, when energy is narrow, the energy routes towards the biological processes that confers the survival of the individual rather than promoting growth, longevity and reproduction. The physiological processes like foraging, storage of food and ingestion become the topmost priorities over reproduction because reproductive processes are energetically expensive and can be carried later on (Bronson, 1985). During low food availability when the thermoregulatory demands are high, generally gonadal regression occurs and species undergoes sexual rest. Even in species that breed year round, reproductive processes are curtailed under low food availability (Bronson, 1998). Further, in case of goats the sources of energy are very much uncommon and striking in nature. They can store energy in forms of glycogen and fat. But, under favorable and unfavorable conditions of foraging they are able to get energy from glucose and protein. In almost all of the animals glucose is most important ready source of energy. But, for protein the explanation is different. Under malnutrition (which may be caused due to scarcity of food, heavy competition and obviously due to huge parasitic load) proteins are used by them. But, under favorable and well nourished condition goats generally consume protein and use it for the maintenance of basal metabolism. Explanation may come from the evolution of goats from wild ones to domesticated animal. Under wild conditions they are thrived to migrate from one portion of grazing land to another one. At that time fat may be used as direct source of energy or by producing glucose (by gluconeogenesis). This is the only reason that under restricted conditions (food and water deprivation for prolonged time) they are never under "deprived stress" (Machen, 1981).

During the period of energetic challenges, ingestive behavior is favored over other activities. The ingestive process is triggered by a variety of metabolic sensory stimuli i.e. low glucose in blood and its metabolites, low level of peripheral hormone like insulin and leptin along with central feeding-stimulatory circuits involving neuropeptide Y (NPY) and agoutirelated protein (AgRP). These processed are of highly adaptive to the survival, bringing metabolic fuels, nutrients, water, salt and minerals for the maintenance of cell structure and function as survival of the organism is the utmost for species perpetuation. This is the reason the neuropeptides that stimulate ingestion during energetic challenges, inhibit the hypothalamic–pituitary–gonadal (HPG) system. When food becomes available ad libitum creating low energy demands, eating becomes minor and reproduction becomes the topper. Natural selection always favors the population whose individuals give priority to reproduction over feeding (Schneider et al., 2000).

The mechanisms that integrate the energy to reproduction might be important in clinical as well as agricultural aspects. Human female having the body weight at both extremes and also the one who is diabetic have more chances of reproductive abnormalities (Loucks et al., 1994). These abnormalities are mainly due to low circulating levels of ovarian steroids especially estrogens those causes osteoporosis and other impaired cognitive function (Misra, 2008). In addition to its clinical relevance, this area of research is now focused to improve breeding and lactation performance in dairy and meat animals. More over this beautiful adaptive strategy has made this group of ruminants a successful survivor under harsh conditions, environmental stresses and has made this animal a short day breeder when immune modulation and reproduction are occurring simultaneously under cold stress.

2. ENERGETIC EFFECTS ON THE HPG SYSTEM

There are many chemical messengers and metabolic processes that influence the reproduction acting through HPG system. Gonadotrophin releasing hormone (GnRH) being the master control of HPG and whose cell bodies are located in the area spanning from the preoptic area (POA) to the arcuate nucleus (Arc) of the hypothalamus. The neurohormone, GnRH has two modes of secretion. The pulse mode occurs during the follicular phase, when low concentration of estradiol causes hyposecretion of GnRH and LH. It means that in pulse mode the estradiol exerts negative feedback and limits the secretion of GnRH and LH. The other mode of GnRH secretion is surge mode which occurs during the periovulatory phase when high concentrations of estradiol exert positive feedback effects on GnRH. The GnRH pulse generator is a little-understood oscillating neural circuit that results in the pulsatile secretion of GnRH from terminals in the median eminence into the pituitary portal system. Each pulse of GnRH leads to the release of a pulse of LH from leuteotrophs in the anterior pituitary (Qarke and Cummins, 1982) along with FSH. These LH and FSH pulses are critical for follicle development and steroid secretion. The rising levels of estradiol have positive feedback on GnRH and LH, and these actions of estradiol are required for the LH surge, which triggers ovulation.

Metabolic challenges, such as food deprivation, inhibit the HPG system at many levels. The primary target is the GnRH pulse generator, and these effects are similar in males and females (Bronson, 1985). Pulsatile LH secretion, follicle development and ovulation can be reinstated by pulsatile treatment with GnRH in food-deprived rats, sheep, pigs, cows, monkeys and women (Armstrong and Britt, 1985; Bronson, 1985; Cameron, 2004; Cameron and Nosbisch, 1991). In addition, metabolic challenges alter the GnRH, LH and FSH surge, independent of their effects on pulsatile LH secretion (Crump et al., 1982; Medina et al., 1993, 1998). In sheep, metabolic challenges, such as food deprivation, inhibit GnRH secretion into the pituitary portal circulation (Foster and Bucholtz, 1995; Ansar-Ahmed et al., 1985). GnRH gene expression, immune reactivity or content are either unchanged or increased by energetic challenges, perhaps reflecting inhibited GnRH release from neurons (Ebling et al., 1995; McShane et al., 1992). Inhibition of GnRH secretion leads to a cascade of inhibitory effects, including decreased gonadotrophin secretion, retarded follicle development, inhibited synthesis of gonadal steroids and, in rodents and non-human primates, decreases in steroid-induced reproductive behaviors. GnRH transcription or translation might be important in those mammals that are induced ovulators. In musk shrews, the inhibitory effects of food restriction on reproduction are rapidly reversed by refeeding (Gill and Rissman, 1997; Temple and Rissman, 2000). These results emphasize a role for direct effects of metabolic challenges on GnRH neurons.

3. ENERGETIC EFFECTS ON MOTIVATION *VS* PERFORMANCE OF INGESTIVE AND REPRODUCTIVE BEHAVIOR

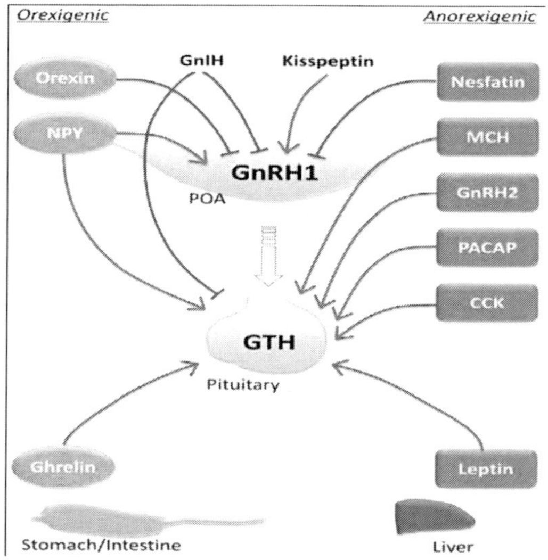

Schematic representation showing different factors affecting reproductive energy demand (Photo courtesy: www. google.co.in)

The metabolic signals, hormones and neuropeptides optimize reproductive success by prioritizing behavioral options, i.e., by changing the motivation to engage in either reproductive or ingestive behaviors. For example, in rats, food restriction increases the tendency to eat and decreases the tendency to engage in sex behavior. It is important to note that this process does not always include laboratory measures of behavioral performance, such as the amount of food eaten in a restricted time period. Brain mechanisms that control sex behavior and eating behavior do so by altering motivational aspects of the behavior as well as performance, and the neuroendocrine mechanisms that govern motivation are only partially representative of those that determine performance. For example, in some species, metabolic signals, peripheral hormones and central neuropeptides influence food procurement behaviors without necessarily influencing the amount of food ingested (Schneider et al., 2002). In Syrian hamsters, a period of food deprivation fails to influence subsequent food intake. In fact, hamsters rarely change meal size and frequency in response to a variety of stimuli that influence these parameters in rats (Buckley and Schneider, 2001, 2003; Schneider et al., 2000), and yet, these same metabolic stimuli increase hunger motivation, as measured by the tendency to eat an unpalatable substance (DiBattista and Bedard, 1987., Schneider et al., 1988). In nature, Syrian hamsters live in burrows where they are known to hoard large quantities of food. In the laboratory, when their nest boxes are connected to artificial burrows leading to an external food source, Syrian hamsters exhibit hoarding behavior. After a period of food deprivation, Syrian hamsters show significant increases in the amount of food hoarded, and treatment with leptin during food deprivation significantly attenuates the food-deprivation-induced increase in hoarding (Buckley and Schneider, 2003). Similarly, in Siberian hamsters, hoarding behavior is increased by food deprivation (Bartness and Clein, 1994) and by central treatment with AgRP (Day et al., 2005), NPY or NPY agonists, and hoarding is decreased by treatment with NPY antagonists (Schneider et al., 2002). It would be interesting to examine whether the reverse is true; that is, when food is plentiful and energy demands are low, does the increased availability of energy, elevated concentrations of plasma leptin, and decreased central NPY secretion decrease the motivation to engage in hoarding and increase the motivation to engage in sex? These results emphasize that effects of leptin or other hormones and neuropeptides and metabolic sensory stimuli on the motivational aspects of sex behavior should be examined separately from the performance of the behaviors. Leptin treatment increases performance; that is, lordosis duration in OVX, steroid-primed hamsters fed ad libitum (Wade et al., 1997) and food deprivation-induced decreases in lordosis duration are prevented with NPY antagonists (Keen-Rhinehart and Bartness, 2007). It would be interesting to examine the motivational aspects of sex behavior, such as courtship and solicitation. It is plausible that neuropeptides, hormones and metabolic signals do not universally influence the ingestion of food or the performance of sex, but rather, they influence the motivation to engage in species-specific behaviors that ensure survival and reproductive success. So far, it appears that in hamsters, these factors influence food hoarding, which allows hamsters to ingest the food at a constant rate in the relative safety and more temperate environment of their burrow, and might also modulate fertility, sexual motivation and performance.

In summary, there is evidence that metabolic challenges inhibit reproduction at many levels, including the GnRH pulse generator, the GnRH, LH and FSH surges, and finally, by direct effects of metabolic fuel availability on the brain mechanisms that prioritize courtship, mating and ingestive behavior. This review will summarize evidence that metabolic and hormonal signals are detected in the caudal brain stem and periphery. Peripheral neural signals from the vagus nerve are relayed in the nucleus of the solitary tract (NTS), and metabolic signals are detected in the area postrema (AP) or medial NTS and relayed to hypothalamic areas, such as the paraventricular nucleus (PVN), or to the POA via the parabrachial nucleus. From the PVN, neuronal projections are purported to influence GnRH secretion via contact with GnRH axons in the Arc. GnRH neurons also receive input from the anteroventral nucleus (Gu and Simerly, 1997). Other hormones and neuropeptides might act more directly in the hypothalamus to influence GnRH neurons in the Arc or POA. This neural pathway will be discussed in detail in the Central effectors section. A system for organizing the factors that control energy balance and reproduction

Hormones inform the brain about the energetic status and reproductive state of the animal, and thereby enhance the occurrence of behavioral and metabolic adjustments that are appropriate for the environmental, reproductive and energetic conditions. For example, leptin secretion is related to adipocyte number and the flux of oxidizable metabolic fuels into adipocytes (Levy et al., 2000; Wang et al., 2007). In most cases, the animal has the option to ignore the hormonal information if more urgent energetic needs arise related to survival and reproduction. Hormones can alter or modulate the metabolic stimulus. For example, some hormones alter the availability of oxidizable fuels (the primary metabolic sensory stimulus), and thus, have indirect effects on the central effector systems. For example, leptin increases energy expenditure, thermogenesis and fuel oxidation, and thus, has the capacity to influence food intake and reproduction indirectly by making more fuels available for oxidation, thereby changing the primary sensory metabolic stimulus (Bai et al., 1996). In food-restricted ewes, pulsatile LH secretion was reinstated within an hour of refeeding and restored to the pulse frequency of fed ewes by 9 days of refeeding, despite the fact that adipose tissue and plasma leptin concentrations did not increase significantly above the levels of the food-restricted groups (Szymanski et al., 2007). Together, these results show the GnRH pulse generator and sex behavior are influenced by the minute-to-minute availability of oxidizable metabolic fuels.

According to the metabolic hypothesis, body fat content, caloric intake and energy expenditure control reproductive function by acting through a common sensory stimulus, the general availability of oxidizable metabolic fuels. Body fat can buffer against food deprivation by virtue of the fuels that can be hydrolyzed and mobilized from lipids. Factors that increase energy expenditure, such as cold exposure, can inhibit estrous cyclicity, but only when the increase in energy expenditure is not offset by utilization of oxidizable fuels mobilized from adipose tissue or by increased food intake. For example, prolonged housing at cold ambient temperatures inhibits estrous cycles in Syrian hamsters only when the increased energy expenditure is not offset by increased intake (Schneider and Wade, 1990).

4. CONCLUSION

It is evident from very brief review presented in the chapter that, there are different factors play significant roles both in regulation and modulation of reproductive energy budget. The pathway(s) involved are however still unclear and hence, in depth extensive molecular studies are warranted to encompass this issue.

ACKNOWLEDGEMENT

Financial support to SG by Council for Scientific and Industrial Research (CSIR), New-Delhi, India is gratefully acknowledged.

REFERENCES

Ansar-Ahmed, S., Penhale, WJ, and Talal, N, Sex hormones, immune responses, and autoimmune diseases, Mechanism of sex hormone action, Am J Pathol, 121, 531-551, 1985.

Armstrong, D and Britt, JH, Pulsatile administration of gonadotropin-releasing hormone to anestrous sows Endocrine changes associated with GnRH-induced and spontaneous oestrus, Biol Reprod, 5, 33-37, 1985.

Bai, YL, Zhang, SY, Kim, KS, Lee, JK and Kim, KH, Obese gene expression alters the ability of 30A5 preadipocytes to respond to lipogenic hormones, J Biol Chem, 271, 13939-13942, 1996.

Bartness, TJ and Clein, MR, Effects of food deprivation and restriction, and metabolic blockers on food hoarding in Siberian hamsters, Am J Physiol Regul Integr Comp Physiol, 266, R1111–R1117, 1994.

Bronson, FH and Heideman, PD, Seasonal regulation of reproduction in mammals, The Physiology of Reproduction, 2nd Ed. (eds E. Knobil & J.D. Neill), Raven Press, New York, 541–584, 1994.

Bronson, FH, Energy balance and ovulation: small cages versus natural habitats, Reprod, Fert Dev, 10, 127–137, 1998.

Bronson, FH, Mammalian reproduction: an ecological perspective, Biol Reprod, 32,

1–26, 1985.

Buckley, CA and Schneider, JE, Food hoarding, but not food intake is increased by food deprivation and decreased by leptin treatment in Syrian hamsters, Am J Physiol, 285, R1021-1029, 2003.

Buckley, CA and Schneider, JE, Leptin treatments that decrease food intake do not support a conditioned taste aversion in Syrian hamsters (Abstract), Horm Behav, 39, 326, 2001.

Buijs, RM, Hermes, MH and Kalsbeek, A, The suprachiasmatic nucleus paraventricular nucleus interactions: a bridge to the neuroendocrine and autonomic nervous system, Prog Brain Res, 119, 365–382, 1998.

Cameron, JL and Nosbisch, C, Suppression of pulsatile luteinizing hormone and testosterone secretion during short term food restriction in the adult male rhesus monkey (Macaca mulatto), Endocrinol, 128, 1532–1540, 1991.

Cameron, JL, Interrelationships between hormones, behaviour, and affect during adolescence: complex relationships exist between reproductive hormones, stress-related hormones, and the activity of neural systems that regulate behavioural affect, Comments on part III, Ann N Y Acad Sci, 1021, 134-42, 2004.

Crump, AD, Lomax, MA and Rodway RG, Oestradiol- induced luteinizing hormone CLH) release is inhibited by 2-deoxyglucose infusion in sheep, J Physiol, 330, 94, 1982.

Day, DE, Keen-Rhinehart, E and Bartness, TJ, Role of NPY and its receptor subtypes in foraging, food hoarding, and food intake by Siberian hamsters, Am J Physiol Regul Integr Comp Physiol, 289, R29–R36, 2005.

DiBattista, D and Bedard, M, Effects of food deprivation on hunger motivation in golden hamsters *(Mesocricetus auratus)*, J Comp Psychol, 101, 183–189, 1987.

Ebling, FJP, Alexander, IHM, Urbanski, HF and Hastings, MH, Effects of N-methyl-D-aspartate (NMDA) on seasonal cycles of reproduction, body weight and pelage colour in the male Siberian hamster, J Neuroendocrinol, 7, 555–566, 1995.

Foster, DL and Bucholtz, DC, Metabolic signals and the timing of puberty in sheep, In The Neurobiology of PubertyEds, TM Plant and PALee, Society for Endocrinol, Bristol, 243–257, 1995.

Gill, CJ and Rissman, EF, Female sexual behaviour is inhibited by short and long-term food restriction, Physiol Behav, 61, 387-394, 1997.

Gu, GB and Simerly, RB, Projections of the sexually dimorphic anteroventral periventricular nucleus in the female rat, J Comp Neurol, 384, 142-164, 1997.

Keen-Rhinehart, E and Bartness, TJ, NPY Y1 receptor is involved in ghrelin- and fasting-induced increases in foraging, food hoarding, and food intake, Am J Physiol Reg Integr Comp Physiol, 292, R1728–R1737, 2007.

Levy, JR, Gyarmati, J, Lesko, JM, Adler, RA and Stevens, W, Dual regulation of leptin secretion: intracellular energy and calcium dependence of regulated pathway, Am J Physiol Endocrinol Metab, 278, E892-E901, 2000.

Loucks, AB, Heath, EM, Verdun, M and Watts, JR, Dietary restriction reduces luteinizing hormone (LH) pulse frequency during waking hours and increases LH pulse amplitude during sleep in young menstruating women, J Clin Endocrinol Metab, 78, 910–915, 1994.

Machen, R, Nutrient Requirements of Goats: Angora, Dairy, and Meat Goats in Temperate and Tropical Countries, 6th Ed, Washington, DC, National Academy Press, 2–4,1981.

McShane, TM, May T, Miner, JL and Keisler, DH, Central actions of neuropeptide-Y may provide a neuromodulatory link between nutrition and reproduction, Biol Reprod, 46, 1151-1157, 1992.

Medina, CL, Nagatani, S, Darling, TA, Bucholtz, DC, Tsukamura, H, Maeda, K and Foster, DL, Glucose availability modulates the timing of the luteinizing hormone surge in the ewe, J Neuroendocrinol, 10, 785–792, 1998.

Medina, VM, Dawson-Basoa, ME and Gintzler, AR, 17 beta-estradiol and progesterone positively modulate spinal cord dynorphin: relevance to the analgesia of pregnancy, Neuroendocrinol, 58, 310–315, 1993.

Misra, M, Bone density in the adolescent athlete, Rev Endocr Metab Disord,9, 139–144, 2008.

Nelson, RJ, Demas, GE, Klein, SL and Kriegsfeld, LK, The influence of season, photoperiod, and pineal melatonin on immune function, J Pineal Res, 19, 149–165, 1995.

Nelson, RJ, Demas, GE, Klein, SL and Kriegsfeld, LK, The influence of season, photoperiod, and pineal melatonin on immune function, J Pineal Res, 19, 149–165, 1995.

Qarke, JJ and Cummins, JT, The temporal relationship between gonadotropic releasing hormone (GnRH) and luteinizing hormone (LH) secretion in ovariectomized ewes, Endocrinol, 111, 1737-1739, 1982.

Schneider, JE and Wade, GN, Effects of diet and body fat content on cold-induced anoestrusin Syrian hamsters, Am J Physiol, 259, R1198-R1204, 1990.

Schneider, JE, Blum, RM and Wade, GN, Metabolic control of food intake and estrous cycles in Syrian hamsters. I Plasma insulin and leptin, Am J Physiol Reg Integr Comp Physiol, 278, R476–R485, 2000.

Schneider, JE, Buckley, CA, Blum, RM, Zhou, D, Szymanski, L, Day, D and Bartness, TJ, Metabolic signals, hormones and neuropeptides involved in the control of energy balance and reproductive success in hamsters, Eu J Neurosci, 16, 377–379, 2002.

Schneider, JE, Lazzarini, SJ, Friedman, MI and Wade, GN, Role of fatty acid oxidation in food intake and hunger motivation in Syrian hamsters, Physiol Behav, 43, 617-623, 1988.

Szymanski, LA, Schneider, JE, Friedman, MI, Kurose, JHY, Blache, D, Rao, A, Dunshea, FR and Clarke, IJ, Changes in insulin, glucose and ketone bodies, but not leptin or body fat content precede restoration of luteinising hormone secretion in ewes. J Neuroendocrinol, 19, 449–460, 2007.

Temple, JL and Rissman, EF, Acute refeeding reverses food-restriction induced hypothalamic-pituitary-gonadal axis deficits, Biol Reprod, 63, 1721-1726, 2000.

Wade, GN and Schneider, JE, Metabolic fuels and reproduction in female mammals, Neurosci Biobehav Rev, 16, 235–272, 1992.

Wade, GN, Lempicki, RL, Panicker, AK, Frisbee, RM and Blaustein, JD, Leptin facilitates and inhibits sexual behaviour in female hamsters, Am J Physiol Reg Integr Comp Physiol, 272, R1354–R1358, 1997.

Wang, C, Mao, X, Wang L, Liu, M, Wetzel, MD, Guan, Kl, Dong, LQ and Liu, F, Adiponectin sensitizes insulin signalling by reducing p70 S6 kinase-mediated serine phosphorylation of IRS-1, J Biol Chem, 282, 7991-7996, 200.

❐

Chapter-10

EXPERIMENTS WITH GOATS: LIMITATIONS AND CHALLENGES

Somenath Ghosh[1], Sunil Kumar Rai[2], Ravi S. Pandey[1] and Chandana Haldar[3*]

[1] *Biochemistry Laboratory, Department of Zoology, University of Allahabad, Allahabad-21102, India.*

[2]*Post-Doctoral Fellow, Department of Biochemistry and Molecular Medicine, School of Medicine, University of California, Davis, CA, USA*

[3]*Pineal Research Laboratory, Department of Zoology, Banaras Hindu University, Varanasi-221005, India.*

**Email of Corresponding Author: chaldar2011@gmail.com*

ABSTRACT

Goat farming is one of the most ancient house hold business for poor and rural persons around the globe. Goats are reared for their meat, milk and fur purposes for house hold as well as trading purposes. Goat milk is so nutritive that it is called as "Swiss-Baby's Foster Mother" in Sweden. In India, goats are reared in almost all rural families due to similar purposes. Hence, goats are called as "Poor Man's Cow" in India. The main problem goat farming in India is their reduced natality rate. To prevent the immature death of kids, goats are generally cross-bred and thus, the purity of breed is lost. Lack of purity in breeds is the key research problem while experimenting on goats. Due to peculiar reproductive nature and reduced natality goats are emerging as new experimental model not only for veterinary clinicians but also for scientists and even for immunologists also. Research on goats is beneficial also from socio-economic point of view. In present chapter we will address only those problems which are mainly associated with the rearing and caring of goats taking their purity of breed and socio-economic importance in consideration.

Keywords: Challenges; Experiments; Goats; Limitations.

1. INTRODUCTION

Vaccinations are an integral part of the livestock health management program. They provide cheap insurance against diseases that commonly affect sheep and goats. Probably, the only universally recommended vaccine for goats is CD-T. CD-T toxoid that provides three-way protection against enterotoxaemia (overeating disease) caused by *Clostridium perfringins* types C and D and tetanus (lockjaw) caused by *Clostridium tetani*. Seven and 8-way combination vaccines for additional *clostridia* diseases such as blackleg and malignant edema are available, but generally not necessary for small ruminants.

2. VACCINATION IN GOATS

For betterment of does and to provide passive immunity to their off springs through the colostrums; does are vaccinated 2 to 4 weeks prior to parturition. However, females giving birth for the first time are vaccinated twice in late pregnancy, about four weeks apart. Maternal antibodies will protect kids for about two months, if offspring have ingested adequate colostrums. Kids should receive their first CD-T vaccination when they are 6 to

8 weeks old followed by a booster 2 to 4 weeks later. If pastured animals are later placed in a feed lot for concentrate feeding, producers should consider re-vaccinating them for enterotoxaemia type D. If the kids whose mother was not vaccinated for C and D, kids are required to be vaccinated with some success at two to three days of age and again in two weeks. However, vaccinations in later age will be more successful since during much young stage, cloistral antibodies interfere with vaccinations and get failed. A better way is to vaccinate the offspring from non-vaccinated dams at 1 to 3 weeks, with a booster 3 to 4 weeks later. Anti-toxins can provide immediate short-term immunity if dams were not vaccinated or in the event of disease outbreak or vaccine failure. For the sore mouth (contagious etyma), a viral skin disease commonly affecting sheep and goats, live vaccines are effective. Since vaccines against sore and mouth diseases are live vaccine, handling must be proper with use of gloves. Foot rot and foot scald is one of the most ubiquitous diseases in goat. It causes considerable economic loss due to the costs associated with treating it and the premature culling of affected animals. Foot rot vaccines should be administered every 3 to 6 months and especially prior to anticipated outbreaks of hoof problems (i.e. prior to the wet/rainy season). Abortion is when a female loses her offspring during pregnancy or gives birth to weak or deformed babies. There are vaccines (individual and combination) for several of the agents that cause abortion in sheep: enzootic *(Chlamydia sp.)* and vibriosis (*Campylobacter fetus*). Abortion vaccines should be administered prior to breeding. Risk factors for abortion include an open flock and a history of abortions in the flock. Since the disease-causing organism is carried by domestic cats, the best protection is to control cat population by and keeping cats from contaminating feed sources. Rabies is one of the deadly diseases though the risk to goats is usually minimal and hence, rabies vaccination may be considered if the flock is located in a rabies belt. The cost of the rabies vaccine relative to the value of the animals should be considered as well. The large animal rabies vaccine has been approved to use for sheep but currently there is no licensed rabies vaccine for goat. Hence, the best way is to vaccinate all the dogs and cats. In order for vaccination programs to be successful, label directions must be carefully followed and vaccines need to be stored, handled, and administered properly. Only healthy livestock should be vaccinated. It is also important to note that vaccines have limitations and that the immunity imparted by vaccines can sometimes by inadequate or overwhelmed by disease. Scientists are currently working on vaccines to protect small ruminants against worms.

3. LIMITATION OF IMMUNIZATION IN GOATS

In India and almost all over the world the most extensive study has till date being conducted with special emphasis on immunization in ruminants and goats in particular. Thus, the veterinary significance of goat immunity can be regarded as partially explored. But, in rural India, goat farmers are not familiar with the vaccination against diseases and hence the goats' mortality is quite high due to several lethal bacterial and viral diseases along with pneumonia in various forms (particularly pulmonary adenomatosis). Enterotoxaemia, anthrax, pneumonia, clostridia diseases and lumbar paralysis in goats result in high mortality. Most mortality is due to liver-fluke infestation due to which animals become quite weak and died of gastro-intestinal problems. There are several diseases that cause their untimely death in goats. Thus, it is need of time, context and obviously of economy to explore out the most details of immune organs of goats and to pin point the factors which regulates the multi-factorial regulation of open circuit of immune system.

4. GOATS AS AN EXPERIMENTAL MODEL

Goats are large mammals which have many similarities to humans in terms of physiology. They are easy to handle and suffer from many diseases which affect humans. They also have short gestation periods yet give birth to young of a similar weight to human babies, making them excellent for studying development and genetics. They are also used extensively in veterinary research, studies of digestion in ruminants, and research on the impact of farming on the environment. Goats are frequently used as a model for cattle and other large mammals, as they are smaller, less expensive to keep and easier to breed.

5. IDENTIFICATION OF THE ANIMAL

Identification:

Each goat in a herd should be marked in the same manner by using some identification, such as tattooing, metal labels or notching the ears. With the tattoo method, combinations of four letters or numbers can be used. Before applying the tattoo ink, the ear must be cleaned with soapy water.

Branding:

It involves shearing a number, letter, design or a combination of these on the skin with a hot iron or chemicals. It is most commonly applied on the horse, camel, cattle and buffaloes.

Tattooing:

It consists of piercing the outlined of desired number or letters on the skin inside the ear and then incorporating a black vegetable pigment into these punctures.

Tagging:

Tags or the labels are made of the light metal or strong plastic with the number stamped on them and then fixed generally to the ear with a special tagging forceps.

Dentition and ageing of the animal

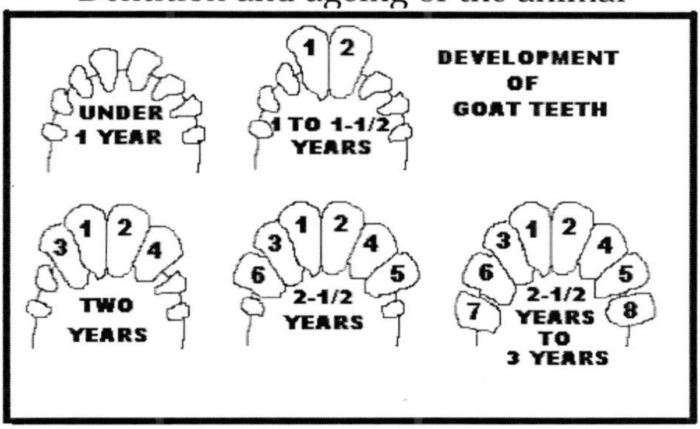

(Photo courtesy: www.google.co.in)

The approximate age of the animal can be determined within the certain limits by the teeth. One can note the time of appearance and the degree of wear of the temporary and permanent teeth. In goats, by the first year the full set of temporary incisors are present. Six molars are present in each jaw.

At the age of 14 months, the two central temporary incisors are shed and are soon replaced by two permanent incisors which are broader than that of the previous one. The second, third and the corner pair are replaced by the permanent teeth by 3rd, 4th and 5th years of the age respectively. The goat ages rapidly after 5 years and both incisors and molars become more worn (Fandos et al., 1993).

6. CONCLUSION

From the small review we may say that limitations in goat research do exists in goat research but experiments on this direction is on its way for soon relief.

ACKNOWLEDGEMENT

Financial support to SG by Council for Scientific and Industrial Research (CSIR), New-Delhi, India is gratefully acknowledged.

REFERENCES

Fandos P, Orueta JF and Olanda A, Tooth wear and its relation to kind of food: the repercussion on age criteria in Capra pyrenaica, Acta Theriol, 38, 93-102, 1993. *www.google.co.in*

Chapter-11

SYNTHESIS, CHARACTERIZATION AND EVALUATION OF PHYSIOLOGICAL IMPACTS OF MELATONIN LOADED POLY (D, L-LACTIC ACID) NANO-PARTICLES (MEL-PLA-NANO-PARTICLES) AND ITS PUTATIVE USE IN OSTEOPOROSIS.

Somenath Ghosh[1], Sunil Kumar Rai[2] and Ravi S. Pandey[1*]

1Biochemistry Laboratory, Department of Zoology, University of Allahabad, Allahabad-21102, India.

[2] Post-Doctoral Fellow, Department of Biochemistry and Molecular Medicine, School of Medicine, University of California, Davis, CA, USA

**Email of Corresponding Author: rspandey2004@yahoo.com*

ABSTRACT

Melatonin loaded PLA-Nano particles are nowadays important in biological system for it's biodegradable nature in the mammalian system for systemtic/targetted drug delivery system. The aim of the present study is note the effect of Mel-PLA-Nano particles and toxicity in osteoprosis which is a lesser explored area of research. Different concentrations of commercially available melatonin and PLA were prepared by dissolving in dichloromethane (DCM). The final dried nano-paticles were stored at 4°C for future use and a part was used for structural analysis by Scanning Electron Microscopy (SEM), Transmission Electron Microscopy (TEM), Fourier Transform Infrared Spectroscopy (FTIR). Further, to evaluate the toxicity and immunological impact of nano-particles on rats; animals were divided into 2 groups (control and nano-particle treated, n=5/group) for 7 days. After treatment animals were sacrificed and blood, liver and kidney were collected. A fraction of blood was processed for assessment of cell mediated immune parameters (TLC, DLC and % LC) and remaining blood was centrifuged at 3000 x g at 4°C for 30 min. Plasma was seperated and used for evaluation of cytokines, (IL-2, IL-6, TNF-α, IFN-γ, IL-1β), makers of renal function test (urea, creatinine and BUN). Both plasma and tissue homogenates were used for estimations of AST, ALT, ACP and ALP. The nano-particle size was evaluated by SEM and TEM and was further confirmed by FTIR and in vitro controlled drug release efficasy. We noted signifacntly high ($p < 0.05$) levels of cell mediated immune parameters (TLC, DLC and %LC) and cytokine parameters (IL-2 and TNF-α) upon nano-particle treatment. But rest of the parametrs were found to be significantly low (like IL-6, IL-1β, AST, ALT; $p > 0.05$) or un-affected (like IFN-γ, ACP and ALP). From our preliminary study we may conclude that we have successfully synthesized Mel-PLA-nano-particles and their physilogical effects were non-toxic to the animal.

Keywords: Mel-PLA-nano-particles, Characterization, Immune effect, Synthesis, Toxicity.

Abbreviations:

ACP: Acid Phosphatase; AFM: Atomic Force Microscopy; ALP: Alkaline Phosphatase; ALT: Alanine Aminotransferase; AST: Aspartate Aminotransferase; DCM: Dichloromethane; DSC: Differential Scanning Calorimetry FTIR: Fourier-transform Infrared Spectroscopy; IL-1β: Interleukin-1 Beta; IL-2: Interleukin-2; IL-6: Interleukin-6;

IFN-γ: Interferon Gamma; LFT: Liver Function Test; RFT: Renal Function Test; Mel: Melatonin; PLA: Poly Lactic Acid; SEM: Scanning Electron Microscopy; TEM: Transmission Electron Microscopy; TNF-α: Tumor Necrosis Factor-Alpha.

1. INTRODUCTION

Melatonin (N-acetyl-5-methoxytryptamine) is a hormone produced and released by the pineal gland, which has potential to regulate neuro-immune-modulation, anti-cancerous, apoptosis and cell-division by scavenging free radicals (Tan et al., 2007; Hardeland et al., 1993). The immune system being an open and multidirectional system has greatly evolved to interact with the endocrine system to combat with the environmental threats. Recently the important role of melatonin has been carried out into the area of neuro-endocrine-immunology in reference to bone deformities (Maestroni et al., 2002). Melatonin receptors are present on the Lymphocytes (Drazen and Nelson, 2001). This hormone is well established in the regulation of apoptosis and cell-division by enhancing the secretion of mitogens (Xu et al., 2007). The effect of melatonin as an anti-oxidant and anti-apoptotic had been proved in a seasonal breeder *Funambulus pennanti* (Ahmad & Haldar, 2010). The potential clinical benefit of the use of melatonin as an anti-oxidant and anti-apoptotic will be helpful for treatment of many degenerative diseases, such as osteoporosis (a bone degenerative disease) and Alzheimer (a neurodegenerative disease). Considering the important role of melatonin, a proposal has planned to encapsulate the melatonin with biodegradable polymeric nanoparticles and its controlled delivery with desired release kinetics. Biodegradable polymeric nanoparticles have been extensively used with great interest in the area of nano-biotechnology as delivery systems for active molecules and drugs for therapeutic use (Pandey et al., 2013), due to their controlled and sustained-release properties, and biocompatibility with tissue and cells (Cheng et al., 2007; Kim and Kang, 2008). The current chapter is based on the controlled release of melatonin by encapsulating with biodegradable polymeric nano-particles. The polymeric nanoparticles have the potential to act as a carrier of active molecules and drugs at target sites, protecting them from physiological environment and increase their biological activity. So we have planned a Melatonin and melatonin loaded polymeric biodegradable nano-particles in regulation of osteoporosis and bone deformities in rodents.

Why polymeric nano-particles?

The polymeric composition (hydrophobicity, surface charge, and biodegradation profile) of the nano-particles, any adjuvant substances, and the associated active molecules and drugs (molecular weight, charge, localization in the nano-spheres by adsorption or incorporation) have a great influence on the active molecules and drugs absorption, bio-distribution pattern and elimination. The polymeric nano-particles technology used in the recent years has great significance in improving the efficacy of the active molecules and drugs. Nano-particles can be prepared from a variety of biodegradable polymers such as synthetic as well as natural polymers. The choice of materials depends on several factors including (i) size and morphology of the nano-particles (ii) surface charge and permeability of the nano-particle (iii) degree of biodegradability, biocompatibility and cytotoxicity (iv) drug loading and desired release profile.

Rationale of the study:

The main aim of this proposal is to improve the osteoporosis conditions (either post-oestrous or induced) in rodents by developing melatonin loaded polymeric nano-particles and their controlled delivery in rodents as requirement. (i) Osteoporosis is not only an indigenous problem but also it is well prevalent throughout the world. Except for Ca^{+2} supplementation there is no cure is suggested for the treatment of this clinical condition. (ii) Till date, age factor and menopause are regarded as main cause of osteoporosis particularly in humans. But, in this regard, further literature and clinical investigations are completely lacking. (iii) The improvement of osteoporosis by controlled delivery of melatonin loaded nano-particles is the most untouched area of research and during the literature survey we have found very few literatures.

2. CURRENT STATUS OF RESEARCH

National status:

In recent years, rapid scientific and technological advancement have been made in controlled release of active molecules and drugs using biodegradable polymer based nano-particles. To overcome the undesirable side effects, and to protect the efficacy of active molecules and drugs from physiological environment, controlled delivery approach is necessary in order to get optimum therapeutic results. At present, scientists are involved to develop biodegradable polymeric nano-particles by using various biodegradable polymers such as poly- (lactic-co-glycolic acid; PLGA), poly (D,L-lactic acid; PLA), poly (ε-Caprolactone; PCL), chitosan gelatin, and poly-alkyl-cyanoacrylates (Gou et al., 2009; Zheng, et al., 2010; Jain et al., 2006; Kumari et aI., 2011; Pandey et al., 2015) for target delivery of active molecules and drugs related to cancer, diabetes, malaria and other harmful diseases. Our lab is engaged in exploring immunomodulatory role of melatonin considering the bone marrow functioning in rodents and birds (Vishwas et al., 2013, 2014). Recently we have published our findings in direction of melatonin's immune protection because of its anti-oxidative and anti-apoptotic action via inhibiting caspase-3 activity in spleen lymphocytes of seasonal breeders under radiation induced stress (Sharma et al. 2008) and also by bone marrow functioning (Rastogi et al., 2015; personal communication). In India very few groups are involved in bone deformity study (Compston, 2002). The reasons behind the mechanism of melatonin action in bone anomalies (as suggested by Jager and Kuchroo, 2010) have never been explored at National/International level till date. To the best of our knowledge there is no any information available in India concerning synthesis of melatonin loaded polymeric nano-particle and its delivery approach in comparison to melatonin except for the report of Pandey et al., 2015.

International status:

The role of melatonin in execution of different physiological functions like modulation of reproduction (Haldar et al., 1992), immunity (Reiter et al., 2000) is very much well evident. Reports are available on melatonin functioning in bone marrow functioning in modulation of immunity (Vishwas et al., 2013). But, unfortunately, till date a single literature is not available even at international level, regarding the therapeutic use of melatonin or melatonin rich compound in therapeutic use of melatonin in bone deformities study. Further, internationally, very few groups are working in the area of melatonin loaded polymeric nano-particles. Melatonin loaded polysorbate 80-coated Eudragit S100 nanopanicles provided an increase in the m-eitro effect of melatonin against lipid peroxidation in comparison to the melatonin in aqueous solution (Schaffazick et al., 2005). Hafner et al (2009) developed melatonin loaded lecithin/chitosan nano-particles for the transmucosal delivery of melatonin and observed that nano-particles enhance melatonin transport across the epithelial barrier.

3. HYPOTHESIS

It is evident from previous literature (Tamini et al., 2008) that melatonin may impair bone softening by activating Bone Morphogenic Protein 15 (BMP-15). In India (about 50%) and almost 45% population of post-menopausal women of the world are suffering from impaired bone functioning and Fragile Bone Disease (FBD; WHO, 2009). This is due to low level of estrogen in circulation which in turn affects the Ca^{+2} absorption from blood. Further, the cesarean deliveries of mothers also lead to impaired bone functions just after menopause. Till date, there is no literature available regarding the proper management of bone fragility in females particularly after menopause except for some commercially available Ca^{+2} supplementation. Taking this background the present dissertation is divided into different parts which are in progress. Since, the studies are underway (unpublished data) we will be providing the glimpses of results in terms of levels of significance wherever necessary.

4. EXPERIMENTAL APPROACHES

To investigate the possibility of developing and optimization of all conditions for preparations of melatonin loaded polymeric nano-particles:

Depending on the physicochemical characteristics of melatonin it is essential to choose the polymer and the method of preparation to achieve an efficient entrapment and controlled delivery of the melatonin. Emulsion polymerization is one of the fastest methods for nano-particle preparation and is readily, scalable. The method is based on the use of on organic or amt.., continuous phase. The continuous organic phase methodology involves the dispersion of monomer into an emulsion or inverse micro-emulsion. During preparation of nano-panicles, surfactants or protective soluble polymers may be used to prevent aggregation in the early stages of polymerization. To avoid toxicity, it is necessary to eliminate the organic solvents and surfactants from the prepared nano-panicles. Emulsification-solvent evaporation method may be used for emulsification or the polymer solution into an aqueous phase and evaporation of the organic solvent by inducing polymer precipitation as nano-spheres. The solvent is subsequently, evaporated by increasing the temperature under pressure with continuous stirring. This can be controlled by adjusting the stir rate, type and amount of dispersing agent, viscosity of organic and aqueous phases, and temperature.

Nano-particles preparation:

The most suitable biodegradable polymer based nano-particles containing melatonin was prepared by emulsification and solvent evaporation/nano-precipitation method. Briefly, known amounts polymer and melatonin was added into the organic solvent (based on the nature of dispersed phase) and co-dissolved by stirring. The resulting, organic solution was added drop wise into the aqueous phase containing emulsifier under high speed stirring. Nano-particles suspension was collected using an ultra-centrifuge (1, 50, 000 g, 30 min, 4^0C) and washed with distilled water at least three times. Organic solvent was removed from nano-particles suspension by the rotor evaporation technique under reduced pressure at 40^0 C. and finally dried powder of nano-particles obtained. The final dried nano- particles was stored at 4^0 C until further use.

Nano-particles characterizations:

Characterization and evaluation of particle size, morphology, structure and stability of melatonin loaded nano-particles was carried out by using SEM, AFM, FTIR and particle size analyzer. Thermal characterization of pure polymeric nano- particles, melatonin and melatonin loaded nano- particles was carried out using differential scanning calorimetry (DSC).

Evaluation of toxicity of nano-particles in animal models:

Administration of melatonin loaded PLA nanoparticles in animals:

After synthesis of melatonin loaded PLA nanoparticles the therapeutic efficacy toxicity (if any) of nanoparticles were evaluated in rats. Animals were divided into two groups (control and nanoparticle treated; n=5/group) with ad libitum access to commercially available animal palette and water. In the control group, animals were administrated with normal saline where the treated animals received melatonin loaded PLA nanoparticles. After 7 days, animals were sacrificed and blood and desired tissues (liver and kidney) were collected and were homogenized in respective buffers and were processed for evaluation of different immunological parameters and evaluation of toxicity in terms of Liver Function Test (LFT) and Renal Function Test (RFT).

5. RESULTS AND DISCUSSION

Evaluation of toxic impacts of melatonin loaded PLA nanoparticle:

Any kind of foreign chemical introduced in body can evoke different physiological side-effects which are mainly of two types. 1. Toxic effects. 2. Immunological side effects. Thus, to evaluate the same we assessed the LFT and RFT in both control and nanoparticle treated groups.

Effect on Liver Function Test (LFT):

Serum Glutamic oxaloaceate Transaminase (SGOT or Aspartate Aminotransferase; AST) and Serum Glutamic Pyruvic Transaminase (SGPT or Alanine Aminotransferase) are stored in liver under normal conditions. But, upon clinical conditions where the liver is injured (which may be induced due to any kind of drug treatment or due to alcoholic shock) the level of both SGOT and SGPT may be elevated. Thus, assessment of SGOT and SGPT are regarded as the universal markers of liver function. Upon nanoparticle treatment we noted significantly low levels ($p > 0.05$) of SGOT and SGPT as compared to control in plasma and liver. This provided us the first clue that the nanoparticle treatment is not going to affect the most biochemically efficient organ of the body; i.e. the liver. Further, we were interested in evaluating the functioning of another physiologically important organ i.e. the kidney which was assessed by Renal Function Test (RFT).

Effect on Renal Function Test (RFT):

Kidney is one of the most important organs of body actively and effectively participating in detoxification. Any kind of drug administration may hamper renal function as well as glomerular filtration rate (GFR), hence our next aim was to investigate the Renal Function Test (RFT) by assessing plasma urea, Blood Urea Nitrogen (BUN), creatinine, Alkaline Phosphatase (ALP) and Acid Phosphatase (ACP). We noted significantly low levels ($p > 0.05$) of plasma urea, BUN and creatinine levels in nanoparticle treatment groups; however the ACP and ALP levels were found to be unaffected as compared to control. Thus, our results have suggested that the nanoparticle treatment neither had negative impact on liver function nor had negative effect on kidney function. But, any kind of drug treatment (including the nanoparticles) can evoke immunological anomaly in body. Hence, our next aim was to investigate the status of cell mediated and humoral immune parameters upon nanoparticle treatment.

Evaluation of immunological effects of melatonin loaded PLA nanoparticles:

(i) ***Effect on Cell mediated Immune Parameters :*** The cell mediated immune parameters are the first marker for any kind of immunogenic or hypersensitivity reaction. To assess the same, we noted significantly high ($p < 0.05$) levels of Total Leukocyte Count (TLC), Differential Leukocyte Count (DLC) and % Lymphocyte Count (%LC) upon nanoparticle treatment as compared to control. These results suggested that, the elevated levels of peripheral cellular components of immunity may be due to some kind of hypersensitivity reaction because of nanoparticle treatment. To investigate the same, we noted humoral immune parameters i.e. cytokines.

(ii) ***Effect on Humoral Immune Parameters:*** The open circuit of blood immune parameters is chemically coordinated by lymphokines, chemokines and cytokines. Among them, the most effective are the cytokines which are producing the broadest spectrum coordinating a number of immunologically active cells. The cytokines are mainly of three types pro-inflammatory (e.g. IL-2), anti-inflammatory (e.g. IL-6) and switch between pro-and anti-inflammation (e.g. TNF-α). We noted significantly high ($p < 0.05$) levels of IL-2 and TNF-α and significantly low ($p > 0.05$) levels of IL-6 and IL-1β (a cytokine directly associated with plasma cortisol) in nanoparticle treatment group in comparison to control. Further, IFN-γ (a particular cytokine marker for viral infection) was fond unaffected both in control and nanoparticle treatment group.

6. CONCLUSION

From our preliminary study, we may conclude that we have successfully synthesized Melatonin loaded PLA-nanoparticles which are neither having toxic (in terms of LFT and RFT) nor having immunological (in terms of cell mediated and humoral) side effects. Thusa our preliminary work is based on the synthesis and optimization of melatonin loaded polymeric nano-particles with desired panicle size, increased entrapment efficiency, and controlled release kinetics. The optimized formulation will be applied for treatment of constitutive and induced bone anomalies as well as a special focus on osteoporosis in rodents with relevance to post menopausal mothers as they are backbone of our family and society.

ACKNOWLEDGEMENT

Financial support to SG by Council for Scientific and Industrial Research (CSIR, New Delhi, India) as Research Associate (RA) is gratefully acknowledged.

REFERENCES

Ahmad R, Haldar C. (2010). Melatonin and androgen receptor expression interplay modulates cell-mediated immunity in tropical rodent *Funambulus pennanti:* an *in-vivo* and *in-vitro* study. Scan J Immunol. 71: 420-430.

Compston JE. (2002). Bone marrow and bone: a functional unit. J Endocrinol. 173: 387-394.

Cheng J, Teplya BA, Sherifia I et al. (2007). Formulation of functionalized PLGA-PEG nanoparticles for *in vivo* targeted drug delivery. Biomat. 28: 869–876.

Drazen DL, Nelson RJ. (2001). Melatonin receptor subtype MT2 (Mel 1b) and not MT1 (Mel 1a) is associated with melatonin-induced enhancement of cell mediated and humoral immunity. Neuroendocrinol. 74: 178–184.

Gou M, Zheng X, Men K et. al. (2009). Poly (epsilon-caprolactone)/poly (ethyleneglycol)/poly(epsilon-caprolactone nanoparticles: preparation, characterization, and application in doxorubicin delivery. J Phys Chem B. 113:12928-12933.

Hafner A, Lovrić J, Voinovich D et al. (2009). Melatonin-loaded lecithin/chitosan nanoparticles: physicochemical characterisation and permeability through Caco-2 cell monolayers. Int J Pharm. 381: 205-213.

Haldar C, Shavali SS, Singh S. (1992). Photoperiodic response of pineal-thyroid axis of the female Indian palm squirrel, *Funambulus pennant.* J Neural Transm Gen Sect. 90: 45-52.

Hardeland R, Reiter RJ, Poeggeler B et al. (1993). The significance of the metabolism of the neurohormone melatonin: antioxidative protection and formation of bioactive substances. Neurosci Biobehav Rev. 17: 347–357.

Jäger A, Kuchroo VK. (2010). Effector and regulatory T-cell subsets in autoimmunity and tissue inflammation. Scand J Immunol. 72: 173-184.

Jain PK, Lee KS, El-Sayed IH et al. (2006). Calculated absorption and scattering properties of gold nanoparticles of different size, shape, and composition: applications in biological imaging and biomedicine. J Phys Chem B. 110: 7238-7248.

Kim BG, Kang IJ. (2008). Evaluation of the effects of biodegradable nanoparticles on a vaccine delivery system using AFM, SEM, and TEM. Ultramicroscopy. 108: 1168–1173.

Kumari A, Yadav SK, Pakade YB et al. (2011). Nanoencapsulation and characterization of Albizia chinensis isolated antioxidant quercitrin on PLA nanoparticles Colloids Surf B. Biointerfaces. 82: 224-232.

Maestroni GJ, Sulli A, Pizzorni C et al. (2002). Melatonin in rheumatoid arthritis: synovial macrophages show melatonin receptors. Ann N Y Acad Sci. 966: 271–275.

Pandey SK, Ghosh S, Maiti P, et al. (2015) Therapeutic efficacy and toxicity of tamoxifen loaded PLA nanoparticles for breast cancer. Int J Biol Macromol. 72: 309-319.

Pandey SK, Haldar C, Patel DK et al. (2013). Biodegradable polymers for potential delivery systems for therapeutics. Adv Polym Sci. 254: 169–202.

Reiter RJ, Tan D, Osuna C et al. (2000). Actions of melatonin in the reduction of oxidative stress. A review. *J Biomed Sci.* 7: 444-458.

Sharma S, Haldar C, Chaube SK. (2008). Effect of exogenous melatonin on X-ray induced cellular toxicity in lymphatic tissue of Indian tropical male squirrel *Funambulus pennant*. Int J Radiat Biol. 84: 363-374.

Schaffazick SR, Pohlmann AR, de Cordova CA et al. (2005). Protective properties of melatonin-loaded nanoparticles against lipid peroxidation. Int J Pharm. 289 (1-2): 209-213.

Tamimi F, Torres J, Kathan C et al. (2008). Bone regeneration in rabbit calvaria with novel monetite granules. J Biomed Mater Res A. 87: 980-985.

Tan DX, Manchester LC, Terron MP et al. (2007). One molecule, many derivatives: a never-ending interaction of melatonin with reactive oxygen and nitrogen species? J Pineal Res. 42: 28–42.

Vishwas DK, Mukherjee A, Haldar C. (2013). Improvement of oxidative stress and immunity by melatonin: An age dependent study in golden hamster. Exp Gerontol. 48: 168–182.

Vishwas DK, Mukherjee A, Haldar C. (2014). Melatonin improves humoral and cell-mediated immune responses of male golden hamster following stress induced by dexamethasone. J Neuroimmunol. 259: 17-25.

World Health Organization. (2009). Annl Rep. (www.who.org. retrieved on 01/03/2018).

Xu L, Kitani A, Fuss I et al. (2007) Cutting edge: regulatory T cells induce CD4+CD25-Foxp3-T cells or are self-induced to become Th17 cells in the absence of exogenous TGF-beta. J Immunol. 178: 6725–6729.

Zheng X, Kan B, Gou M et al. (2010) Preparation of MPEG-PLA nanoparticle for honokiol delivery *in vitro*. Int J Pharm. 386 (1-2): 262-267.

❐

Chapter-12

NANOTIZED BIOCOMPATIBLE ANTICANCEROUS COMPOUND(S) AND THEIR IMPLICATION TO PERSONALIZED NANOMEDICINE

Neera Yadav[1], Shama Parveen[1] and Monisha Banerjee[1]

[1]*Molecular and Human Genetics Lab, Department of Zoology, University of Lucknow, Lucknow*
Email of Corresponding Author: monishabanerjee30@gmail.com

ABSTRACT

Nanoparticles have been in use of medical sciences since ancient times in India as an integral part of Ayurveda. Earlier, gold and silver nanoparticles were in use but later, scientists tried to prepare other metallic and synthetic nanoparticles. These particles are of sub-micron size ranging from 1-1000 nm. It was observed that metallic nanoparticles other than silver and gold were not that much suitable for medical applications as the compounds produce toxicity inside the body of the patient. It was need of the time to search for biocompatible materials to be used in place of harmful metals. Gradually, other non-toxic materials such poly (d, l-lactide-co-glycolic) acid (PLGA), polyglycolic acid (PGA), polylactic acid (PLA), polyethylene glycol (PEG) and poly-l-lysine (PLL) were used to prepare nanoparticles. These compounds are colloidal in nature containing an active drug encapsulated within the polymer with no or less toxicity. The detection and monitoring of target expression before therapy initiation and during the whole treatment will help more in disease diagnosis of specific individual. Ligand-functionalized therapeutic nanomaterials are new hope for individualized treatment of diseases.

Keywords: Nanoparticles, Biocompatible, Cervical cancer, Personalized nano-medicine

1. INTRODUCTION

Cervical cancer is the third most common cancer with about one lac new cases in 2016 and estimated to be about 1.04 lacs during 2020. It is common with an annual incidence of more than 4,50,000 cases worldwide and is the third most common cause of cancer related death in females. The high mortality could be the cause of the fact that most cervical cancer cases are diagnosed at advanced stage. Cervical cancer is caused by human papilloma viruses (HPVs). There are two main types of cervical cancer named for the type of cell where the cancer started: (1) Squamous cell carcinoma makes up about 80% to 90% of all cervical cancers. These cancers arise in cells on the outer surface covering of the cervix. (2) Adenocarcinoma makes up 10% to 20% of all cervical cancers. These cancers arise in the glandular cells that line the lower birth canal. There are 5,27,624 new cases with a mortality of 2,65,672 worldwide while in India 1,22,844 new cases with a mortality of 67,477 every year has been recorded. (Burni *et al, 2017,* ICO Report).

Treatment approaches have not achieved satisfactory outcomes despite rapid progress in understanding the biology of HPVs and cervical cancer cells. The failure of anticancer therapies is mainly because of the side effects of most anticancerous drug potency, together with the development of drug resistance which leads to decreased drug efficacy. Nanotechnology has found various ways to improve the management of cancer. In the last few years, drug delivery research has witnessed remarkable growth due to utilization of nanoparticles as controlled release reservoirs for the targeted delivery of drug for combating many diseases to ensure an effective and safe use of nanomaterials for medical application. The interaction between a material and the biological system of interest must be studied with particular focus on the environment in which the biomaterial will be placed in. In drug delivery, to evaluate a nanoparticle's biocompatibility is central and is a possible first key step for the safe delivery of drug. The biocompatible nanoparticle has ability to perform with an appropriate host response in a specific situation. The high degree of biocompatibility is achieved when a material interacts with the cell without inducing unacceptable toxic, immunogenic, thrombogenic and carcinogenic response.

Biocompatibility is one of the most important characteristics of biomedical polymer material whose surface is required to interact with a biological system. Since proteins are viewed as the primary and the most important player in mediating polymer organism interaction, the status of the proteins on material surface is believed to determine the ultimate biocompatibility of a given polymer. In order to achieve specific response between polymer surfaces and the adjacent cells, non-specific interaction should be minimized. It is important to design a biocompatible polymer surface so that their interaction with biomolecules result into induction of specific protein to protein adsorption and cellular response. The application of biocompatible nanoparticle to personalized medicine provides an unprecedented opportunity. Biocompatible nanoparticles offer several advantages as to design flexibility, small size, and large surface to volume ratio and increase avidity for target molecules. Within the nascent but rapidly growing field of nanomedicine personalized medicine application are among the most promising and exciting innovation. Personalized medicine consist of healthcare strategy where specific therapeutics are prescribed to patients on the basis of genetic, phenotypic and environmental factors that influence the response to therapy. It has long been recognized that individual patients respond differently to the same drug in terms of efficacy and safety due to complexity and heterogeneity of disease and patients' genomic structure. The goals for biocompatible nanoparticles in drug delivery include-

- **a)** more specific drug targeting and delivery
- **b)** reduction in toxicity
- **c)** faster development of new safe medicines

In principle, nanomedicine could deliver a drug (or even a combination of drugs) precisely where it is needed in the body. This increases a drug's effectiveness while avoiding side effects caused by flooding the entire body with a conventional chemotherapy, all this by virtue of their size. One of the field's core assumptions is that nanoparticles are small enough to leak through shoddily built blood vessels around tumors, but too big to pass through normal vasculature into healthy tissues.

2. Wnt SIGNALING IN CERVICAL CANCER

The wingless-type (Wnt) signaling pathway involves different proteins that are needed for cell proliferation and differentiation in different tissue (Bahrami *et al,* 2017). Wnt pathway initiates by binding proteins to the frizzled (Fz) family receptor when fz receptor activated a signal is forwarded to the phosphoprotein Dishevelled. The Wnt signaling could be categorized into 2 groups the canonical and non-canonical pathway. In the canonical Wnt pathway phosphorylation of Dishevelled led to translocation of β-catenin to the nucleus. Afterwards, it induces a cellular response through gene transduction together with TCF/LEF transcription factors (Moradi *et al,* 2017). Increasing evidence indicates that Wnt signaling plays substantial role in tumorigenesis and tumor propagation. Activation of Wnt signaling activates proliferation and metastasis of cancer cell, whereas inhibition of Wnt signaling blocks cancer stemness tumor propagation (Ramchandran *et al,* 2014). Also, Wnt signaling pathway has been suggested to be contributed in chemo sensitivity (Dandekar *et al,* 2014) and radio resistance. In cervical carcinogenesis aberrant activation of the Wnt pathway is a phenomenon that has been observed since the first stage of neoplastic development. Overexpression of the Fz receptor has been shown and the inhibition of negative regulators are found to be hyper methylated during malignant progression of cervical cancer. Coordinated inactivation of negative regulators permits the Wnt/B catenin pathway to keep the cell replication and transformation signals activated, then promoting the tumoral phenotype. Cytoskeletal reorganization is an important process for cell motility, invasion and metastasis. P^{120} is a member of a family of proteins that have an armadillo repeat domain which regulates the activity of small GTPases, in particular, the Rho, Rac and cdc42 families that participate in the Wnt/PCP pathway. In a study p120 gene was found to be overexpressed in early stage in patients with lymph node metastasis. The evidence suggested that Wnt/PCP pathway promotes cell migration by means of Rho family small GTPase activation. Because of the notorious importance the Wnt pathway in the development and its promotion of the tumoral phenotype of cervical cancer, it has been proposed that pathway becomes the second hit toward carcinogenesis. Although the behavior of the Wnt7a protein in human cancer appears to be tissue-specific (Ramos-Solano *et al,* 2015) demonstrated, for the first time, its tumor suppressor properties in CC-derived cell lines. Down regulation of Wnt7a could comprise an important step during cervical carcinogenesis. However, future experiments with samples from patients with or without cervical lesions must be performed in order to fully understand the participation of Wnt7a in CC progression.

3. USE OF PLGA NANOPARTICLES IN CERVICAL CANCER

Poly (lactic-co-glycolic acid) PLGA is one of the most effective biodegradable polymers for the development of nanomedicines because it undergoes hydrolysis in the body to produce the body biodegradable metabolite monomers, lactic acid and glycolic acid (Acharya and Sahoo, 2011). It has been approved by the US FDA to use in drug delivery systems due to controlled and sustained release properties, low toxicity and biocompatibility with tissue and cells. PLGA nanoparticles are internalized in cells partly through fluid phase pinocytosis and also through clathrin-mediated endocytosis (Alimohammadi *et al,* 2014). PLGA- nanoparticles rapidly escape the endo-lysosome and enter the cytoplasm. This facilitates interaction of nanoparticles with the vesicular membranes leading to transient and localized destabilization of the membrane resulting in the escape of nanoparticles into the cytosol. A novel biodegradable docetaxel-loaded PLGA-TPGS, a random copolymer can be synthesized from lactide, glycolide and D-alfa-tocopheryl Polyethylene glycol 1000 succinate (TPGS) nanoparticles. *In vitro* cellular uptakes of such nanoparticles were investigated in human cervix carcinoma cells (HeLa) by using CLSM (confocal laser scanning microscopy) demonstrating fluorescence. The developed nanoparticles had significant cytotoxicity against HeLa cells, which was time and concentration dependent. The study concludes that PLGA-TPGS random copolymer-based nanoparticles are efficient drug delivery systems for cancer chemotherapy (Yuandong *et al,* 2010).

4. PLGA WORK IN INDIA

Nanoparticle synthesis technology provides an effective way to deliver anti-cancer drug(s) into tumors. Curcumin nanoparticle formulation (Nano-CUR) with PLGA is a well-known example of PLGA associated nanomedicine. This formulation has shown to be effective for improved therapeutic effects in metastatic ovarian and breast cancer cells. To implement this efficient Nano-CUR formulation to achieve improved anti-cancerous effects on cervical cancer, several studies have been already conducted. Results obtained from these experiments have shown inhibited growth of cervical cancer cells, induced apoptosis, and arrested cell cycle in the G1-S transition phase. Moreover, Nano-CUR formulation effectively reduced the tumor burden in NOD SCID gamma (NSG) orthotopic xenograft mouse model. Additionally, Nano-CUR treatment caused a marked decrease in the levels of miRNA-21 (an onco-miRNA associated with chemo-resistance), in *in vitro* and *in vivo* models and enhanced the expression of miRNA-214 (a tumor suppressor). Altogether, these results suggest that Nano-CUR provides a beneficial approach for a rational strategy to widen the chemo-preventive and therapeutic modality for the overall management of cervical cancer (Zaman *et al, 2016*).

5. NECESSITY OF PERSONALIZED MEDICINE

In the near future, the availability of ligand-functionalized therapeutic nanomaterials will have a clear impact on the individualized treatment of diseases. In the detection and monitoring of target expression before initiating therapy and during the whole treatment will clearly be of utmost importance. Similarly, multivalent nanoparticles are complex objects in which behavior depends on a variety of physicochemical properties (Xue-Qing Zhang *et al,* 2012). Presently, efforts should be made to better understand how ligand-functionalized nanomaterials interact with their targets. In parallel, a better comprehension of the correlations between target expression patterns and cancer prognosis is also required. When both of these aspects are addressed, the therapeutic targets to select for the rational design of nanomedicine will become clearer.

REFERENCES

Alimohammadi YH, Joo SW. PLGA-based nanoparticles as cancer drug delivery systems. Asian Pac J Cancer Prev. 2014; 15:517-35.

Bahrami A, Amerizadeh F, ShahidSales S, Khazaei M, Ghayour Mobarhan M, Sadeghnia HR, Maftouh M, Hassanian SM, Avan A. Therapeutic potential of targeting wnt/β catenin pathway in treatment of colorectal cancer: Rational and progress. Journal of cellular biochemistry. 2017 Aug;118(8):1979-83.

Bruni L, Barrionuevo-Rosas L, Albero G, Serrano B, Mena M, Gómez D, Munoz J, Bosch FX, de Sanjose S. Human papillomavirus and related diseases in the world. Summary report 27 July 2017.

Dandekar S, Romanos Sirakis E, Pais F, Bhatla T, Jones C, Bourgeois W, Hunger SP, Raetz EA, Hermiston ML, Dasgupta R, Morrison DJ. Wnt inhibition leads to improved chemosensitivity in paediatric acute lymphoblastic leukaemia. British journal of haematology. 2014 Oct;167(1):87-99.

Ma Y, Zheng Y, Liu K, Tian G, Tian Y, Xu L, Yan F, Huang L, Mei L. Nanoparticles of poly (lactide-co-glycolide)-da-tocopheryl polyethylene glycol 1000 succinate random copolymer for cancer treatment. Nanoscale research letters. 2010 Jul;5(7):1161.

Moradi A, Ghasemi F, Anvari K, Hassanian SM, Simab SA, Ebrahimi S, Hesari A, Forghanifard MM, Boroushaki MT, Shahid Sales S, Avan A. The cross regulation between SOX15 and Wnt signaling pathway. Journal of cellular physiology. 2017 Dec;232(12): 3221-5.

Ramachandran I, Ganapathy V, Gillies E, Fonseca I, Sureban SM, Houchen CW, Reis A, Queimado L. Wnt inhibitory factor 1 suppresses cancer stemness and induces cellular senescence. Cell death & disease. 2014 May;5(5):e1246.

Ramos-Solano M, Meza-Canales ID, Torres-Reyes LA, Alvarez-Zavala M, Alvarado-Ruíz L, Rincon-Orozco B, Garcia-Chagollan M, Ochoa-Hernández AB, Ortiz-Lazareno PC, Rösl F, Gariglio P. Expression of WNT genes in cervical cancer-derived cells: Implication of WNT7A in cell proliferation and migration. Experimental cell research. 2015 Jul 1;335(1):39-50.

Zaman MS, Chauhan N, Yallapu MM, Gara RK, Maher DM, Kumari S, Sikander M, Khan S, Zafar N, Jaggi M, Chauhan SC. Curcumin nanoformulation for cervical cancer treatment. Scientific reports. 2016 Feb 3; 6:20051.

Zhang XQ, Xu X, Bertrand N, Pridgen E, Swami A, Farokhzad OC. Interactions of nanomaterials and biological systems: Implications to personalized nanomedicine. Advanced drug delivery reviews. 2012 Oct 1;64(13):1363-84.

Acharya S, Sahoo SK. PLGA nanoparticles containing various anticancer agents and tumour delivery by EPR effect. Advanced drug delivery reviews. 2011 Mar 18;63(3):170-83.

❐

Chapter-13

NANOTECHNOLOGY STRATEGIES IN CANCER THERAPY: INNOVATION AND FUTURE PERSPECTIVES

Saima Amjad, A.K. Sharma and M. Serajuddin*

Fish Biology Research Lab

Department of Zoology, University of Lucknow, Lucknow-226007

Email of Corresponding Author: lu.fisheries@gmail.com

ABSTRACT

Nanotechnology in medicine has provided new opportunities to emerge latest avenues to treat incurable diseases through targeted drug delivery. It is trying to find something new and enhance the properties of a drug by employing every branch of science including physics, biology, chemistry, engineering, biotechnology. The uncontrolled division in the abnormal cells of the body causes cancer, and chemo drugs kill uncontrolled growth of cell, but it also damages the healthy cells. Due to these side-effects, targeted drug delivery is used to increase the efficacy and reduced the dose by delivering the drug at a specific location. Nanotechnology has been utilized several strategies of drug delivery to solid tumors in order to improve the efficacy of the drug by using nanodevices and nanoparticles. Nanodevices that are used as a carrier for drugs to target diseased cell are dendrimer, micelles, nanorobots, nanocapsules, fullerenes, nanotubes and liposomes. Quantum dots are used for imaging and diagnosis, because of their better diagnostic property, especially for a tumor in comparison of traditionally used fluorescence labeling probe. Miniature nanodevices are also developed for providing high-quality images that are not possible with traditional tools. However, the use of nanotechnology in medicine needed attention on safety and toxicological issues. Therefore, more studies needed to understand the biological responses of nanomaterials to develop more biocompatible nanodevices and nanoparticle for safe treatment of cancer.

Keywords: Nanodevices, Nanoparticles, Cancer, Nanotechnology

1. INTRODUCTION

Cancer is known to develop via a multistep carcinogenesis process entailing numerous cellular physiological systems such as cell signaling and apoptosis, making it a highly incomprehensible and complex disease (Reichert and Wenger, 2008; Zou, 2005). Cancer is a cluster of diseases involving abnormal cell growth with the potential to spread to other parts of the body (WHO, 2018 and NCI, 2018). More than 100 types of cancer affect the human (NCI, 2018). Cancer ranks second leading cause of death globally. The major causes of cancer are the genetic mutation from environmental factors (90-95%) and the rest are caused by genetic inherited (5-10%). The other reasons besides pollution which caused cancer such as economic, lifestyle and behavioral factors (Manton et al., 2008). The other common causes of cancer include diet and obesity (30–35%), tobacco (25–30%), infections (15–20%), radiation (up to 10%), stress, lack of physical activity (Anand et al., 2008; Islami et al., 2018).

The most common therapy used in cancer treatments is restricted to radiation, chemotherapy, and surgery (Singhal, 2010). The drugs used in chemotherapy are cytotoxic chemicals which kills cancer cells simultaneously it also kills healthy cells (Grossman and Scott, 2012). The other common challenges in cancer treatment are insufficient drug concentration reaching the tumor site, nonspecific systemic distribution of anticancer agents, challenges in the development of multiple drug resistance and limited diagnosis (Das et al., 2009 and Parveen and Sahoo, 2008).

Nobel Laureate physicist Richard P. Feynman at an American Physical Society meeting at Caltech on 29 December 1959 delivered a talk on "There's Plenty of Room at the Bottom" It provides an inspiration for the field of nanotechnology (Feynman, 1960). The term nanotechnology was first proposed by Norio Taniguchi in 1974. Nanotechnology has the potential to create materials and devices at atomic, molecular and supramolecular levels. Nanotechnology application used in the field of health and medicine is known as Nanomedicine (Freitas, 1999). Nanomedicine application has the widespread approach in medical application ranges from nanodevices, nanomaterials and nanoparticles are used for drug delivery and diagnosis of cancer treatment to overcome the side effects of a chemotherapy drug and for better therapeutic approaches. But the current problem of nanomedicine is that nanoparticles and nanodevices are intentionally introduced into the human body and there is a gap of knowledge in understanding the biological interaction of these nanomaterials inside the body. This book chapter classifies the nanotechnology tools such nanodevices and intelligent materials are currently applied in diagnosis and treatment of cancer.

2. NANOSIZED TECHNOLOGIES FOR TARGETED DRUG DELIVERY AND DIAGNOSIS OF CANCER

Nanoparticles and nanodevices used for drug delivery provide the opportunity to encapsulate poorly soluble drugs (Kipp, 2004; Zhang et al., 2008), modify nano drug blood circulation and their distribution to targeted organ (Alexis et al., 2008; Bertrand and Leroux, 2012) and provide shield to therapeutic molecules (Whitehead et al., 2009). Nanoparticles used for drug delivery are various types on the basis of size, design, and materials. The individuality of each particle differs in terms of drug loading capacity, drug stability, drug release rates, and targeted drug delivery ability. Nanoparticles are particles with size ranges between 1-100 nanometers (nm). In nanomedicine, nanoparticles are specifically engineered by attaching the different elements and drugs and make it biocompatible for targeted drug delivery. Nanoparticles will usually be taken up by the liver, spleen, and other parts of the reticuloendothelial system (RES) when injected into the human body (Brigger et al., 2002). Nanoparticles coated with a substance such as poly (vinyl pyrrolidone) prevent to destroy inside the body and after injection 5-10% of the nanoparticles still circulating in the blood for eight hours (Gaur, 2000). Hence, the nanoparticles have the ability to target only the cancerous cell due to the long circulation time period and reduce the dose and severity of side effects of a chemotherapeutic drug. Hence, some important nanomaterials used in nanomedicine is discussed in this review.

Micelles:

A micelle is cumulative of surfactant molecules dispersed in a liquid colloid. A typical micelle structure in aqueous solution contains forms an aggregate with the hydrophilic "head" regions part in contact with adjoining solvent and appropriately the hydrophobic single-tail regions in the micelle center. This form is caused by the packing behavior of single-tail lipids in a bilayer. Micelles usefulness is in delivering of poorly water-soluble drugs (Saha et al., 2013) especially in chemotherapeutic treatment (Qiu et al., 2013) withdrawing the attention of researchers from last several years.

Liposomes:

Liposomes are artificial spherical vesicle shape may have one or bilayer membranes. Vesicle size is a significant factor in defining the half-life circulation of liposomes both number and size of bilayers affects the quantity of pharmaceutical drug loading in the liposomes. Size of the liposome can vary from 25 nm to 2.5 µm vesicles (Vahed et al., 2017). Currently, liposomes are used in a different variety of human malignancies treatment and currently it is the best well-studied drug delivery system (Gabizon, 1998; Martin, 1998; Drummond, 1999; Gelmon, 1999). Solid tumors have "leaky" microvasculature supporting structure which permits the uptake of large approximately 100 nm size drug-loaded carriers (Hobbs et al., 1998; Huang et al., 1993; Yuan et al., 1994; Yuan and Dellian et al., 1995) and their successful interaction with cancerous cells, or diffusion of encapsulated chemotherapy drug to near the tumor to easily diffuse into the cancer cell in free form. In addition liposomes have several other advantages over the subsequent free drugs, have favorable pharmacokinetics properties where encapsulation of a drug usually rapid which results to increase the time duration of circulation of drug in blood (Hwang, 1987; Allen et al., 1995; Allen and Stuart, 1999) Moreover, drug loaded liposome also increases effectiveness in tumor when compared to the free chemotherapy drug (Gabizon et al., 1994; Northfelt et al., 1998; Stewart et al., 1998).

Dendrimers:

Dendrimers are artificial, branched with an average diameter of 1.5-14.5 nm, three-dimensional, mono-disperse macromolecules having a central core and several functionalities at the surface periphery of nanometer dimensions (Patri et al., 2002; Andresen et al., 2005; Perumal et al., 2008; Theoharis et al., 2009; Agrawal and Jain 2007). Due to their unique architectural structure, inner core work as a "host" have a capacity to entrap the drug molecules (guest) and the surface functionality provides the attachment site for the gene, several types of drugs and targeting agent (Perumal et al., 2008; Gupta et al., 2006). Biodegradable polyester dendrimers (2,2-bis(hydroxymethyl)-propanoic acid monomers) have been synthesized for the release of doxorubicin drug inside the cell after hydrolysis of the hydrazone linkage (Padilla et al., 2002). The comparative study of polyester dendrimer-drug conjugate in C-26 colon carcinoma-bearing mice shows similar efficacy to doxorubicin-encapsulated liposome (Lee et al., 2006). Another chemotherapeutic drug Cisplatin (CDDP) which is used in treatment of various types of cancers, including; ovarian, head, non-small cell lung, cancers, testicular and bladder was also used in drug delivery in conjugation of PAMAM (polyamidoamine) dendrimers (Dasari and Tchounwou, 2014; Kirkpatrick et al., 2011).

Nanorobots:

The most advanced and emerging technology in nanomedicine is to create miniature machines or surgeons whose constituents are at or near the scale of a nanometer (10^{-9} meters) (Sierra et al., 2005; Vaughn, 2006; Ghosh and Fischer, 2009). The potential uses of nanorobots in nanomedicine include early detection and targeted drug delivery for chemotherapeutic drug (Zyga, 2007; Lavan et al., 2003), diabetes monitoring (Melki,

2007; Shanthi and Sravan, 2007), biomedical instrumentation, pharmacokinetics (Patel et al., 2006), surgery (Leary et al., 2006) and health care. The expected future advancement in the field of medical nanotechnology is to inject the nanorobots at a cellular level into the patient body.

Researchers at Harvard and Massachusetts Institute of Technology synthesized a nanostructure by attaching a special RNA strand (approximately 10 nm size) to nanoparticles loaded with a chemotherapy drug. The attached RNA strands on drug loaded nanoparticle targeted the cancerous cells and adhere at the surface to release the drug into the cancer cell (Bullis, 2008). Zhang et al., (2014) also reported the therapeutic efficacy of antitumor drug doxorubicin loaded inside the noncovalently intercalated self-assembled DNA origami nanostructures. The DNA origami nanostructures have a potential for targeted drug delivery and safe of cancer therapeutics *in vivo.*

Fullerenes:

Fullerene (C60) is a molecule of carbon consists of a hollow sphere, ellipsoid or tube, and the most common one is the spherical fullerene also named buckyball. Fullerene and its derivatives are used in various types of cancer treatment it is hypotoxic and through injection, it can distribute all tissues quickly (Qingnuan et al., 2002; Wang et al., 2006). The unique physiochemical properties of fullerenes are extensively used for investigation of tumor imaging and therapy (Chen et al., 2012).

Quantum dots and encapsulated fluorophores:

Quantum dots (QD) are very small semiconductor nanostructure size ranges between 2-10 nm diameters. Due to its quantum confinement property they will emit fluorescence from visible to infrared wavelengths on excitation and these light frequencies can be tuned with the changing dot's size (Sabaeian and Khaledi-Nasab, 2012; Khaledi-Nasab et al., 2014), material and shape giving rise to potential applications in medical diagnosis (Ramírez, 2015). QDs cancer therapy is a new type of fluorescent probe for cellular and molecular imaging. In comparison to conventional organic dyes and fluorescent proteins used for cancer diagnosis QDs have the improved brightness of the signal, photostable, wavelength-tunable emission, and different colors of QDs can be excited simultaneously by a single light source (Hoult and Phil, 2000; Caravan et al., 1999). CA125 is a protein which is known as tumor marker was detected in ovarian cancer specimen with QD605 at maximum emission wavelength 605 nm and it showed a brighter more specific image and stable than fluorescein isothiocyanate (Wang et al, 2004). CdSe/CdS/ZnS QDs were used as an optical imaging in pancreatic cancer cells using transferring and anti-Claudin-4 antibody (Qian et al., 2007).

QDs has been used as a multifunctional nanoparticle conjugated with A10 RNA, aptamer which recognizes prostate specific membrane antigen (PSMA) to target and drug delivery to cancer cells (Bagalkot et al., 2007). QD-aptamer with Doxorubicin (anthracycline drug with fluorescent properties) conjugate provides better imaging to target cancer cell (Gao et al., 2004).

Encapsulated fluorophores are used for diagnosis of the cancer cell. These smart nanoscale fluorescent materials can be synthesized by traditional organic fluorophores encapsulation inside the nanoparticle carrier. Encapsulation of fluorophore increase the signal strength and sensitivity of an assay and resistance the dye from photobleaching and also control the biodistribution and clearance rate compared to free organic fluorophores (Wang et al., 2006; Yan et al., 2007).

Nanoshell:

A nanoshell is a metallic nanoparticle consisting of a dielectric core like silica covered with an ultrathin metallic layer, which is usually gold (Loo et al., 2004). These silica-gold nanoshell have been used as a targeted therapy probe in vitro experiment for human prostate (Gobin et al., 2008; Strern et al., 2007), liver (Liu et al., 2010), brain (Bernardi et al., 2008) and breast (Loo et al., 2004; Loo et al., 2005; Lowery et al., 2006) cancer treatment and bio-imaging enhancement. Scientist Shunying Liu from East China Normal University describes, how the nanoshell target cancerous cell, small peptides are attached on the surface of nanoshell which increased the targeting ability to reach the specific cancerous cell and attach to the marker of the cell and acidic environment of the cancer cell stimulate the unloading of antitumor drugs. The specific gold nanoshells are also used in targeted hyperthermia treatment. In this process nanoshells could allow to an absorbed light near-infrared region and when it reached to the targeted site, slightly increased the temperatures than normal to destroy the cancerous cell.

Nanoscale Cantilever:

Nanoscale cantilever work as a diagnostic tool for cancer detection, it resembles comb-like structure having equally distributed teeth. It is coated with specific antibodies provide rapid and selective detection of cancer proteins (NCI, 2006). When the cancerous protein attaches to the antibodies attach on the cantilever, it changes its conformational shape (Labuda and Grütter, 2011; Sansoz and Gang, 2010; Lee et al., 2010). The change in confirmation is due to change in conductivity indicates the positive result for a cancer patient. Nanoscale cantilever provides early detection of cancer before tumor formation in the cancer patient and helps in prevention and treat the disease at earliest on the appropriate time.

3. CHALLENGES AND FUTURE PERSPECTIVES

In last few years' nanotechnologies open new avenues through the multi-directional approach to cancer treatment. The advantage of nanomedicine over conventional drugs includes, the reduced negative impact of high doses of chemotherapy drugs, long circulation time, biocompatibility, have an ability to target specific tissues with controlled release of drugs. Apart from many advances and benefits of nanomedicine, there is some drawback like the major concern is associated with toxicity of nanoparticles. So, there is a need to biosafety evaluation of nanomaterials before intravenous and subcutaneous injections and there is also a chance of accumulation of these nanomaterials inside the human tissues or organs (Zhao and Castranova, 2011). Hence, nanomedicine applications in cancer treatment have some drawback but it still brings new hope to revolutionize drug delivery therapy.

ACKNOWLEDGEMENT

The first author Saima Amjad acknowledges Department of Science and Technology for providing DST INSPIRE Fellowship to bring out the research work and extends their thanks to Head, Department of Zoology for providing necessary facilities.

REFERENCES

"Cancer Fact sheet N°297". (February 2018. Retrieved 21 March (2018). World Health Organization.

Agrawal, P., Gupta, U., & Jain, N. K. (2007). Glyco-conjugated peptide dendrimers-basednanoparticulatesystemforthedeliveryofchloroquinephosphate. Biomaterials, 28(22),

3349-3359.

Alexis, F., Pridgen, E., Molnar, L. K., & Farokhzad, O. C. (2008). Factors affecting the clearance and biodistribution of polymeric nanoparticles. Molecular pharmaceutics, 5(4), 505-515.

Allen, T. M. (1999). Liposome pharmacokinetics classical, sterically stabilized, cationic liposomes and immunoliposomes. Liposome Rational design.

Allen, T. M., Hansen, C. B., & de Menezes, D. E. L. (1995). Pharmacokinetics of long-circulating liposomes. Advanced Drug Delivery Reviews, 16(2-3), 267-284.

Anand, P., Kunnumakara, A. B., Sundaram, C., Harikumar, K. B., Tharakan, S. T., Lai, O. S., ... & Aggarwal, B. B. (2008). Cancer is a preventable disease that requires major lifestyle changes. Pharmaceutical research, 25(9), 2097-2116.

Andresen, T. L., Jensen, S. S., & Jørgensen, K. (2005). Advanced strategies in liposomal cancer therapy: problems and prospects of active and tumor specific drug release. Progress in lipid research, 44(1), 68-97.

Bagalkot, V., Zhang, L., Levy-Nissenbaum, E., Jon, S., Kantoff, P. W., Langer, R., & Farokhzad, O. C. (2007). Quantum dot– aptamer conjugates for synchronous cancer imaging, therapy, and sensing of drug delivery based on bi-fluorescence resonance energy transfer. Nano letters, 7(10), 3065-3070.

Bernardi, R. J., Lowery, A. R., Thompson, P. A., Blaney, S. M., & West, J. L. (2008). Immuno-nanoshells for targeted photothermal ablation in medullo-blastoma and glioma: an *in vitro* evaluation using human cell lines. Journal of neuro-oncology, 86(2), 165-172.

Bertrand, N., & Leroux, J. C. (2012). The journey of a drug-carrier in the body: an anatomo-physiological perspective. Journal of controlled release, 161(2), 152-163.

Brigger, I., Dubernet, C., & Couvreur, P. (2012). Nanoparticles in cancer therapy and diagnosis. Advanced drug delivery reviews, 64, 24-36.

Bullis, Kevin (April 29, 2008). Nano RNA Delivery. MIT Technology Review.

Caravan, P., Ellison, J. J., McMurry, T. J., & Lauffer, R. B. (1999). Gadolinium (III) chelates as MRI contrast agents: structure, dynamics, and applications. Chemical reviews, 99(9), 2293-2352.

Chen, Z., Ma, L., Liu, Y., & Chen, C. (2012). Applications of functionalized fullerenes in tumor theranostics. Theranostics, 2(3), 238.

Das, M., Mohanty, C., & Sahoo, S. K. (2009). Ligand-based targeted therapy for cancer tissue. Expert opinion on drug delivery, 6(3), 285-304.

Dasari, S., & Tchounwou, P. B. (2014). Cisplatin in cancer therapy: molecular mechanisms of action. European journal of pharmacology, 740, 364-378.

Defining "Cancer". (Retrieved 28 March 2018). National Cancer Institute.

Drummond, D. C., Meyer, O., Hong, K., Kirpotin, D. B., & Papahadjopoulos, D. (1999). Optimizing liposomes for delivery of chemotherapeutic agents to solid tumors. Pharmacological reviews, 51(4), 691-744.

Feynman, Richard, P. (1960). There's plenty of room at the bottom. Engineering Science. 22, 22-36.

Freitas, R. A. (1999). Nanomedicine, volume I: basic capabilities (Vol. 1). Georgetown, TX: Landes Bioscience.

Gabizon, A. A. (1998). Clinical trials of liposomes as carriers of chemotherapeutic agents: synopsis and perspective. In Medical applications of liposomes (pp. 625-634).

Gao, X., Cui, Y., Levenson, R. M., Chung, L. W., & Nie, S. (2004). *In vivo* cancer targeting and imaging with semiconductor quantum dots. Nature biotechnology, 22(8), 969

Gaur, U., Sahoo, S. K., De Tapas, K., Ghosh, P. C., Maitra, A., & Ghosh, P. K. (2000). Biodistribution of fluoresceinated dextran using novel nanoparticles evading reticuloendothelial system. International journal of pharmaceutics, 202(1-2), 1-10.

Gelmon, K. A., Tolcher, A., Diab, A. R., Bally, M. B., Embree, L., Hudon, N., ... & Burge, C. (1999). Phase I study of liposomal vincristine. Journal of clinical oncology, 17(2), 697-697.

Ghosh, A., & Fischer, P. (2009). Controlled propulsion of artificial magnetic nanostructured propellers. Nano letters, 9(6), 2243-2245.

Gobin, A. M., Moon, J. J., & West, J. L. (2008). EphrinAl-targeted nanoshells for photothermal ablation of prostate cancer cells. International journal of nanomedicine, 3(3), 351.

Grossman, J. H., & McNeil, S. E. (2012). Nanotechnology in cancer medicine. Physics Today, 65(8), 38.

Gupta, U., Agashe, H. B., Asthana, A., & Jain, N. K. (2006). Dendrimers: novel polymeric nanoarchitectures for solubility enhancement. Biomacromolecules, 7(3), 649-658.

Hobbs, S. K., Monsky, W. L., Yuan, F., Roberts, W. G., Griffith, L., Torchilin, V. P., & Jain, R. K. (1998). Regulation of transport pathways in tumor vessels: role of tumor type and microenvironment. Proceedings of the National Academy of Sciences, 95(8), 4607-4612.

Hoult, D. I. (2000). Sensitivity and power deposition in a high-field imaging experiment. Journal of Magnetic Resonance Imaging, 12(1), 46-67.

Huang, S. K., Martin, F. J., Jay, G., Vogel, J., Papahadjopoulos, D., & Friend, D. S. (1993). Extravasation and transcytosis of liposomes in Kaposi's sarcoma-like dermal lesions of transgenic mice bearing the HIV tat gene. The American journal of pathology, 143(1), 10.

Hwang, K, J. (1987) Liposome pharmacokinetics, in Liposomes: from biophysics to therapeutics (Ostro MJ, ed). Marcel Dekker, New York, pp 109–156.

Islami, F., Goding Sauer, A., Miller, K. D., Siegel, R. L., Fedewa, S. A., Jacobs, E. J., ... & Flanders, W. D. (2018). Proportion and number of cancer cases and deaths attributable to potentially modifiable risk factors in the United States. CA: a cancer journal for clinicians, 68(1), 31-54.

Kipp, J. E. (2004). The role of solid nanoparticle technology in the parenteral delivery of poorly water-soluble drugs. International journal of pharmaceutics, 284(1-2), 109-122.

Kirkpatrick, G. J., Plumb, J. A., Sutcliffe, O. B., Flint, D. J., & Wheate, N. J. (2011). Evaluation of anionic half generation 3.5–6.5 poly (amidoamine) dendrimers as delivery vehicles for the active component of the anticancer drug cisplatin. Journal of inorganic biochemistry, 105(9), 1115-1122.

Labuda, A., & Grütter, P. H. (2011). Exploiting cantilever curvature for noise reduction in atomic force microscopy. Review of scientific instruments, 82(1), 013704.

Lavan, D. A., McGuire, T., & Langer, R. (2003). Small-scale systems for in vivo drug delivery. Nature biotechnology, 21(10), 1184.

Leary, S. P., Liu, C. Y., Apuzzo & M. L. J. (2006). Toward the Emergence of Nanoneurosurgery: Part III. Nanomedicine: Targeted Nanotherapy, Nanosurgery, and Progress Toward the Realization of Nanoneurosurgery. *Neurosurgery*. **58** (6): 1009–1026.

Lee, C. C., Gillies, E. R., Fox, M. E., Guillaudeu, S. J., Fréchet, J. M., Dy, E. E., & Szoka, F. C. (2006). A single dose of doxorubicin-functionalized bow-tie dendrimer cures mice bearing C-26 colon carcinomas. Proceedings of the National Academy of Sciences, 103(45), 16649-16654.

Lee, H., Yu, A., & Flores-Mir, C. (2010). Limited evidence that cantilevers are associated with slightly lower survival rates of implant-supported fixed partial dentures. The Journal of the American Dental Association, 141(11), 1371-1372.

Liu, S. Y., Liang, Z. S., Gao, F., Luo, S. F., & Lu, G. Q. (2010). *In vitro* photothermal study of gold nanoshells functionalized with small targeting peptides to liver cancer cells. Journal of Materials Science: Materials in Medicine, 21(2), 665-674.

Loo, C., Hirsch, L., Lee, M. H., Chang, E., West, J., Halas, N., & Drezek, R. (2005). Gold nanoshell bioconjugates for molecular imaging in living cells. Optics letters, 30(9), 1012-1014.

Loo, C., Lin, A., Hirsch, L., Lee, M. H., Barton, J., Halas, N., ... & Drezek, R. (2004). Nanoshell-enabled photonics-based imaging and therapy of cancer. Technology in cancer research & treatment, 3(1), 33-40.

Lowery, A. R., Gobin, A. M., Day, E. S., Halas, N. J., & West, J. L. (2006). Immunonanoshells for targeted photothermal ablation of tumor cells. International journal of nanomedicine, 1(2), 149.

Manton, K. G., Akushevich, I., & Kravchenko, J. (2008). Cancer mortality and morbidity patterns in the US population: an interdisciplinary approach. Springer Science & Business Media.

Martin, F. J. (1998). Clinical pharmacology and antitumor efficacy of DOXIL (pegylated liposomal doxorubicin). In Medical applications of liposomes (pp. 635-688).

Melki, Benjamin (January 31, 2007) Nanorobotics for Diabetes. nanovip.com

National Cancer Institute. 11 Feb. 2006 <http://nano.cancer.gov/news_center/monthly_feature_2006_jan.asp>.

Northfelt, D. W., Dezube, B. J., Thommes, J. A., Miller, B. J., Fischl, M. A., Friedman-Kien, A., ... & Henry, D. H. (1998). Pegylated-liposomal doxorubicin versus doxorubicin, bleomycin, and vincristine in the treatment of AIDS-related Kaposi's sarcoma: results of a randomized phase III clinical trial. Journal of clinical oncology, 16(7), 2445-2451.

Padilla De Jesús, O. L., Ihre, H. R., Gagne, L., Fréchet, J. M., & Szoka, F. C. (2002). Polyester den dritic systems for drug delivery applications: *in vitro* and *in vivo* evaluation. Bioconjugate chemistry, 13(3), 453-461.

Parveen, S., & Sahoo, S. K. (2008). Polymeric nanoparticles for cancer therapy. Journal of drug targeting, 16(2), 108-123.

Patel, G. M., Patel, G. C., Patel, R. B., Patel, J. K., & Patel, M. (2006). Nanorobot: a versatile tool in nanomedicine. Journal of Drug Targeting, 14(2), 63-67.

Patri, A. K., Majoros, I. J., & Baker Jr, J. R. (2002). Dendritic polymer macromolecular carriers for drug delivery. Current opinion in chemical biology, 6(4), 466-471.

Perumal, O. P., Inapagolla, R., Kannan, S., & Kannan, R. M. (2008). The effect of surface functionality on cellular trafficking of dendrimers. Biomaterials, 29(24-25), 3469-3476.

Qian, J., Yong, K. T., Roy, I., Ohulchanskyy, T. Y., Bergey, E. J., Lee, H. H., ... & Prasad, P. N. (2007). Imaging pancreatic cancer using surface-functionalized quantum dots. The Journal of Physical Chemistry B, 111(25), 6969-6972.

Qingnuan, L., Xiaodong, Z., Ruili, L., Xiaoguang, S., Shaoliang, C., & Wenxin, L. (2002). Preparation of 99mTc-C60 (OH) x and its biodistribution studies. Nuclear medicine and biology, 29(6), 707-710.

Qiu, J. F., Gao, X., Wang, B. L., Wei, X. W., Gou, M. L., Men, K., ... & Huang, M. J. (2013). Preparation and characterization of monomethoxy poly (ethylene glycol)-poly (ε-caprolactone) micelles for the solubilization and *in vivo* delivery of luteolin. International journal of nanomedicine, 8, 3061.

Ramírez, H. Y., Flórez, J., & Camacho, Á. S. (2015). Efficient control of coulomb enhanced second harmonic generation from excitonic transitions in quantum dot ensembles. Physical Chemistry Chemical Physics, 17(37), 23938-23946.

Reichert, J. M., & Wenger, J. B. (2008). Development trends for new cancer therapeutics and vaccines. Drug discovery today, 13(1-2), 30-37.

Sabaeian, M., & Khaledi-Nasab, A. (2012). Size-dependent intersubband optical properties of dome-shaped InAs/GaAs quantum dots with wetting layer. Applied optics, 51(18), 4176-4185.

Saha, S. C., Patel, D., Rahman, S., & Savva, M. (2013). Physicochemical characterization, solubilization, and stabilization of 9-nitrocamptothecin using pluronic block copolymers. Journal of pharmaceutical sciences, 102(10), 3653-3665.

Sansoz, F., & Gang, T. (2010). A force-matching method for quantitative hardness measurements by atomic force microscopy with diamond-tipped sapphire cantilevers. Ultramicroscopy, 111(1), 11-19.

Shanthi, V., & Musunuri, S. (2007). Prospects for Medical Robots. AZo Journal.

Sierra, D. P., Weir, N. A., & Jones, J. F. (2005). A review of research in the field of nanorobotics (No. SAND2005-6808). Sandia National Laboratories.

Singhal, S., Nie, S., & Wang, M. D. (2010). Nanotechnology applications in surgical oncology. Annual review of medicine, 61, 359-373.

Stern, J. M., Stanfield, J., Lotan, Y., Park, S., Hsieh, J. T., & Cadeddu, J. A. (2007). Efficacy of laser-activated gold nanoshells in ablating prostate cancer cells *in vitro*. Journal of Endourology, 21(8), 939-943.

Stewart, S., Jablonowski, H., Goebel, F. D., Arasteh, K., Spittle, M., Rios, A., ... & Dezube, B. J. (1998). Randomized comparative trial of pegylated liposomal doxorubicin versus bleomycin and vincristine in the treatment of AIDS-related Kaposi's sarcoma. International Pegylated Liposomal Doxorubicin Study Group. Journal of Clinical Oncology, 16(2), 683-691.

Theoharis, S., Krueger, U., Tan, P. H., Haskard, D. O., Weber, M., & George, A. J. (2009). Targeting gene delivery to activated vascular endothelium using anti E/P-Selectin antibody linked to PAMAM dendrimers. Journal of immunological methods, 343(2), 79-90.

Vahed, S. Z., Salehi, R., Davaran, S., & Sharifi, S. (2017). Liposome-based drug co-delivery systems in cancer cells. Materials Science and Engineering: C, 71, 1327-1341.

Vaughn JR (2006). Over the Horizon: Potential Impact of Emerging Trends in Information and Communication Technology on Disability Policy and Practice. National Council on Disability, Washington DC: 1–55.

Wang, H. Z., Wang, H. Y., Liang, R. Q., & Ruan, K. C. (2004). Detection of tumor marker CA125 in ovarian carcinoma using quantum dots. Acta Biochimica et Biophysica Sinica, 36(10), 681-686.

Wang, L., Wang, K., Santra, S., Zhao, X., Hilliard, L. R., Smith, J. E., ... & Tan, W. (2006). Watching silica nanoparticles glow in the biological world. Analytical Chemistry 78: 646-654.

Whitehead, K. A., Langer, R., & Anderson, D. G. (2009). Knocking down barriers: advances in siRNA delivery. Nature reviews Drug discovery, 8(2), 129.

Yan, J., Estévez, M. C., Smith, J. E., Wang, K., He, X., Wang, L., & Tan, W. (2007). Dye-doped nanoparticles for bioanalysis. Nano today, 2(3), 44-50.

Yuan, F., Dellian, M., Fukumura, D., Leunig, M., Berk, D. A., Torchilin, V. P., & Jain, R. K. (1995). Vascular permeability in a human tumor xenograft: molecular size dependence and cutoff size. Cancer research, 55(17), 3752-3756.

Yuan, F., Leunig, M., Huang, S. K., Berk, D. A., Papahadjopoulos, D., & Jain, R. K. (1994). Mirovascular permeability and interstitial penetration of sterically stabilized (stealth) liposomes in a human tumor xenograft. Cancer research, 54(13), 3352-3356.

Zhang, L., Chan, J. M., Gu, F. X., Rhee, J. W., Wang, A. Z., Radovic-Moreno, A. F., ... & Farokhzad, O. C. (2008). Self-assembled lipid– polymer hybrid nanoparticles: a robust drug delivery platform. ACS nano, 2(8), 1696-1702.

Zhang, Q., Jiang, Q., Li, N., Dai, L., Liu, Q., Song, L., ... & Du, Y. (2014). DNA origami as an *in vivo* drug delivery vehicle for cancer therapy. ACS nano, 8(7), 6633-6643.

Zhao, J., & Castranova, V. (2011). Toxicology of nanomaterials used in nanomedicine. Journal of Toxicology and Environmental Health, Part B, 14(8), 593-632

Zou, W. (2005). Immunosuppressive networks in the tumour environment and their therapeutic relevance. Nature Reviews Cancer, 5(4), 263.

Zyga, L. (2007). Virtual 3D nanorobots could lead to real cancer-fighting technology. Nanotechnology News Physorg, 116071209, 1-3.

❒

Chapter-14

NUTRIGENOMICS: AN OVERVIEW

Yashika Awasthi*, Arun Ratn, Anubha Shukla, Manoj Kumar and Sunil P. Trivedi

Environmental Toxicology & Bioremediation Laboratory, Department of Zoology, University of Lucknow, Lucknow-226007
**Email of Corresponding Author: yashikaawasthi.ya@gmail.com*; sat060523@gmail.com*

ABSTRACT

Nutrigenomics is the future of nutritional science. This will give the direction of our knowledge towards how it will influence diet-related issue in animals. The structure of nutrigenomics is based on certain points like diet can act on the human genome and food components can reduce the risk of disease. Dietary components have ability to alter various genetic events, like transcription and translation which cause alters biological responses such as cell growth, cell division, apoptosis, cell differentiation and other which are important in diseases. This heterogeneity could only show the way towards new research aims. Nutrigenomics was born due to the need to move from Epidemiology and Physiology to Molecular Biology and Genetics. Currently number of steps that need to understand like gene/protein, how nutrient effect the gene/protein expression. The proper use of these approach may help in prevention of diseases related to diet.

Keywords: Nutrigenomics, nutrients, human genome, gene/protein expression.

1. INTRODUCTION

Nutrigenomics is one of the sensitive genomic tools in nutrition research. This is mainly used to promote knowledge of nutrition in direction of how this influences metabolic pathways, homeostatic control and prevention of diet-related diseases (Garg et al., 2014). Nutrition research has started to recognize nutrient can be dietary signals that influence many processes like cell growth, cell differentiation and other epigenetic changes (Murdoch et al., 2016). In future, nutrigenomics hopefully shows organ-specific effects of bioactive dietary components responses at metabolite profiles (metabolome), organization of the chromatin (epigenome); protein-expression patterns, including posttranslational modifications (proteome) and gene-expression patterns (transcriptome).

The term nutrigenomics was first coined by Pelegrin in 2001 (Kaur et al., 2013). Nutrigenomics is an emerging science which investigates the relationship between nutrition and expression of genes and to provide scientific basis for controlling the worldwide epidemic of specific diseases, mostly notably cardiovascular diseases and cancer (Elsamanoudy et al., 2016). Nutrigenomics attempts to gain knowledge to study the genome wide influence of nutrition. Furthermore, nutrigenomic come in focus after the completion of the human genome project which made attention to the importance of genes which is directly related to human nutrition. Since, then there has been an advance identification of nutrients as ligand molecule influencing gene expression. Moreover, the involvement of the molecular biology and bioinformatics tools also allow measuring the subtle change in gene expression. Nutrigenomics is concern with the study of genome-wide influences of nutrition (Mead, 2007). From a nutrigenomics view, nutrients are dietary signals that are detected by the cellular sensor systems which influence gene/genes and proteins expression. So, patterns of gene expression and protein synthesis in response to particular nutrients or nutritional regimes can be viewed as 'dietary signatures' (Fenech et al., 2011). The concept of nutrigenomics can be seen is analogous to that of pharmacogenomics with important difference that pharmacogenomics is concerned with the effects of drugs that are pure compounds-administered in precise (usually small) doses-whereas in case of nutrigenomics it must encompass the complexity and variability of nutrition (Surendiran et al., 2008). Nutrigenomics seeks to examine this dietary impact in specific cells, tissues and organisms to understand how nutrition influences homeostasis.

2. CONCEPT OF NUTRIGENOMICS

The concept of nutrigenomics builds on the premises that:

- food components can reduce the risk of disease by causing alteration of multiple processes involved with onset, incidence, progression, and/or severity;
- diet can act on the human genome and have potential to alter the expression of genes and gene products;
- diet could potentially compensate for or accentuate effects of genetic polymorphisms; and
- nutrient play key role in balancing of health and disease states and on an individual's genetic background.

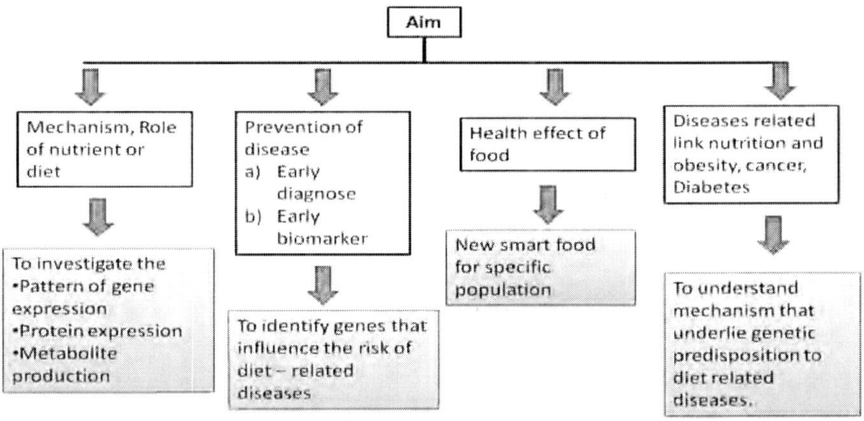

Fig.1. Represent different aim of nutrigenomics.

3. NUTRIGENOMICS: DIETARY SIGNAL FROM NUTRIENTS TO GENES

Nutrigenomics potentially involve the consumer's benefits like health maintenances and improvement and also generate consumer's knowledge for its genetic association with nutrient.

Numerous dietary components have ability to alter various genetic events, thereby affect the health. In addition to nutrient such as calcium, zinc, selenium, folate and vitamin there are also bioactive food component that have significant role in influence health (Milner, 2004). Bioactive food component simultaneously modified several processes i.e. it is very difficult to identify which process bring phenotypic change.

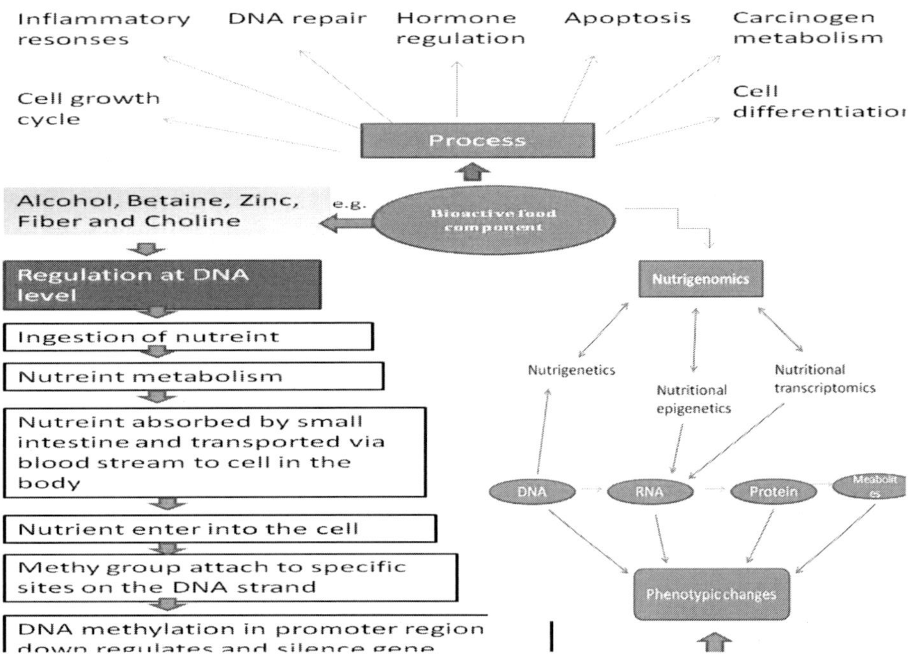

Fig. 2. Bioactive food components can influence genetic and epigenetic events associated with change in phenotypes (Trujillo et al., 2006).

Nutritional Epigenetics:

Epigenetic events can be modified by bioactive food components and various proteins are involved in overall process like DNA methyltransferase, methyl cytosine guanine dinucleotide and histone modifying enzymes (Meeran et al., 2010). Dietary factors shift the activity of DNA methyltransferase by change the methyl group. Studies show that DNA methylation is dependent on bioactive food component and indirectly it controls the expression of genes (Stefanska et al., 2012). Diets on its own shift the DNA hypomethylation and tumor suppressor gene silencing with age, leads to alteration in genotype, gene expression profile, cellular phenotype and an increased risk of cancer (Lomberk, 2011; Stefanska et al., 2012).

Using The "Omics" of Nutrition to Identify How Dietary FactorsContribute to Establishing A Phenotype:

DNA modification alone will not give full concept related to dietary factor may influence to a person phenotype. Various dietary components can modify the gene at transcriptional and translational level which cause alters biological responses such as cell growth, cell division, apoptosis, cell differentiation and other which are important in diseases.

Studies of nutrient at protein level can be used to identify and quantify bioactive proteins and peptides and show how they affect biology and respond to diet (Daliri et al., 2017). In the animal model, proteomic analysis found to play a key role in solving major nutrition related problems e.g. obesity, diabetes, Cardiovascular diseases and cancer etc. (Wang et al., 2006). Proteomics is a primordial resource for nutrigenomics e.g. activity of butyrate, which can alter the expression of several protein from the ubiquitin proteasome system and regulate key proteins which control cell cycle regulation, cell growth and cell differentiation (Sales et al., 2014).

Metabolomics, newest "omics" in nutrition, can be defined as screening of dose and temporal changes in cellular small molecular weight compound in response to dietary treatment. Metabolomic profiles are used for comparing patterns of variations between different groups i.e. health versus diseases and control versus treated (Hadi et al., 2017). In addition, metabolites can be used to monitor the outcome of treatment stratagies, such as pharmacological and dietary intervention (Van Gaal and Dirinck, 2016).

Molecular Tools for Nutrition Research:

Various techniques are available for the screening of genetic background, gene at transcriptional level, proteomics and metabolome ultimately result found by these are used for developing strategies targeted to provide optimal nutrition to individual. The use of microarray technologies – allow generating new information related to (i) physiological effect of different dietary protein, (ii) Omega 3 polyunsaturated fatty acids, (iii) dietary condition in cancer.

Other tool RT-PCR successfully used to evaluate the interaction of gene and diet by the quantification of change in gene expression. For the proteomic analysis one of the molecular techniques is 2D Gel electrophoreses. Metabolomics tools are now being applied to the analysis of:

1. Food components,
2. The identification of their metabolites in body fluids and biological tissues,
3. The evaluation of their bioavailability and metabolism,
4. The role of gut microflora, and
5. The physiological responses to a particular bioactive component (Astarita and Langridge, 2013).

Tools Used for Making Strategies for Interaction Between Gene and Nutrition:

Genomics tools can be used in two different strategies in molecular nutrition research. The first strategy approach will provide detailed molecular data on the interaction between nutrition and the genes and proteins, the expression of which is influenced by nutrients, are identified using genomics tools such as transcriptomics, proteomics and metabolomics. The first strategy, typically applied by smaller research groups, will reveal the identification of transcription factor that regulate pathways which influences homeostasis (Table 1). The second strategies are based on SYSTEMS BIOLOGY approach: to develop biomarkers of early metabolic dysregulation and susceptibility (stress signatures) that areinfluenced by diet and this approach might be play important role in for human nutrition studies (Müller and Kersten, 2003; Afman and Müller, 2006).

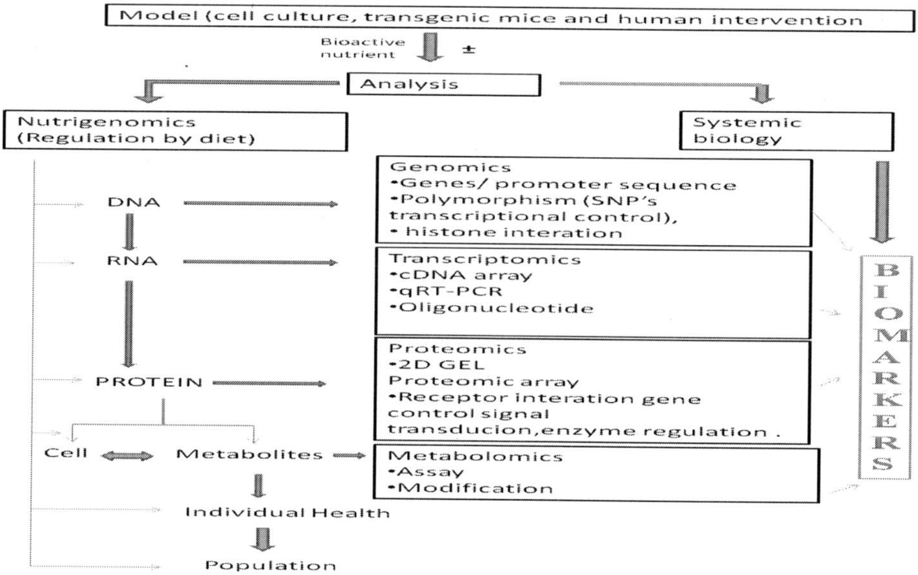

Fig.3. Using the "omics" of nutrition to identify dietary biomarkers by using two different tools based on two strategies (Modified from Van Ommen and Stierum, 2002).

4. CONCLUSION AND FUTURE PERSPECTIVES

Food is basically a complex mixture of nutrients and other components. Most dietary signals are weak and must be considered for longer exposure period. Nutrigenomics offers the tools to measure weak signals and elucidating nutrient signaling pathways that contribute in diet related diseases, such as cancer, obesity and cardiovascular diseases.

Nutritional strategies for health in the future will require more attention for building up desired outcomes on nutritional signals related to human health (Allison et al., 2015). So, the strategy here is based on following areas: to identify the dietary signals; to elucidate the underlying mechanism of dietary sensor; to find out the target genes of these sensors; to recognize the change in phenotype due to initiation of cascade of signaling pathway and for the search of exact ligand; and to find 'signatures' (gene/protein expression and metabolite profiles) that allow the discrimination of healthy versus unhealthy individuals to enable early dietary intervention.

The use of such an approach may allow for prevention and protection from disease by blocking or suppress the initiation, promotion, and progression of pathways that lead to a lethal phenotype. The success of nutritional approaches will depend on the ability to identify and validate nutrigenetic, epigenetic changes due to bioactive food component, proteomic, and metabolomic biomarkers to determine cause, effect, and susceptibility to disease.

Nutrient	Compound	Transcription factor	Gene Alteration	Deficient Diet-Disease Potential	Food Intake
Vitamins	Folic acid (Vitamin B9) Vitamin B12 (Cobalamin) Vitamin B6 (Pyridoxine) Niacin (Vitamin B3) Vitamin E (Tocopherols) Vitamin D (Calciferol)	PXRVD RRAR, RXR	Chromosome break and hampers DNA repair/ methylation. Chromosome break and hampers DNA repair/ methylation Hampers DNA repair Mimics radiation damage Prevent gene variation	Cancer, heart disease, brain dysfunction, male infertility, leukemia Same as folic acid Same as folic acid Colon cancer, heart disease, immune dysfunction Colon, breast, prostate cancer	Pork, tuna, prawns, kidney, liver, poultry, carrots, turnips and celery, mushrooms, beans, almonds, wheat products, rice bran, as well as milk and other dairy products Tomato, spinach, broccoli, blueberries, mangoes, kiwi, papaya, almonds, hazelnuts, peanuts, wholegrain cereals and vegetable oils
Proteins	Amino acids	C/EBPs	Alters gene expression	Kwashiorkor, marasmus	Egg, milk, soya milk, tofu, yoghurt, cheese, broccoli, almonds, peanuts, cashew, poultry
Fats	Fatty acids Cholesterol	PPARs, SREBPs, LXR, HNF4, ChREBP SREBPs, LXRs, FXR	Alters gene expression	Obesity, CVD, Diabetes	Salmon, sardines, herring, mackerel, soyoil, sunflower oil, palm oil, flaxseeds, rapeseeds, peanuts, walnuts, almonds, mustard seeds, cloves, oregano, cauliflower, broccoli
Minerals	Zinc	MTF1	Chromosome breaks	Brain and immune dysfunction	Oysters, beef, crab, pork, lobster, chicken, spinach, broccoli, cashew nuts, almond, milk, cheese, yogurt
Other Food components	Flavonoids	ER, NFκB, AP1	Alters gene expression	Cancer	Onion, green bean, broccoli, curly kale, endive, celery, cranberry, orange juice, grape fruits, lemons, red, blue and purple berries, peppers, tomatoes and eggplants

Table 1. Transcription-factor pathways mediating nutrient–gene interactions and diseases (Modified from Michael Müller and Sander Kersten, 2003; V. S. Neeha & Priyamvadah Kinth, 2013).

REFERENCES

Afman, L., & Müller, M. (2006). Nutrigenomics: from molecular nutrition to prevention of disease. Journal of the American dietetic association, 106(4), 569-576.

Dang, T. S., Walker, M., Ford, D., & Valentine, R. A. (2014). Nutrigenomics: the role of nutrients in gene expression. Periodontology 2000, 64(1), 154-160.

Gladine, C., & Mazur, A. (2014). Nutrigenomic effects of omega-3 fatty acids. Lipid Technology, 26(10), 227-229.

Massaro, M., Scoditti, E., Carluccio, M. A., Montinari, M. R., & De Caterina, R. (2007). Omega–3 Fatty Acids, Inflammation and Angiogenesis: Nutrigenomic Effects as an Explanation for Anti-Atherogenic and Anti-Inflammatory Effects of Fish and Fish Oils. Journal of nutrigenetics and nutrigenomics, 1(1-2), 4-23.

Müller, M., & Kersten, S. (2003). Nutrigenomics: goals and strategies. Nature Reviews Genetics, 4(4), 315-322.

Neeha, V. S., & Kinth, P. (2013). Nutrigenomics research: a review. Journal of food science and technology, 50(3), 415-428.

Ronteltap, A., Van Trijp, J. C. M., Renes, R. J., & Frewer, L. J. (2007). Consumer acceptance of technology-based food innovations: lessons for the future of nutrigenomics. Appetite, 49(1), 1-17.

Trujillo, E., Davis, C., & Milner, J. (2006). Nutrigenomics, proteomics, metabolomics, and the practice of dietetics. Journal of the American dietetic association, 106(3), 403-413.

Van Ommen, B., & Stierum, R. (2002). Nutrigenomics: exploiting systems biology in the nutrition and health arena. Current opinion in biotechnology,13(5), 517-521.

❒

Chapter-15

LONG CHAIN UNSATURATED FATTY ACIDS AS POTENTIAL ANTIMICROBIAL AGENTS

Priyanka Agarwal, Farheen Jameel and M. Serajuddin*

Department of Zoology, University of Lucknow, Lucknow 226007, India
**Email of Corresponding Author: lu.fisheries@gmail.com*, priagar.87@gmail.com*

ABSTRACT

The pervasiveness of drug-resistant microbial strains and undesirable side effects of currently used antibiotics, accounts for the discovery of alternative antimicrobial compounds with enhanced therapeutic properties for microbial infections. In the present scenario, Long chain unsaturated fatty acids are creating interests as potential remedial antimicrobial agents owing to their efficacy, broad spectrum action and the lack of long-established resistance mechanisms antagonistic to microbes. Predominantly, fatty acid susceptivity is distinctive to gram-positive bacteria, with few exceptional gram-negative species. Long chain unsaturated fatty acids kill microbes by disrupting microbial cell membranes, generating free radicals, augmenting the formation of lipid peroxides that are cytotoxic, activating PMNs, macrophages and T cells to produce other bioactive metabolites, such as prostaglandins, lipoxins, resolvins, protectins and maresins that enhance the phagocytic action of leukocytes and macrophages. Combination therapies are often prescribed during the treatment of bacterial infections because this approach can improve clinical outcome, reduce undesirable side effects of traditional antibiotics and may also reduce the opportunity to select for drug-resistant strains. The most significant role of these long chain unsaturated fatty acids is that, they are growth inhibitory to harmful bacteria but not to commensals.

Keywords: long chain unsaturated fatty acids, antimicrobial activity, free radicals, lipid peroxides.

1. INTRODUCTION

It's a noticeable fact that humans are constantly exposed to harmful pathogenic microorganisms and so the human body must be equipped with certain defensive molecules to safeguard and defend against the potentially harmful microorganisms. The known endogenous defensive molecules include: interferon, cytokines, free radicals, etc. Apart from these molecules, it's also noteworthy that certain lipids could have antimicrobial properties and thus can act as endogenous antimicrobe-like molecule. These antimicrobial lipids have the advantage of being present in all tissues of the body.

The application of certain antibiotics in microbial infections have a decreased clinical efficacy in the present treatment regime which accounts to the pervasiveness of drug-resistant microbial strains (Upton et al., 2003; Zimaity et al., 2004; Moon et al., 2012; Simonart et al., 2005) and some undesirable side effects of the same(Joint Formulary Committee, 2011). This implies for a clinical need to discover alternative antimicrobial compounds with enhanced therapeutic properties for microbial infections. Most investigations too far into the sensitivity of bacteria to fatty acids relies on the study of effects of short chain fatty acids and long-chain saturated and monounsaturated fatty acids and paid less attention to the toxicity of polyunsaturated fatty acids (PUFAs)(Nieman, 1954; Kodicek, 1949; Kabara,

1978; Kanai & Kondo, 1979).

ω-3 Fatty acids are essential dietary PUFAs, including α-linolenic acid (C18:3 n-3, LNA), i.e. abundantly available in some edible plants whereas various long-chain PUFAs (LC-PUFAs), found naturally at high levels in fish and marine vertebrates (Leaf at al., 2003; Berge & Barnathan, 2005)are known to exert highly vigorous effect against Gram-positive bacteria including eicosapentaenoic acid (EPA; C20:5n-3) (Desbois et al., 2009; Desbois, 2013), docosahexaenoic acid (DHA), γ-linolenic acid (GLA; C18:3n-6) (Feldlaufer et al., 1993; Butcher et al., 1976; Asthana et al., 2006; Huang et al., 2010; Zhang et al., 2012) and dihomo-γ-linolenic acid (DGLA; C20:3n-6) (Feldlaufer et al., 1993). Investigations reveal that fatty acid sensitivity is a characteristic of gram-positive bacteria, with exception of few gram-negative species (Nieman, 1954; Kodicek, 1949; Kabara, 1978; Kanai & Kondo, 1979)as is evident from the fact that Gram-negative bacteria possess an impermeable outer membrane, which acts as barricade against many hydrophobic substances including long-chain fatty acids (Galbraith & Miller, 1973; Sheu& Freese, 1973; Sheu et al.,1975).

The anti-bacterial activity of PUFAs against *Staphylococci, streptococci, Mycobacteria, Helicobacter, Bacilli,* enveloped viruses and fungi is well known (Das, 1985; Das, 2006).

Both linoleic and oleic acids form an important antibacterial component in the herbs (*Helichrysum pedunculatum* and *Schotia brachypetala*) used for dressing wounds in South Africa (Dilika et al., 2000; McGaw et al., 2002).Fatty acid derivatives found in microorganisms, algae, or plants, also show effective antimicrobial activities that can be exploited for chemical defense against microorganisms (Pfefferle et al., 1996; Lopez & Gerwick, 1988; Dellar et al., 1996; Lacey & Lord, 1981; McDonald et al., 1981). This implies that naturally occurring free fatty acids may have a therapeutic role.

2. PUFAs AND BACTERIAL STRAINS

The action of some skin surface lipids could act as the first line of defense in humans as is evident from the observation that *Staphylococcus aureus* and coagulase-negative *staphylococci*, group A *streptococci* are present on normal human skin but do not cause any harm. This is supported by the observation that group A streptococcus exposed to oleic acid (OA,18:1n-9) showed decreased survival rates, resulting in the condensation of the nucleoid and distortion of the *streptococcal surface* indicating the abilityof this fatty acid to alter the integrity of the cell membrane (Speert et al., 1979). Moreover, the long-chain unsaturated fatty acids such as linoleic, linolenic and oleic acids are bactericidal to important pathogenic microorganisms, including methicillin-resistant *Staphylococcus aureus* (McDonald and colleagues, 1981; Farrington et al., 1992; Kabara et al.,1972; Knapp &Melly, 1986), *Helicobacter pylori* (Hazell & Graham, 1990; Sun, O'Connor, &Roberton, 2003), and *Mycobacteria* (Seidel & Taylor, 2004).

Shin et al., 2006 reported that EPA significantly reduced the growth of four Gram-positive bacteria, *B. subtilis, L. monocytogenes, S. aureus* ATCC 6538, *S. aureus* KCTC 1916 and one Gram-negative bacterium, *P. aeruginosa,* which is in line with other studies that show long-chain unsaturated fatty acids, including linoleic acid do not inhibit Gram-negative bacteria such as *Escherichia coli* (Dilika et al.,2000; Freese et al., 1973; Kabara et al., 1972; Sun et al., 2003).

ω-6 PUFAs (Arachidonic acid, AA) affect the bacterial cells resulting in peroxidation and generation of free radicals, which in turn, invade the bacterial outer membrane and facilitate the action of antimicrobials (Giamarellos-Bourboulis et al., 1998).Similar mechanism is exhibited by EPA action on *S. aureus* ATCC 6538 and *P. aeruginosa* (Shin et al., 2006).This has been shown by SEM study of EPA treated cells, which demonstrated that both *S. aureus* ATCC 6538 and *P. aeruginosa* has an altered cell morphology in comparison to their control. Such morphological features may be the consequence of lysis of outer membrane followed by loss of cellular electron dense material (indicates the loss of cell constituents and breakdown of the cell wall)(Koyama et al., 1997).

Human lymphocytes contain large amounts of esterified AA (and possibly, EPA and DHA and other PUFAssuch as LA, GLA, DGLA, ALA and their metabolites) that can be released with appropriate stimulation (like cell membrane perturbation due to invading microorganisms)to inactivate viruses and activate PMNs (polymorphonuclear leukocytes), macrophages and T cells to produce other antimicrobial substances such as lipid peroxides, PGs, LTs, lipoxins, resolvins, protectins and maresins, that are critical to their antibacterial, antifungal, antiviral and immunomodulatory actions (Walker et al., 2011; Wu et al., 2016; Sordi et al., 2013; Gobbetti et al., 2014; Das, 1985; Das, 2006;Kohn et al., 1980; Das, 1985; Das et al., 1987a;Das et al., 1987b; Das, 2005; Sangeetha et al., 1990; Das et al., 1990; Das, 1991; Sangeetha et al., 1992.)

The most significant role of these PUFAs is that, theyare able to selectively enhance the growth of useful bacteria (commensals) and, possibly, prevent the proliferation of harmful microbiota. This imply that dietary supplementation of these fatty acids does not cause any deleterious effect on gut microflora (Yuet al., 2014; Pusceddu et al., 2015; Patterson et al., 2014; Costantini et al., 2017; Das, 2002).

3. SYNERGISTIC ACTION

Fatty acids are antimicrobial components of the innate immune system found on mammalian skin and the application of exogenous LC-PUFAs during therapy would augment these natural defenses (Georgel et al., 2005; Drake et al., 2008; Brogden et al.,2012; Thormar & Hilmarsson, 2007; Fluhr et al., 2001). Combination therapies are often prescribed during the treatment of bacterial infections (Bojar & Holland, 2004; Harper, 2004; Joint Formulary Committee, 2011) because this approach can improve clinical outcome, reduce undesirable side effects due to the application of lower doses of individual agents and may also reduce the opportunity to select for drug-resistant strains in target bacteria (Kalan & Wright, 2011) as well as in the non-target skin microbiota (Bojar & Holland, 2004; Moon et al., 2012). Few studies have examined the interactions of fatty acids and LC-PUFAs with clinical antibiotics against bacteria (Clarke et al., 2007; Giamarellos-Bourboulis et al., 2000; Kitahara et al., 2006).

Desbois and Lawlor, 2013 showed that against *S. aureus,* three of the six LC-PUFAs (15-OHEPA, DGLA and HETrE) acted in synergy with the anti-acne agent benzoyl peroxide (BPO), while five of the LC-PUFAs(DGLA, EPA, GLA, 15-hydroxyeicosatrienoic acid [HETrE], and 15-hydroxyeicosapentaenoic acid [15-OHEPA]) acted synergistically with the aminoglycoside antibiotic neomycin (NEO). At this stage, the nature of the synergy between the LC-PUFAs with BPO or NEO is unknown, but one may speculate that they improve the penetration of the clinical agents into the cell because fatty acids can increase cell membrane permeability (Kravchenko et al., 2003).

4. CONCLUSION AND FUTURE PROSPECTS

Long chain unsaturated fatty acids now-a-days are attracting attention as potential therapeutic antimicrobial agents due to their potency, broad spectrum of activity and the lack of classical resistance mechanisms against the actions of microbes. AA and other PUFAs have been shown to possess significant anti-microbial properties for a variety of organisms including bacteria, viruses and fungi. AA seems to possess the ability to enhance immune response (both cellular and humoral),directly inhibit fatty acid synthesis that is critical for bacteria to survive, inactivate enveloped viruses. PUFAs and their metabolites may be used in combinationwith currently available traditional antibiotics to preventand manage various infections. It is also likely that these bioactive lipids could be used to reverse/overcome antibiotic resistance thatis assuming a major issue in fighting many infections. The present study is limited to *in vitro* observations and *in vivo* studies are needed to understand the true potential of applying LC-PUFAs in combination with clinical antimicrobial drugs for treating microbial infections.

ACKNOWLEDGEMENT

The authors like to thank The Head, Department of Zoology, University of Lucknow, Lucknow for the administrative support and UGC, for the award of fellowship.

REFERENCES

Asthana R.K., Srivastava A., Kayastha A.M., Nath G. and Singh S.P. (2006). Antibacterial potential of γ-linolenic acid from *Fischerella* sp. colonising Neem tree bark. World J. Microbiol. Biotechnol. 22: 443–448.

Berge J.P. and Barnathan G. (2005). Fatty acids from lipids of marine organisms: molecular biodiversity, roles as biomarkers, biologically active compounds, and economical aspects. Adv. Biochem. Eng. Biotechnol. 96: 49–125.

Bojar R.A. and Holland K.T. (2004). Acne and *Propionibacterium acnes*. Clin. Dermatol. 22: 375–379.

Brogden N.K., Mehalick L., Fischer C.L., Wertz P.W. and Brogden K.A. (2012). The emerging role of peptides and lipids as antimicrobial epidermal barriers and modulators of local inflammation. Skin Pharmacol. Physiol. 25: 167–181.

Butcher G.W., King G. and Dyke K.G.H. (1976). Sensitivity of *Staphylococcus aureus* to unsaturated fatty acids. J. Gen. Microbiol. 94: 290–296.

Clarke S.R., Mohamed R., Bian L., Routh A.F., Kokai-Kun J.F., Mond J.J., Tarkowski A. and Foster S.J. (2007). The *Staphylococcus aureus* surface protein isdA mediates resistance to innate defenses of human skin. Cell Host Microb. 1: 1–14.

Costantini L., Molinari R., Farinon B. and Merendino N. (2017). Impact of omega-3 fatty acids on the gut microbiota. Int J Mol Sci 18. doi: https://doi.org/10.3390/ijms18122645. pii: E2645.

Das U.N. (1985b). Antibiotic-like action of essential fatty acids. Can Med Assoc J; 132: 1350.

Das U.N. (2006). Do unsaturated fatty acids function as endogenous anti-bacterial and anti-viral molecules? Am J Clin Nutr. 83: 390–1.

Das U.N., Begin M.E., Ells G., Huang Y.S. and Horrobin D.F. (1987b). Polyunsaturated fatty acids augment free radical generation in tumor cells in vitro. Biochem. Biophys. Res. Comm. 145: 15–24.

Das U.N., Huang Y.S., Begin M.E., Ells G. and Horrobin D.F. (1987a). Uptake and distribution of cis-unsaturated fatty acids and their effect on free radical generation in normal and tumor cells *in vitro.* Free Rad. Biol. Med. 3, 9–14

Das U.N., Padma M., Sangeetha P., Ramesh G. and Koratkar R. (1990). Stimulation of free radical generation in human leukocytes by various stimulants including tumor necrosis factor is a calmodulin dependent process. Biochem Biophys Res Commun 167: 1030–6.

Das U.N. (1985a). Can essential fatty acid deficiency predispose to AIDS? Can Med Assoc J 132: 900–2.

Das U.N. (2005). Essential fatty acids and acquired immunodeficiency syndrome. Med Sci Monit 11: RA206–11.

Das U.N. (2002). Essential fatty acids as possible enhancers of the beneficial actions of probiotics. Nutrition 18: 786–9.

Das U.N. (1991). Tumoricidal action of cis-unsaturated fatty acids and its relationship to free radicals and lipid peroxidation. Cancer Lett 56: 235–43.

Dellar J.E., Cole M.D. and Waterman P.G. (1996). Unusual antimicrobial compounds from *Aeollanthus buchnerianus*. Experientia 52: 175–9.

Desbois A.P., Mearns-Spragg A. and Smith V.J. (2009). A fatty acid from the diatom *Phaeodactylum tricornutum* is antibacterial against diverse bacteria including multi-resistant *Staphylococcus aureus* (MRSA). Mar. Biotechnol. 11: 45–52.

Desbois A.P. and Lawlor K.C. (2013). Antibacterial Activity of Long-Chain Polyunsaturated Fatty Acids against *Propionibacterium acnes* and *Staphylococcus aureus*. Mar. Drugs 11; 4544-4557; doi:10.3390/md11114544.

Desbois, A.P. (2013). Antimicrobial properties of eicosapentaenoic acid (C20: 5n-3). In Marine Microbiology: Bioactive Compounds and Biotechnological Applications; Kim, S.-K., Ed.; Wiley-VCH Verlag GmbH & Co. KGaA: Weinheim, Germany; 351–367.

Dilika F., Bremner P.D. and Meyer J.J.M. (2000). Antibacterial activity of linoleic and oleic acids isolated from *Helichrysum pedunculatum*: a plant used during circumcision rites. Fitoterapia 71: 450–2.

Drake D.R., Brogden K.A., Dawson D.V. and Wertz P.W. (2008). Antimicrobial lipids at the skin surface. J. Lipid Res. 49: 4–11.

El-Zimaity D., Kearns A.M., Dawson S.J., Price S. and Harrison G.A.J. (2004). Survey characterisation and susceptibility to fusidic acid of *Staphylococcus aureus* in the Carmarthen area. J. Antimicrob. Chemother. 54: 441–446.

Farrington M., Brenwald N., Haines D. and Walpole E. (1992). Resistance to desiccation and skin fatty acids in outbreak strains of methicillin-resistant *Staphylococcus aureus*. Journal of Medical Microbiology 36: 56–60.

Feldlaufer M.F., Knox D.A., Lusby W.R. and Shimanuki H. (1993). Antimicrobial activity of fatty acids against *Bacillus* larvae, the causative agent of American foulbrood disease. Apidologie 24: 95–99.

Fluhr J.W., Kao J., Jain M., Ahn S.K., Feingold K.R. and Elias P.M. (2001). Generation of free fatty acids from phospholipids regulates stratum corneum acidification and integrity. J. Invest. Dermatol. 117: 44–51.

Freese E., Shew C. W. and Galliers E. (1973). Function of lipophilic acids as antimicrobial food additives. Nature 241: 321–325.

Galbraith H. and Miller T. B. (1973). Effect of long-chain fatty acids on bacterial respiration and amino acid uptake. Journal of Applied Bacteriology 36: 659–675.

Georgel P., Crozat K., Lauth X., Makrantonaki E., Seltmann H., Sovath S., Hoebe K., Du X., Rutschmann S., Jiang Z., Bigby T., Nizet V., Zouboulis C.C. and Beutler B. (2005). A toll-like receptor 2-responsive lipid effector pathway protects mammals against skin infections with Gram-positive bacteria. Infect. Immun. 73: 4512–4521.

Giamarellos-Bourboulis E. J., Grecka P., Dionyssiou-Asteriou A. and Giamarellou H. (2000). Impact of n-6 polyunsaturated fatty acids on growth of multidrug-resistant *Pseudomonas aeruginosa:* Interactions with amikacin and ceftazidime. Antimicrobial Agents Chemotheraphy 44: 2187–2189.

Gobbetti T., Coldewey S.M., Chen J., McArthur S., le Faouder P., Cenac N., et al. (2014). Nonredundant protective properties of FPR2/ALX in polymicrobial murine sepsis. Proc Natl Acad Sci U S A 111: 18685–90.

Harper J.C. (2004). An update on the pathogenesis and management of *acne vulgaris.* J. Am. Acad. Dermatol. 51: S36–S38.

Hazell S. L. and Graham D. Y. (1990). Unsaturated fatty acids and viability of *Helicobacter (Campylobacter) pylori.* Journal of Clinical Microbiology 28: 1060–1061.

Huang C.B., George B. and Ebersole J.L. (2010). Antimicrobial activity of n-6, n-7 and n-9 fatty acids and their esters for oral microorganisms. Arch. Oral Biol. 55: 555–560.

Joint Formulary Committee. British National Formulary, 61st ed. (2011). British Medical Association and Royal Pharmaceutical Society of Great Britain: London, UK.

Kabara J. J., Swieczkowski D. M., Conley A. J. and Truant J. P. (1972). Fatty acids and derivatives as antimicrobial agents. Antimicrobial Agents Chemotheraphy 2: 23–28.

Kabara J.J. (1978). Fatty acids and derivatives as antimicrobial agents. A review. In: Kabara J.J., ed. The pharmacological effect of lipids. Champaign, Ill: American Oil Chemist Society: 1-14.

Kalan L. and Wright G.D. (2011). Antibiotic adjuvants: multicomponent anti-infective strategies. Expert Rev. Mol. Med. 13: 1–17.

Kanai K. and Kondo E. (1979). Antibacterial and cytotoxic aspects of long-chain fatty acids as cell surface events: selected topics. Japan J Med Sci Biol32: 135-74.

Kitahara T., Aoyama Y., Hirakata Y., Kamihira S., Kohno S., Ichikawa N., Nakashima M., Sasaki H. and Higuchi S. (2006). *In vitro* activity of lauric acid or myristyl amine in combination with six antimicrobial agents against methicillin-resistant *Staphylococcus aureus* (MRSA). Int. J. Antimicrob. Agents 27: 51–57.

Knapp H. R. and Melly, M. A. (1986). Bactericidal effects of polyunsaturated fatty acids. Journal of Infectious Diseases 154: 84–94.

Kodicek E. (1949). The effect of unsaturated fatty acids on gram-positive bacteria. Society for Experimental Biology Symposia 3: 218-31.

Kohn A., Gitelman J. and Inbar M. (1980). Unsaturated free fatty acids inactivate animal enveloped viruses. Arch Virol 66: 301–7.

Koyama S., Yamaguchi Y., Tanaka S. and Motoyashiya J. (1997). A new substance (yoshixol) with an interesting antibiotic mechanism from wood oil of Japanese traditional tree (kiso-hinoki), *Chamaecyparis obtuse*. General Pharmacology 28: 797–804.

Kravchenko I.A., Golovenko N.Y., Larionov V.B., Aleksandrova A.I. and Ovcharenko N.V. (2003). Effect of lauric acid on transdermal penetration of phenazepam *in vivo*. Bull. Exp. Biol. Med., 6: 579–581.

Lacey R.W. and Lord V.L. (1981). Sensitivity of staphylococci to fatty acids: novel inactivation of linolenic acid by serum. J Med Microbiol 14: 41–9.

Leaf A., Kang J. X., Xiaoand Y. F. and Billman G. E. (2003). Clinical prevention of sudden cardiac death by n-3 polyunsaturated fatty acids and mechanism of prevention of arrhythmias by n-3 fish oils. Circulation 107: 2646–2652.

Lopez A. and Gerwick W.H. (1988). Ptilodene, a novel icosanoid inhibitor of 5-lipoxygenase and Na+/K+ ATPase from the red marine alga *Ptilotafilicina*. J Agardh Tet

Lett 29: 1505–6.

McDonald M.I., Graham I., Harvey K.J. and Sinclair A. (1981). Antibacterial activity of hydrolysed linseed oil and linolenic acid against methicillin-resistant *Staphylococcus aureus*. Lancet 2: 1056.

McGaw L.J., Jäger A.K. and van Staden J. (2002). Isolation of antibacterial fatty acids from *Schotiabrachypetala*. Fitoterapia 73: 431–3.

Moon S.H., Roh H.S., Kim Y.H., Kim J.E., Ko J.Y. and Ro Y.S. (2012). Antibiotic resistance of microbial strains isolated from Korean acne patients. J. Dermatol. 39: 833–837.

Nieman C. (1954). Influence of trace amounts of fatty acids on the growth of microorganisms. Bacteriol Reviews18: 147-63.

Patterson E., O' Doherty R.M., Murphy E.F., Wall R., O' Sullivan O., Nilaweera K., et al. (2014). Impact of dietary fatty acids on metabolic activity and host intestinal microbiota composition in C57BL/6J mice. Br J Nutr 111:1905–17.

Pfefferle C., Kempter C., Metzger J.W. and Fiedler H.P. (1996). E-4-oxonon-2-enoic acid, an antibiotically active fatty acid produced by *Streptomyces olivaceus* Tu 4018. J Antibiot 49: 826–8.

Pusceddu M.M., El Aidy S., Crispie F., O'Sullivan O., Cotter P., Stanton C., et al. (2015). N-3 Polyunsaturated fatty acids (PUFAs) reverse the impact of early-life stress on the gut microbiota. PLoS ONE;10. e0139721.143;

Shina S.Y., Bajpaia V.K., Kimb H.R. and Kanga S.C. (2007). Antibacterial activity of eicosapentaenoic acid (EPA) against foodborne and food spoilage microorganisms. LWT 40: 1515–1519.

Sangeetha P., Das U.N. and Koratkar R. (1990). Free radical generation in human leukocytes by cis unsaturated fatty acids is a calmodulin dependent process. Prostaglandins Leukot Essen Fatty Acids 39: 27–32.

Sangeetha Sagar, Das U.N., Koratkar R., Ramesh G., Padma M. and Kumar G.S. (1992). Cytotoxic action of cis-unsaturated fatty acids on human cervical carcinoma (HeLa) cells: relationship to free radicals, and lipid peroxidation and its modulation by calmodulin antagonists. Cancer Lett; 63: 189–98.

Seidel V. and Taylor, P. W. (2004). *In vitro* activity of extracts and constituents of Pelagonium against rapidly growing mycobacteria. International Journal of Antimicrobial Agents 23:613–619.

Sheu C. W. and Freese E. (1973). Lipopolysaccharide layer protection of Gram-negative bacteria against inhibition by long-chain fatty acids. Journal of Bacteriology, 115: 869–875.

Sheu C. W., Salomon D., Simmons J. L., Sreevalson T. and Freese E. (1975). Inhibitory effects of lipophilic acids and related compounds on bacteria and mammalian cells. Antimicrobial Agents Chemotheraphy 7: 349–363.

Simonart T. and Dramaix M. (2005). Treatment of acne with topical antibiotics: Lessons from clinical studies. Br. J. Dermatol. 153: 395–403.

Sordi R., Menezes-de-Lima Jr O., Horewicz V., Scheschowitsch K., Santos LF. and Assreuy J. (2013). Dual role of lipoxin A4 in pneumosepsis pathogenesis. Int Immunopharmacol 17: 283–92.

Speert D.P., Wannamaker L.W., Gray E.D. and Clawson C.C. (1979). Bactericidal effect of oleic acid on group A *streptococci*: mechanism of action. Infect Immun 26: 1202–10.

Sun C. Q., O'Connor C. J. and Roberton A. M. (2003). Antibacterial actions of fatty acids and monoglycerides against *Helicobacter pylori*. FEMS Immunology Medical Microbiology 36: 9–17.

Thormar H. and Hilmarsson H. (2007). The role of microbicidal lipids in host defense against pathogens and their potential as therapeutic agents. Chem. Phys. Lipids150: 1–11.

Upton A., Lang S. and Heffernan H. (2003). *Mupirocin* and *Staphylococcus aureus*: A paradigm of emerging antibiotic resistance. J. Antimicrob. Chemother. 51: 613–617.

Walker J., Dichter E., Lacorte G., Kerner D., Spur B., Rodriguez A., et al. (2011). Lipoxin a4 increases survival by decreasing systemic inflammation and bacterial load in sepsis. Shock 36: 410–6.

Wu B., Capilato J., Pham M.P., Walker J., Spur B., Rodriguez A., et al. (2016). Lipoxin A4 augments host defense in sepsis and reduces *Pseudomonas aeruginosa* virulence through quorum sensing inhibition. FASEB J 30: 2400–10.

Yu H.N., Zhu J., Pan W.S., Shen S.R., Wei-Guang Shan, Das U.N. (2014). Effects of fish oil with high content of n-3 polyunsaturated fatty acids on mouse gut microbiota. Arch Med Res 45: 195–202.

Zhang H., Zhang L., Peng L., Dong X., Wu D., Wu V.C. and Feng F. (2012). Quantitative structure-activity relationships of antimicrobial fatty acids and derivatives against *Staphylococcus aureus*. J. Zhejiang Univ. Sci. B13: 83–93.

❒

Chapter-16

ROLE OF FISH SKIN MUCUS IN NATURAL RESISTANCE OF FISH AGAINST PATHOGENS

Farheen Jameel, Priyanka Agarwal and M. Serajuddin*

Department of Zoology, University of Lucknow, Lucknow 226007, India
**Email of Corresponding Author: lu.fisheries@gmail.com*

ABSTRACT

Fish being in direct contact with its aquatic environment, need to develop effective strategies to fight all kinds of challenges. Fish skin mucus, secreted by the goblet cells located in the epidermis of fish is the key innate immune component and is considered to be the first defense blockade of fish, providing mechanical and physical protection to the fish. It can trap particles, viruses and bacteria which are removed from the mucosa by water current. It also contains many immunological defense parameters such as immunoglobulins, antimicrobial peptides (AMPs) and complement factors. Immunoglobulin-like molecules have been found in few fish species and only one functional class of immunoglobulin, IgM is found in teleost's. AMPs rapidly move to the site of infection because of their smaller size, positive charge and amphiphilic structure and interact with negatively charged cytoplasmic membranes of microbes, inhibiting cell wall synthesis leading to destruction of microbes. Complement factors include lectins, like pentraxins which are pathogen recognition proteins and mannose-binding lectins, which act as opsonin's or bacterial agglutinins. From the above-mentioned immunological defense parameters, AMPs can be the potential candidates for developing antimicrobial agents, however human clinical trials are still required. More studies are needed to further evaluate factors involved in mucus microbial colonization and fish disease.

Keywords: Fish skin mucus, Antimicrobial activity, Immunoglobulins, AMPs, Complement factors.

1. INTRODUCTION

All living organisms including fish acquire multiple defense mechanisms to combat with the wide range of pathogenic and non-pathogenic microorganisms. Innate and acquired are the two type of immune systems that an organism possesses which provide resistance against various diseases. Antimicrobial substances belong to the innate class of immune system. In fishes, epidermal mucus is the key innate immune component, secreted by the goblet cells present in the epidermis of fish. It functions as a biological interface between fish and its aquatic environment (Shephard, 1993). This layer is thought to have a role in locomotion, osmoregulation (Cameron and Endean, 1973), lubrication (Rosen and Cornford, 1971), mechanical protection (Cameron and Endean, 1973) and in intra-specific chemical communication (Saglio and Blanc, 1989). The mucus layer on the surface of fish is replaced continuously in order to prevent stable colonization of bacteria, parasites and fungi (Nagashima et al., 2003). Many studies suggest that bactericidal activity of fish epidermal mucus is due to the presence of antimicrobial peptides (AMPs) as they are produced constitutively or induced upon infection in fish skin mucus (Sarmaşik, 2002). AMPs have been studied to display a broad-spectrum antimicrobial activity against wide range of gram-negative and gram-positive bacteria, parasites, fungi, yeast, enveloped viruses', pathogenic microorganisms with little or no toxicity to the host cells (Hancock, 2001; Boulanger et al., 2002).

The initial determination of antibacterial activity of epidermal mucus was done by Austin and McIntosh (1988)in rainbow trout *(Oncorhynchus mykiss)*. The rate of susceptibility to bacterial infection is found to be increased after removal of epidermal mucus in *Cyprinus carpio* (Lemaitre et al., 1996). The loss of skin mucus also increased mortality in several fish species, when challenged with *Listonella anguillarum* (Fouz et al., 1990). Antimicrobial activity seems to differ among different fish species such as tilapia *(Tilapia hornorum)* (Tendencia et al., 2004), rockfish *(Sebastes schlegelii)* (Kitani et al., 2008), rainbow trout *(Oncorhynchus mykiss)* (Raida and Buchmann, 2009) and perhaps specific towards certain bacteria (Noya et al., 1995).

2. SKIN MUCUS AND BACTERIAL POPULATION

Fish being in direct contact with its aquatic environment, need to develop effective strategies to fight all kinds of challenges. Effectively, the fish skin mucus is considered to be the first defence blockade of fish (acts as physical and biological barrier), since it can trap particles, viruses and bacteria which are removed from the mucosa by water current (Mayer, 2003). It can also trap and immobilize pathogens before they can contact the epithelial surfaces (Cone, 2009). The bacterial populations present in the fish and water probably share many similarities and the bacterial genera present in the epidermal mucus seems to come from the environment or diet (Diler et al., 2000; Austin, 2006). Moreover, in farmed fishes, the bacterial population changes depending on the diet, storage conditions (PH, temperature, etc) and pollutants concentration in water and on the season (Bernadsky and Rosenberg, 1992).

The pathogenic bacterial species isolated from fish include those belonging to the genus *vibrio (V. ordalii, V. anguillarum, V. alginoly-ticus, Vibrio fluvialis, V.fischeri, V.* albensis, V. scophthalmi, V. harveyi, V. splendidus biotype I), genus Aeromonas (A. schubertii, A. bestiarum, A. jandaei, Aero-monas hydrophila, A. veronii biovar sobria, A. *Caviae); genus Photobacterium (P. angustum, P. damselae, P. damselae subsp. Piscicida, P. logei); Enterobacter aerogenes, Pseudomonas fluorescens, Micrococcus luteus,* Flavobacterium psychro-philum, Moraxella spp., Alcaligenes faecalis, Psychrobacters, Flexibacter spp., Acinetobacter johnsonii, Acinetobacter calcoaceticus, Escherichia coli, Neisseria, coryneforms, Hyphomicrobium vulgare, Alteromonas, Prosthecomicrobium, Caulobacter, Plesiomonas shigelloides, Bacillus cereus, Yersinia ruckeri, Alcaligenes piechaudii, Moraxella, Cytophaga/ Flexibacter, B. Firmus and Pseudomonas marina (Diler et al., 2000; Gonzalez et al., 2000, 2001; Zmyslowska et al., 2001; Austin, 2006; Anja et al., 2000).

3. IMMUNOLOGICAL ROLE OF FISH SKIN MUCUS

Besides providing mechanical and physical protection, skin mucus of fish contains many immunological defense parameters including immunoglobulins, antimicrobial peptides (AMPs) and complement factors (Ellis, 2001; Magnadottir, 2005; Whyte, 2007).

Immunoglobulin-like molecules have been found in few fish species and only one functional class of immunoglobulin, IgM is found in teleost's. (Wilson et al., 1997; Randelli et al., 2008).

Several AMPs isolated from skin secretions of fish, had shown antimicrobial functions such as epinecidin-1 against *Candida albicans* (diploid fungus) (Pan et al., 2009), *Trichomonas vaginalis* (flagellated protozoan) and nervous necrosis virus (Wang et al., 2010), H2B peptide from Atlantic cod against *Aeromonas hydrophila* and *Saprolegnia* (fish pathogens) (Bergsson, 2005), piscidin-2 from striped bass that disrupts fungal membranes.

After microbial infection, AMPs rapidly move to the site of infection due to their small size, positive charge and amphiphilic structure (Zasloff, 2002; Hancock and Sahl, 2006; Hancock and Lehrer, 1998). Their positive charge and hydrophobic nature induce them to interact with negatively charged cytoplasmic membranes of microbes hindering cytoplasmic membrane septum formation, transmembrane pore formation, inhibiting cell wall synthesis, which leads to the destruction of microbes in the external environment (Matsuzaki et al., 1997; Rathinakumar et al., 2009). Complement factors include lectins, like pentraxins which are pathogen recognition proteins (Bayne and Gerwick, 2001) and mannose-binding lectins, which act as opsonins or bacterial agglutinins (Russell and Lumsden, 2005).

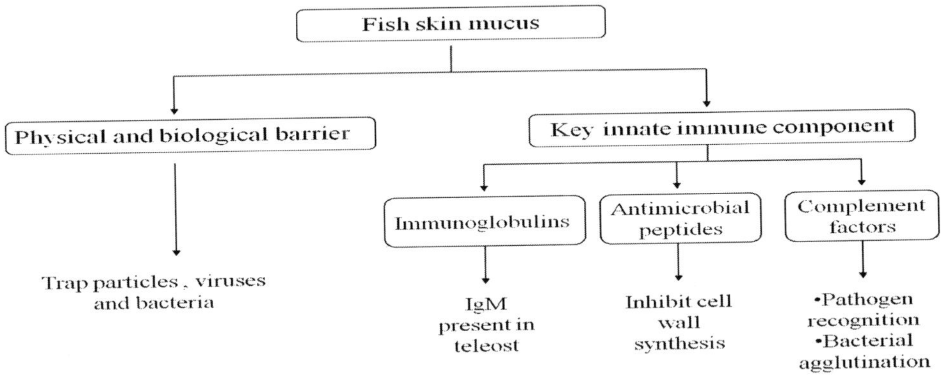

Fig.1. Role of fish skin mucus in defense against microbial pathogens.

4. CONCLUSION

The skin mucus of fish has a critical role in the defense mechanisms of fishes via acting as a mechanical and biological barrier and also containing immunological defense parameters. Concomitantly, AMPs can be the potential candidates for developing antimicrobial agents, however human clinical trials are still required. More studies are needed to further evaluate factors involved in mucus microbial colonization and fish disease and on adhesins which play important role in the establishment of initial interaction of microbes with mucosal surfaces. The physicochemical properties of skin mucus also need to be better understood, in order to design convenient defensive strategies to better defend fish against invading pathogens.

ACKNOWLEDGEMENT

The authors like to thank The Head, Department of Zoology, University of Lucknow, Lucknow for the administrative support and UP-CST, Lucknow for the award of fellowship.

REFERENCES

Anja SS, Morten SB, Inger D, Karl P, Jens LL(2000) Occurrence of antimicrobial resistance in fish-pathogenic and environmental bacteria associated with four Danish rainbow trout farms. Appl Environ Microbiol 66: 4908–4915.

Austin B(2006) The bacterial microflora of fish. Scientific World J 6: 931– 945.

Austin, B. and D. McIntosh (1998) Natural antibacterial compounds on the surface of rainbow trout, *Salmo gairdneri* Richardson. Journal of Fish Diseases 11: 275-277.

Bayne CJ, Gerwick L (2001) The acute phase response and innate immunity of fish. Dev Comp Immunol 25:725–743.

Bergsson G, Agerberth B, Jörnvall H, Gudmundsson GH (2005) Isolation and identification of antimicrobial components from the epidermal mucus of Atlantic cod *(Gadus morhua)*. FEBS J272:4960–9.

Bernadsky G, Rosenberg E(1992) Drag-reducing properties of bacteria from the skin mucus of the Cornetfish *(Fistularia commersonii)*. Microb Ecol 24: 63–76.

Boulanger N, Munks RJL, Hamilton JV, Vovelle F, Brun R, Lehane MJ, Bulet P (2002) A novel antimicrobial peptide with antiparasitic activity in the blood-sucking insect *Stomoxys calcitrans*. J Biol Chem 277:49921–49926.

Cameron A, Endean R (1973) Epidermal secretions and evolution of venom glands in fishes. Toxicon 11: 401-10.

Cone RA(2009) Barrier properties of mucus. Adv Drug Deliv Rev 61: 75– 85.

Diler O, Altun S, Calikusu F, Diler A(2000) A study on qualitative and quantitative bacterial flora of the rainbow trout *(Oncorhynchus mykiss)* living in different fish farms. Turk J Vet Anim Sci 24: 251–259.

Diler O, Altun S, Calikusu F, Diler A(2000) A study on qualitative and quantitative bacterial flora of the rainbow trout *(Oncorhynchus mykiss)* living in different fish farms. Turk J Vet Anim Sci 24: 251–259.

Ellis AE (2001) Innate host defence mechanisms of fish against viruses and bacteria. Dev Comp Immunol 25:827–839.

Fouz B, Devesa S, Gravningen K, Barja JL, Toranzo AE (1990) Antibacterial action of the mucus of turbot. Bull Eur Assoc Fish Pathol 10: 56-59.

Gonzalez CJ, Santos JA, Garcia-Lopez ML, Gonzalez N, Otero A(2001) Mesophilic aeromonads in wild and aquaculture freshwater fish. J Food Prot 64: 687–691.

Gonzalez CJ, Santos JA, Garcia-Lopez ML, Otero A(2000) *Psychrobacters* and related bacteria in freshwater fish. J Food Prot 63: 315– 321.

Hancock REW (2001) Cationic peptides: effectors in innate immunity and novel antimicrobials. Lancet Infect Dis 1:156–164.

Hancock REW, Lehrer R (1998) Cationic peptides: a new source of antibiotics. Trends Biotechnol 16: 82–88.

Hancock REW, Sahl HG (2006) Antimicrobial and host-defence peptides as new anti-infective therapeutic strategies. Nat Biotechnol 24: 1551–1557.

Kitani Y, Kikuchi N, Zhang GH, Ishizaki S, Shimakura K, Shiomi K, Nagashima Y (2008) Antibacterial action of Lamino acid oxidase from the skin mucus of rockfish *Sebastes schlegelii*. Comp Biochem PhysiolB 149: 394-400.

Lemaitre C, Orange N, Saglio P, Saint N, Gagnon T, Molle G (1996)Characterization and ion channel activities of novel antimicrobial proteins from the skin mucosa of carp *(Cyprinus carpio)*. Eur J Biochem 240: 143-149.

Magnadottir B, Lange S, Gudmundsdottir S, Bøgwald J, Dalmo RA (2005) Ontogeny of humoral immune parameters in fish. Fish Shellfish Immunol 19:429–439.

Matsuzaki K, Sugishita KI, Harada M, Fujii N, Miyajima K (1997) Interactions of an antimicrobial peptide, magainin 2, with outer and inner membranes of Gram negative bacteria. Biochim Biophys Acta 1327: 119–130.

Nagashima Y, Kikuchi N, Shimakura K, Shiomi K (2003) Purification and characterization of an antibacterial factor in the skin secretion of rock fish *Sebastes schlegeli*. Comp Biochem Physiol 136C:63–71.

Noya M, Magarinos B, Toranzo AE, and Lamas J (1995) Sequential pathology of experimental Pasteurellosis in Gilthead seabream *Sparus aurata*–a light-microscopic and electron-microscopic study. Diseases of Aquatic Organisms 21: 177-186.

Pan CY, Chen JY, Lin TL, Lin CH (2009) *In vitro* activities of three synthetic peptides derived from epinecidin-1 and an anti-lipopolysaccharide factor against *Propionibacterium acnes, Candida albicans,* and *Trichomonas vaginalis.* Peptides 30:1058–68.

Raida MK, Buchmann K(2009) Innate immune response in rainbow trout *(Oncorhynchus mykiss)* against primary and secondary infections with *Yersinia ruckeri.* Dev Comp Immunol 33: 35-45.

Randelli E, Buonocore F, Scapigliati G (2008) Cell markers and determinants in fish immunology. Fish Shellfish Immunol 25:326–340.

Rathinakumar R, Walkenhorst WF, Wimley WC (2009) Broad-spectrum antimicrobial peptides by rational combinatorial design and high-throughput screening: the importance of interfacial activity. J Am Chem Soc 131: 7609–7617.

Russell S, Lumsden JS (2005) Function and heterogeneity of fish lectins. Vet Immunol Immunopathol 7th Int Vet Immunol Symp 108:111–120.

Saglio P, Blanc J (1989) Intraspecific chemo-communication in immature goldfish, *Carassius auratus* (L): attraction in olfactometer to free amino acid fractions from skin extracts. Biol Behav 14:132-47.

Sarmaşik A (2002) Antimicrobial peptides: a potential therapeutic alternative for the treatment of fish diseases. Turk J Biol 26:201–207.

Shephard KL (1993) Mucus on the epidermis of fish and its influence on drug delivery. Adv Drug Deliv Rev 11:403–417.

Tendencia EA, Dela Pena MR, Fermin AC, Lio-Po G, Casiano H, Choresca JR, Inui Y (2004) Antibacterial activity of tilapia *Tilapia hornorum* against *Vibrio harveyi.* Aquaculture 232: 145-152.

Wang YD, Kung CW, Chen JY (2010) Antiviral activity by fish antimicrobial peptides of epinecidin-1 and hepcidin 1–5 against nervous necrosis virus in medaka. Peptides 31:1026–33.

Whyte SK (2007) The innate immune response of finfish - A review of current knowledge. Fish Shellfish Immunol 23:1127–1151.

Wilson M, Bengten E, Miller NW, Clem LW, Du Pasquier L, Warr GW (1997) A novel chimeric Ig heavy chain from a teleost fish shares similarities to IgD. Proc Natl Acad Sci USA 94:4593–4597.

Zasloff M (2002) Antimicrobial peptides of multicellular organisms. Nature 415: 389–395.

Zmyslowska I, Lewandowska D, Nowakowski T, Kozlowski J (2001) Occurrence of bacteria in water and in vendace *(Coregonus albula)* during rearing in tanks. Pol J Environ Stud10: 51–56.

❐

Chapter-17

SCRUTINIZING OVARIAN CARCINOMA: AN UNWRAPPED GYNECOLOGICAL MALIGNANCY

Juhi Rais, Madhu Tripathi and Md. Arshad*

Molecular Endocrinology Lab
Department of Zoology, University of Lucknow, Lucknow-226007
**Email of Corresponding Author: arshadm123@rediffmail.com*; juhirais44@gmail.com*

ABSTRACT

The Ovary, being an endocrine gland secretes different types of hormones which binds with their respective nuclear receptors and carry out proper functioning of the female reproductive system. However, malfunctioning of any of the receptors or in the development of the ovum can be lethal for the proper functioning of the system and can cause various diseases among which reproductive carcinomas are ranked at the top. Ovarian carcinoma is the 5th most common type of carcinoma that occurs in females and causes more deaths than any other gynecological carcinomas as the early detection of its symptoms are quite subtle and are usually detected at advance grades, thereby shortening the survival rate of an individual. The exact cause of ovarian carcinoma still remains unknown. Multiple cycles of mutation and selection progresses the disease. The experts are adopting many measures for curing the disease but still full control over the problem has not been achieved as the genetic risk factors, mRNA expression profiles and molecular events of oncogenesis are different for each subtype. Therefore, it's necessary to find new ways to develop drugs for this deadly gynecological malignancy.

Keywords: Ovarian Carcinoma, Ovarian Neoplasms, Chemotherapy.

1. INTRODUCTION

The Ovary, one of the principle organs of the female reproductive tract, often found in pairs and produces the female gamete, the Ovum. Ovaries in females are analogous to that of testes in males as both act as gonads as well as endocrine glands. The proper functioning of the female reproductive system is due to the binding of the ovarian hormones with their respective nuclear receptors. However, malfunctioning of any of the receptors or in the development of the ovum can be lethal for the proper functioning of the system and can cause various diseases among which reproductive carcinomas are ranked at the top.

The malignant tumors which arise from the modification of peritoneal mesothelium and cover the surface of the ovary such tumors are termed as Ovarian Carcinoma (Fox, 1993). Ovarian carcinoma causes more deaths than any other gynecological carcinomas as the early detection of its symptoms are quite subtle and are usually detected at the advance grades, thereby shortening the survival rate of an individual (Jayson et al., 2014). The exact cause of ovarian carcinoma still remains unknown.

The Ovarian carcinoma is the 5th most common type of carcinoma that occurs in females. Each year worldwide more than 200000 new cases are diagnosed which constitutes 4% of all carcinomas diagnosed in women and there are 6.6 cases per 100000 women per year (Ferlay, et al., 2010; Hannibal et al., 2008). Over the years the mortality rate of the ovarian carcinoma has not changed having the 5 years of survival rate but from the last 2 decades the small improvements have been recorded as the use of Cisplatin drug has increased the survival rate of the patients when diagnosed in early stage of diagnosis. In more than 80% cases, optimal cytoreductive surgery and a systemic chemotherapy can respond well to a neoplasm which is usually benign (Chan et al., 2006). But in over 50% of women, the Epithelial Ovarian Carcinoma recurs despite of complete remission (CR) and first line chemotherapy. Therefore, experts all over the world are continuously working in finding ways for early detection and treatment of this disease.

2. PRODROMES OF OVARIAN CARCINOMA

In the early stages of ovarian carcinoma, the signs and symptoms are usually absent; if present then it may be very subtle (Jayson et al., 2014). It takes several years to recognize and diagnose in women having ovarian carcinoma. Usually ovarian carcinoma in its early stage is painless but as the mass grows ovarian torsion occurs (Seiden, Michael V, 2012). Some of the common symptoms of ovarian carcinoma are as follows (Seiden, Michael V, 2012; Goff BA, 2012; Schorge, et al., 2008).

- Swelling of the stomach or bloating
- Pelvic or abdominal pain
- Difficulty in eating and indigestion occurs
- Frequently gets fatigue
- Constipation and upset stomach
- Menstrual changes
- Weight loss
- When gets more severe and advanced, accumulates fluid in abdomen and forms ascites

3. RISK FACTORS RELATED TO OVARIAN CARCINOMA

Ovarian carcinoma risk factors are generally hormonal in nature but some of the common risk factors are listed as under:

Age:

Higher the age higher will be the risk of developing ovarian carcinoma. It develops usually after menopause. Half of the diagnosed ovarian carcinomas are developed in women of 63 years of age or older but can affect largely to a population of young women (Schorge et al., 2008).

Obesity:

Various studies have revealed that the ovarian carcinoma and obesity relationship go hand in hand. The women with BMI of at least 30 are at the higher risk of getting ovarian carcinoma (Schorge et al., 2008).

Reproductive History and Gynecological Surgery:

Women who conceive before the age of 26 are at the lower risk of developing ovarian carcinoma whereas the women who never carried a pregnancy after 35 years of age are at higher risk. Oral contraceptives taken by women lower the risk of getting ovarian carcinoma. Prolonged use of Birth control pills even lowers the risk of ovarian carcinoma. Borderline ovarian tumor formation can occur due to the use of fertility medication (Jayson et al., 2014). Those who are successfully treated for infertility and subsequently give birth are at no higher risk stage but those who remain nulliparous besides being treated for infertility will be at higher risk of getting ovarian carcinoma (Schorge et al., 2008; Hoffman et al., 2012). Ovarian Carcinoma is also associated with hormonal conditions such as polycystic ovary syndrome and endometriosis (Jayson et al., 2014). Postmenopausal hormone replacement therapy (HRT) with estrogen increases the risk of ovarian carcinoma (Schorge et al., 2008; Manson and Bassuk, 2005). Tubal ligation may reduce upto two thirds and hysterectomy (removing the uterus) reduces upto one third of the risk of developing ovarian carcinoma.

Family History of Ovarian Carcinoma, Breast Carcinomaor Colorectal Carcinoma:

In 10% of ovarian carcinoma cases, major genetic risk factor for it is a mutation in BRCA1 or BRCA2 DNA mismatch repair genes (Seiden, Michael V, 2012; Schorge, et al., 2008). A strong family history of endometrial carcinoma, colon carcinoma, or other gastrointestinal carcinomas collectively indicates a syndrome known as hereditary nonpolyposis colorectal carcinoma (also known as Lynch syndrome), if this syndrome is present then it confers a higher risk for developing a number of carcinomas, including ovarian carcinoma (Seiden, Michael V, 2012; Schorge, et al., 2008). Other genes that have been associated with ovarian carcinoma are BRIP1, MSH6, RAD51C, RAD51D CDH1, CHEK2, PALB2 and RAD50 are the other genes which are associated with ovarian carcinoma (Norquist et al., 2015; Kuusisto et al., 2011).

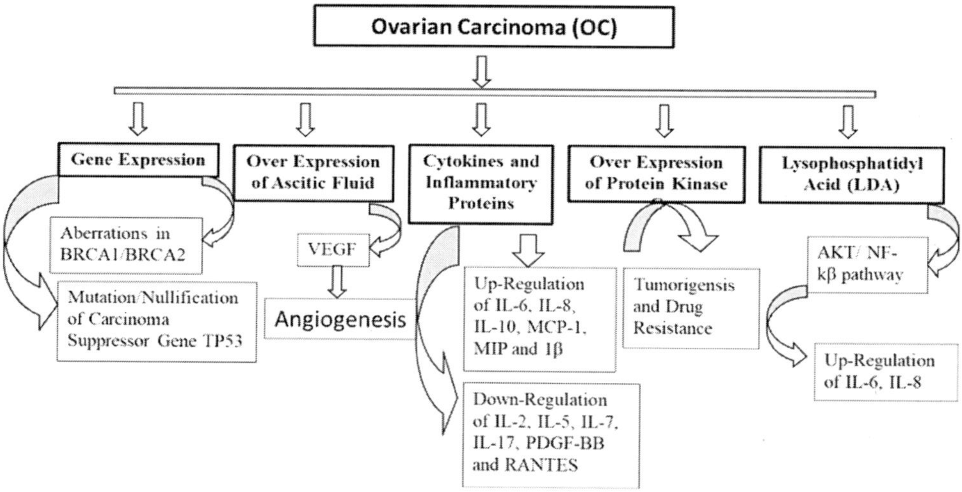

Table 1. The different factors that is responsible for Ovarian Carcinoma

4. OVARIAN CARCINOMA AND ITS TYPES

According to the World Health Organization's classification of ovarian tumors, it recognizes them as serous, mucinous, endometrioid, clear cell and squamous ovarian neoplasms (Fox, 1993). On the basis of histological and genetical features the ovarian carcinomas are divided into 2 types: type I and type II where type I carcinomas are distinguished as low histological grade like endometrioid, mucinous and clear cell carcinomas and the type II are the high histological grade ovarian carcinomas which are generally serous and epithelial ovarian carcinomas (Jayson et al., 2014).

Epithelial Ovarian Carcinoma:

About 90-95% of all ovarian cases the surface epithelial-stromal tumors mark their highest presence. The neoplasms of this group can either be benign or malignant and can arise from any among the three sites: they can be either ovarian surface epithelium (modified peritoneum) or from ectopic endometrial or fallopian tube tissue. It includes serous tumor, endometrioid and mucinous cyst adenocarcinoma, clear cell and Brenner tumors and also a few mixed types of tumors are found (Berek and Robert, 2003).

Serous Tumors:

Approximately 80-85% of all ovarian carcinomas in western industrialized countries are serous (Seidman et al., 2004). The serous carcinomas are morphologically heterogeneous as they are transformed and progressed from other tumor types and thus shows a very broad spectrum of histological appearances. The tumors vary in their sizes as they can be very small or can be too large that can fill the entire abdominal cavity. The serous ovarian carcinoma is lined by tall columnar, ciliated epithelial cells which are filled with serous fluid. Most of the serous carcinoma has papillary architecture but some has glandular, cribiform (Burks et al., 1996; Seidman and Kurman, 1996), solid microcystic (Che et al., 2001) and trabecular architechture.

Mucinous Tumors:

A lining of tall columnar epithelial cells marks the characterstics of muscinous tumors. These are usually bilateral and are closely related to their serous counterparts. Mucinous tumors account for about less than 3% of all ovarian neoplasms (Seidman et al., 2004; Seidman and Kurman, 1996). These tumors can form very large cystic masses. Mullerian mucinous or endocervical mucinous or seromucous or mixed epithelial neoplasms with a mucinous component are the Mucinous borderline tumors lacking goblet cells are classified separately from intestinal mucinous neoplasms and most primary mucinous carcinomas display transitions from intestinal mucinous borderline to carcinoma (Gong et al., 2013; Moreno–Bueno et al., 2001).

Endometrioid Tumors:

Malignant tumors characterizing with atypical cells with poor differentiation and frequent mitosis which makes the tumor more solid are endometrioid tumors which accounts approximately 20% of all ovarian carcinomas (Hoffman et al., 2012). They are beingn as well as malignant and about 40% of the tumors are bilateral which are of malignant type. High levels of microsatellite instability (Moreno–Bueno et al., 2001; Catasus et al., 2004) have been reported in the mutations of CTNNB-1 (A-catenin) (Moreno–Bueno et al., 2001; Catasus et al., 2004; Irving et al., 2005; Wu et al., 2001), PI3CA (encoding phosphotidylinositol 3-kinase [PI3K]) (Wu et al., 2007; Willner et al., 2007) and PTEN (Catasus et al., 2004; Stewart, 2012).

Clear Cell Tumors:

Epithelial cells filled with abundant cytoplasm which may be solid or cystic are known as clear cell tumors (Jayson et al., 2014). They have following characterstics unilateral, low stage, round papillae elongated, hyaline, edematous stroma hobnail cells, cuboidal, monolayer, uniform nuclei and decreased mitotic activity (Yamamoto et al., 2007; Han et al., 2008). In high clear cell carcinoma microsatellite instability (Ueda et al., 2005; Gras et al., 2001) have been reported as well as Mutations in K-ras (Otsuka et al., 2004; Amemiya et al., 2004) and PTEN (Stewart, 2012).

Brenner Tumors:

Brenner tumors are characterized by transitional cells are uncommon. These tumors may be very small or may be very large and can be either solid or cystic (Schorge et al., 2008).

Sex-Cord Stromal Tumors:

Estrogen producing granulose cell tumor, benign thecoma are included under sex-cord stromal tumors which can arise from different cells of mesenchyme, they are the major hormone producing ovarian tumors (Hoffman et al., 2012). They generally include endocrine cells and fibroblasts and occur mostly in women between 50-69 years of age but can occur in young girls too and are usually unilateral and are very aggressive in nature (Schorge et al., 2008).

Germ Cell Tumors:

The tumors which develop from the germ cells are known as Germ cell tumors. Most of the germ cell tumors are benign teratomas and accounts for about 30% of ovarian carcinomas but only 5% of them are malignant when one of the germ cells gets itself converted into squamous cell carcinoma (Schorge et al., 2008; Hoffman et al., 2012). They are generally massive and can occur in young women too.

Secondary Ovarian Carcinoma:

The metastasis from a primary carcinoma apart from ovaries in the body and then affects the ovary, makes the ovarian carcinoma as the secondary carcinoma (Schorge et al., 2008). Some of the common primary carcinomas are colon, breast, appendical and stomach carcinoma and accounts for about 7% of total ovarian carcinoma (Schorge et al., 2008).

5. STAGING OF OVARIAN CARCINOMA

FIGO (International Federation of Gynecology and Obstetrics) has staged the ovarian carcinoma as it has obtained the information after surgery which includes an abdominal hysterectomy, removal of the ovaries and fallopian tubes, the omentum, pelvic (peritoneal) washings and pelvic biopsies (Hoffman et al., 2012; Jayson et al., 2014; Singer et al., 2005). The FIGO stages are as follows: (Hoffman et al., 2012; Singer et al., 2005).

Stage I: In Stage I the carcinoma is completely limited to the ovary. But the Stage I is further divided into 3 main types. They are as follows:

- Stage IA: It involves one ovary with capsule intact. No tumor can be found on ovarian surface and pelvic washing is found to be negative.
- Stage IB: It involves both ovaries with capsule intact but still no tumor can be found on ovarian surface and pelvic washing is found to be negative.
- Stage IC: In this stage tumor involves one or both the ovaries. This stage is divided as:
    - Stage IC1: It shows surgical spill
    - Stage IC2: Capsule has ruptured or tumor gets on ovarian surface
    - Stage IC3: It shows positive pelvic washings

Stage II: It involves one or both the ovaries and pelvic extension of the tumor. It is divided into 2 subtypes:

- Stage IIA: Tumor usually found on uterus or fallopian tubes.
- Stage IIB: Tumor can be found elsewhere in pelvis.

Stage III: In this stage carcinoma can be found outside the pelvis or in the retroperitoneal lymph nodes and usually involves one or both the ovaries. It has following subtypes:

- Stage IIIA: It involves metastasis in retroperitoneal lymph nodes or microscopic extrapelvic metastasis. It has following types:
- ➢ Stage IIIA1: It involves metastasis in retroperitoneal lymph nodes.
    a. Stage IIIA (i): In this the metastasis is less than 10mm in diameter.
    b. Stage IIIA (ii): The metastasis is greater than 10mm in diameter
- ➢ Stage IIIA2: It involves microscopic metastasis in the peritoneum, regardless of retroperitoneal lymph node status.
    - Stage IIIB: In this stage metastasis encroaches peritoneum by less than or equal to 2cm in diameter but metastasis doesn't occur in lymph, liver or spleen capsule.
    - Stage IIIC: In this stage metastasis encroaches peritoneum greater than or equal to 2cm in diameter still metastasis doesn't occur in lymph, liver or spleen capsule.

Stage IV: At this stage distant metastasis occurs i.e. metastasis occurs outside the peritoneum:

- Stage IVA: It involves pleural effusion containing carcinoma cells.
- Stage IVB: This stage involves metastasis occurring to distant organs that includes the parenchyma of the spleen, liver and metastasis also occurs to the inguinal and extra abdominal lymph nodes.

6. TREATMENT AND MANAGEMENT OF OVARIAN CARCINOMA

The management of ovarian carcinoma is done by treating the patients with surgeries, chemotherapies and sometimes radiotherapy is needed. These treatments are performed on the basis of the subtype of the carcinoma the patient is suffering from. Some of the treatments are listed below:

Surgery:

Surgery is the preferred treatment and depends upon how widespread the carcinoma is when diagnosed (the carcinoma stage), as well as its type and grade of carcinoma. Surgery may include removal of one (unilateral oophorectomy) or both ovaries (bilateral oophorectomy), the Fallopian tubes (salpingectomy), the uterus (hysterectomy), and the omentum (omentectomy). For low-grade, unilateral stage-IA carcinomas, only the Fallopian tube or the involved ovary (which must be unruptured) will be removed and if any metastases are found, a second surgery is done to remove the remaining ovary and uterus (Jayson et al., 2014). Debulking surgery, a procedure is performed in advanced carcinomas, where complete removal is not an option. This surgery in women at stage- IV of the disease with extensive metastases in the peritoneum are less likely to be successful (Jayson et al., 2014; Ferlay et al., 2010).

Several different surgical procedures are employed to treat ovarian carcinoma and for stage I and II carcinoma, laparascopic surgery can be used, but in metastases cases it may not be used. For advance stage ovarian carcinoma, lymph-adenectomy should be included in the surgery (Jayson et al., 2014; Ferlay et al., 2010). Some of the procedures like biopsies throughout the peritoneum and abdominal lymphatic system, omentectomy, diaphragm stripping or resection bilateral salpingo-oophorectomy, splenectomy, bowel resection, appendectomy, or a posterior pelvic exenteration can be used to cure the disease depending upon the extent of ovarian carcinoma (Schorge et al., 2008).

Chemotherapy:

After surgery Chemotherapy is used to treat any residual disease and is used to treat ovarian carcinoma since decades. When chemotherapy is performed before a surgery it is called as Neoadjuvant chemotherapy butthis surgery cannot completely remove or optimally debulk a tumor (Jayson et al., 2014). If the tumor is not completely removed during surgery or if the carcinoma is stage IV, Bevacizumab may be used as adjuvant chemotherapy. Approximately 20% of advanced ovarian carcinomas can be cured by chemotherapy (Schorge et al., 2008), malignant germ cell tumors are often more curative than epithelial tumors (Hoffman et al., 2012).

Chemotherapy in ovarian carcinoma usually consists of a group of platinum-based drugs known as platins that are combined with non-platins. Paclitaxel, cisplatin, topotecan, doxorubicin, epirubicin, and gemcitabine are included under common therapies. Carboplatin in combination with either paclitaxel or docetaxel is given as carboplatin is less toxic and has fewer side effects and is superior to cisplatin (Jayson et al., 2014; Berek and Robert, 2003). Chemotherapy can be given in the peritoneal cavity or intravenously (Ferlay et al., 2010). Intraperitoneal chemotherapy causes more adverse side effects than intravenous chemotherapy (Berek and Robert, 2003) and is used when the carcinoma has been optimally debulked.

Radiation Therapy:

In stage 1c and 2 carcinomas, if there is the possibility of residual disease in the pelvis, radiation therapy is used after surgery. In advanced carcinomas Radiotherapy can also be used in palliative care. 5 days a week for 3–4 weeks is a typical course of radiotherapy for ovarian carcinoma. Diarrhea, constipation, and frequent urination are the common side effects of radiotherapy.

Hormonal Therapy:

Ovarian carcinoma is rarely responsive to hormonal treatments though 60% of ovarian tumors have estrogen receptors. Estrogen alone does not have an effect on the carcinoma, whereas tamoxifen and letrozole are rarely effective (Jayson et al., 2014).

7. CONCLUSION AND FUTURE PROSPECTS

Ovarian carcinoma is highly aggressive neoplasm as ovarian carcinoma can arise from any cell type of the ovary by immortalizing and transforming the normal cells into the tumor cell. Ovarian carcinomais often chemo-resistant due to the lack of adequate diagnostic or screening strategies, and thus results in the recurrence of disease. Tumor debulking, chemotherapies, radiotherapies, targeted therapies, immunotherapies, stem cell transplants, photodynamic therapies are some of the measures that are been adopted by the experts for curing the disease but still full control over the problem is not being achieved. Therefore, through translational studies new drugs should be characterized that use several methods like bioinformatics, cell culture system, xenograft and genetic mouse models followed by extended phase II clinical protocols with innovative trial designs. Thus, it's the need of an hour to find better treatment strategies to combat this devastating disease.

ACKNOWLEDGEMENT

Author Juhi Rais acknowledges Department of Science and Technology for providing DST INSPIRE Fellowship to carry out the research work.

REFERENCES

Amemiya S, Sekizawa A, Otsuka J, Tachikawa T, Saito H, Okai T. Malignant transformation of endometriosis and genetic alterations of K-ras and microsatellite instability. International Journal of Gynecology & Obstetrics 2004;86(3):371-76.

Berek, Jonathan S., and Robert C. Bast Jr. "Epithelial ovarian cancer." (2003).

Burks RT, Sherman ME, Kurman RJ. Micropapillary serous carcinoma of the ovary: a distinctive low-grade carcinoma related to serous borderline tumors. Am J Surg Pathol 1996; 20:1319-30.

Catasus L, et al., Molecular genetic alterations in endometrioid carcinomas of the ovary: similar frequency of beta-catenin abnormalities but lower rate of microsatellite instability and PTEN alterations than in uterine endometrioid carcinomas. Human pathology 2004;35(11):1360-68.

Chan JK, et al., Patterns and progress in ovarian cancer over 14 years. Obstet & Gynecol 2006; 108:521-28.

Che M, et al., Ovarian mixed-epithelial carcinomas with a microcystic pattern and signet-ring cells. Int J Gynecol Pathol 2001; 20: 323-28.

Ferlay J, et al., GLOBOCAN 2008 v1.2, Cancer Incidence and Mortality Worldwide: IARC CancerBase No. 10 [Internet]. Lyon, France: International Agency for Research on Cancer; 2010.

Fox H, Pathology of early malignant change in the ovary. Int J Gynecol Pathol 1993; 12:153-55.

Goff BA, Ovarian cancer: screening and early detection. Obstetrics and gynecology clinics of North America 2012; 39(2):183-94.

Gong TT, Wu QJ, Vogtmann E, Lin B, Wang YL. Age at menarche and risk of ovarian cancer: A meta-analysis of epidemiological studies. International Journal of Cancer 2013;132(12): 2894-2900.

Gras E, et al. Microsatellite instability, MLH-1 promoter hypermethylation, and frameshift mutations at coding mononucleotide repeat microsatellites in ovarian tumors. Cancer 2001;92(11):2829-36.

Han G, et al., Mixed ovarian epithelial carcinomas with clear cell and serous components are variants of high-grade serous carcinoma: an interobserver correlative and immunohistochemical study of 32 cases. The American journal of surgical pathology 2008;32(7):955-64.

Hannibal CG, Cortes R, Engholm G, Kjaer SK, Survival of ovarian cancer patients in Denmark: Excess mortality risk analysis of five-year relative survival in the period 1978–2002. Acta obstet et gynecol Scand 2008; 87:1353-60.

Hoffman B, et al., Epithelial Ovarian Cancer, Williams gynecology. McGraw Hill Professional 2012.

Irving JA, et al., Synchronous endometrioid carcinomas of the uterine corpus and ovary: alterations in the β-catenin (CTNNB1) pathway are associated with independent primary tumors and favorable prognosis. Human pathology 2005;36(6):605-19.

Jayson GC, Kohn EC, Kitchener HC, Ledermann JA, Ovarian cancer. The Lancet 2014;384(9951):1376-88.

Kuusisto KM, Bebel A, Vihinen M, Schleutker J, Sallinen SL, Screening for BRCA1, BRCA2, CHEK2, PALB2, BRIP1, RAD50, and CDH1 mutations in high-risk Finnish BRCA1/2-founder mutation-negative breast and/or ovarian cancer individuals. Breast Cancer Res 2011;13(1): R20.

Manson JE, Bassuk SS. The menopause transition and postmenopausal hormone therapy. Harrisons Principles of Internal Medicine 2005;16(2):2209.

Moreno–Bueno G, Gamallo C, Pérez–Gallego L, de Mora JC, Suárez A, Palacios J. β-catenin expression pattern, β-catenin gene mutations, and microsatellite instability in endometrioid ovarian carcinomas and synchronous endometrial carcinomas. Diagnostic Molecular Pathology 2001;10(2):116-22.

Norquist BM et al., Inherited Mutations in Women with Ovarian Carcinoma. JAMA oncology 2015;1-9.

Otsuka J, et al., K-ras mutation may promote carcinogenesis of endometriosis leading to ovarian clear cell carcinoma. Medical Electron Microscopy 2004;37(3):188-92.

Schorge J, et al., Williams gynecology 2008.

Seiden, Michael V, "Gynecologic Malignancies". Williams gynecology. McGraw Hill Professional 2012.

Seidman JD, et al., The histologic type and stage distribution of ovarian carcinomas of surface epithelial origin. Int J Gynecol Pathol 2004; 23:41-44.

Seidman JD, Kurman RJ. Subclassification of serous borderline tumors of the ovary into benign malignant types: a clinicopathologic study of 65 advanced stage cases. Am J Surg Pathol 1996; 20:1331-45.

Singer G, et al., Patterns of p53 mutations separate ovarian serous borderline tumors and low-and high-grade carcinomas and provide support for a new model of ovarian carcinogenesis: a mutational analysis with immunohistochemical correlation. Am J Surg Pathol 2005; 29:218-24.

Stewart SL. Ovarian Cancer Incidence: Current and Comprehensive Statistics. INTECH Open Access Publisher 2012.

Ueda H, Watanabe Y, Nakai H, Hemmi H, Koi M, Hoshiai H. Microsatellite status and immunohistochemical features of ovarian clear-cell carcinoma. Anticancer research 2005; 25(4):2785-88.

Willner J, et al., Alternate molecular genetic pathways in ovarian carcinomas of common histological types. Human pathology 2007;38(4):607-13.

Wu R, et al., Mouse model of human ovarian endometrioid adenocarcinoma based on somatic defects in the Wnt/β-catenin and PI3K/Pten signaling pathways. Cancer cell 2007;11(4):321-33.

Wu R, Zhai Y, Fearon ER, Cho KR. Diverse mechanisms of β-catenin deregulation in ovarian endometrioid adenocarcinomas. Cancer research 2001;61(22):8247-55.

Yamamoto S, et al. Clear cell adenocarcinoma associated with clear cell adenofibromatous components: a subgroup of ovarian clear cell adenocarcinoma with distinct clinicopathologic characteristics. The American journal of surgical pathology 2007;31(7):999-1006.

❐

Chapter-18

ROLE OF INFLAMMATORY BIOMARKERS IN PATHOPHYSIOLOGY OF OSTEOARTHRITIS

Amit Kumar, Md. Arshad and Suchit Swaroop*

Experimental and Public Health Laboratory Department of Zoology
University of Lucknow, Lucknow-226007, India
**Email of Corresponding Author: drsswaroop@gmail.com*

1. INTRODUCTION

Osteoarthritis (OA) is the most prevalent degenerative, (Dominick et al 2014) multifunctional, inflammatory, joint disorder characterized by intra-articular inflammation with synovitis, degeneration of articular cartilage, and change in peri-articular and sub-chondral bone (Steeve et al, 2004). OA is associated with chronic disability between 4th and 5th decade of life (Lutzner et al, 2009) and associated with progressive number of risk factors that are obesity, age, gender, smoking, drug abuse and other systemic disorders. OA is basically a common type arthritis which is considered in the category of polygenic diseases. Unlike other form of arthritis, OA affects only joints and not internal structure of bone or organs. Between the bone joints, the degradation of extra cellular protein matrix such as collagen and aggrecan may lead to develop osteoarthritis (Sandell L and Aigner T, 2005) .This destruction of glycolated proteins from bone joint may cause stiffness, loss of motion in joint, change in joint structure and pain that are symptoms of OA. These kinds of symptoms affect any joint such as hands, hips, spine and knees. Knees joint related osteoarthritis is most common in obese person.

2. MORPHOLOGICAL DAMAGE OF GLYCOLATED EXTRA CELLULAR MATRIX

OA is characterized by progressive catabolism of articular cartilage, sub chondral bone remodeling and joint inflammation. Over catabolism induces the destabilizing joint structure and limiting functional ability of joint (Choi et al, 2013). OA is considered as degenerative disorder because of disruption of matrix equilibrium. Under normal case, articular chondrocytes make a dynamic homeostasis of ECM in the joints (Nakata et al. 1993). The articular chondrocytes maintain the normal level of anabolism and catabolism of ECM such as collagen type-II and aggrecan. These extra cellular components are most abundant proteoglycan (PG) present in articular cartilage of joint (Sandell L and Aigner T, 2005). In osteoarthritic state, the catabolism of ECM increases, whereas synthesis is reduced. Due to this, overall loss of cartilage tissue occurs and thus increases the rate of progression of disease.

In early stage of the disease progression, certain biochemical changes occur, the collagen and aggrecan glycoproteins are disorganized in joints space cause a decrease in proteoglycan content. The disorganization and degradation of proteoglycan cause net increase in water content within the cartilage.

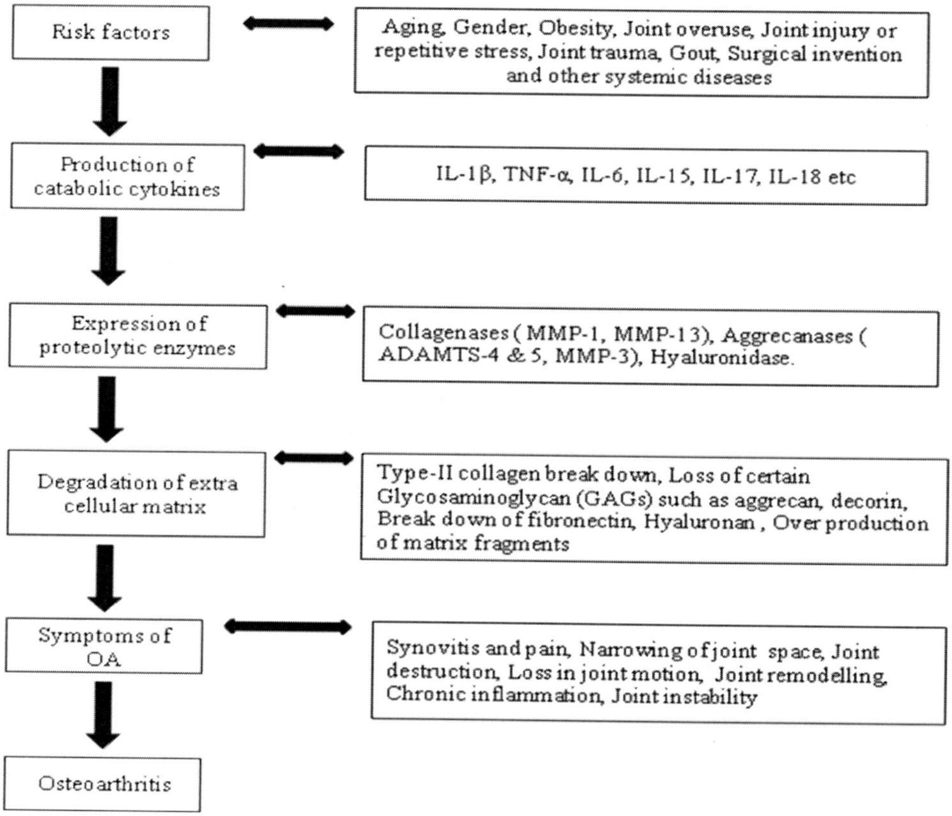

Fig. Schematic representation of overall pathophysiology of the OA.

3. BIOMARKERS INVOLVED IN OSTEOARTHRITIS

Proinflammatory biomarkers:

Proinflammatory cytokines are responsible for disruption of chondrocyte homeostasis. During progression of OA, the large number of cytokines and growth factor production is up regulated in the synovial membrane, cartilage, and subchondral bone (Malemud CJ, 2004). Proinflammatory cytokines seem to be involved in underlying mechanism of cartilage destruction. These cytokines are mainly TNF-α, IL-1, IL-6 and IL-17 (Pelletier *et al,* 2001).

Inteleukin-1 (IL-1):

IL-1 is well studied as pro inflammatory cytokines plays important role in both extracellular matrix degradation as well as induction of nociceptive pathways (Benito *et al,* 2004). IL-1 has been shown to involve in inhibition of ECM synthesis and cartilage break down. Expression of these components in chondrocyte is repressed by IL-1 (Goldring *et al,* 1988). IL-1 induces excessive amount of proteolytic enzymes production such as collagenases and Aggrecanases (ADAMTS-4 &5, MMP-3).IL-1 induces the various cell surface adhesion molecules includes Inter Cellular Adhesion Molecules-1 (ICAM-1/CD54), L-selectin, Vascular endothelial cell adhesion molecule-1(VCAM-1/CD106) (Tsang *et al,* 1997).IL-1 is shown to induce the release of other pro-inflammatory cytokines such as IL-6, IL-8 and other small inflammatory factors, where these have additive or synergistic action on cartilage breakdown.

Interleukin-6 (IL-6):

IL-6 is a glycoprotein of having 184 amino acid residues (Hammacher et al, 1994) pro-inflammatory cytokine and involve in supporting role as destruction of extra cellular matrix and inflammation. It has been demonstrated that, IL-6 is responsible for hypersensitivity in joint cartilage and hyperalgesia (Brenn et al, 2007). IL-6 is known to involve in activation of immune system. IL-1β and TNF-α play a role in the production of IL-6 in the effected joint cartilage tissue and implemented via various joint supporting cells such as chondrocyte, osteoblast cells (Guerne et al, 1990). A role of IL-6 is the induction of pain, interestingly it effects to afferent neurons and pain propagation occur (Obreja et al, 2005). IL-6 is a key mediators that role in to the modification of layer of subchondral bone (Chenoufi et al, 2001).

Tumor Necrosis Factor- alpha (TNF-α):

TNF-α is a pro inflammatory cytokines which plays a role in pathophysiology of OA (Sommer C, 2003). Their action is the catabolic effect on cartilage breakdown and also induces the sensory neurons through the TNF-α receptor 1, 2 (TNFR1 and TNFR2). It causes initiation of events of cascade in inflammatory reactions with other pro-inflammatory cytokines such as IL-1, IL-6 etc (Aoki et al, 2004).

TNF-α is mainly synthesized in synovial chondrocyte cells. Elevated level of TNF-α was also detected in synovial membrane, cartilage tissue, sub-chondral bone layer (Farahat et al, 1993). TNF-α has affinity to binds with two receptor isotypes (TNR-R1 and TNFR2) are present on extra cellular leaflet of about all nucleated cells (Grell et al, 1995). Through the ligand binding to and activation of TNF-R1, there may be two cascade events occurs viz signal transduction through MMPK to the nucleus or via activating caspase-8 toward apoptosis. TNF-α induced neuropathic pain may be blocked through application of anti-inflammatory medication. Ibuprofen and celecoxib are known to reduce the pain through their anti-inflammatory properties (Schafer et al, 2004).

4. ANTI-INFLAMMATORY BIOMARKERS

The most important group of anti-inflammatory cytokines is mainly IL-4, IL-10 and IL-14. These cytokines are belongs to a broad cytokine protein family.

Interleukin-4 (IL-4):

IL-4 behaves like a secreted small ligand that is recognized by IL-4 receptor (IL-4R) (Mueller et al, 2002). The IL-4R is basically divide into two types, Type-I which is formed by dimerization of IL-4Rα and IL-2Ryc receptor chain and recognize by IL-4 whereas Type-II that is composed of two different receptor chains such as IL-4Rα and IL-13Rα1 and recognize by both the ligand such as IL-4 and IL-13 (Laporte et al, 2008).

5. SUMMARY

OA is joint destructive and irreversible disease affecting normal daily life activities and causes negative impact on economics of society. OA is a most complex disorder cannot easily describe the pathophysiological processes involved in disease development. Certain catabolic factors have supplementary role in degeneration of joint supportive muscles and cartilages that lead to joint inflammation, reduction in joint motion and pain. These symptoms irreversibly appear as person gets affected with OA. OA exhibits progressive but gradual loss of ECM. IL-1 has considered as primary catabolic mediator in evolution of the disease by inducing synthesis of ECM degradative enzymes include MMP-1, MMP-13, Aggrecanases (ADAMTS-4 & 5, MMP-3), and hyaluronidase. These enzymes are responsible for degradation of most abundant glycoprotein such as collagen type-II. The production of proteolytic enzymes from chondrocyte is under the influence of IL-1β induced intra cellular signaling down to nucleus. IL-1β primarily induces signal transduction through binding with its type-I receptor whereas binding with type-II receptor leads to inactivation of the receptor and inhibition of further signal transduction. Number of cytosolic kinases and transcription factors are positively associated in signal transmission towards nucleus. Apart from above role of IL-1β, it also induces the nociceptive pathway and leads to pain. Previous studies do not clearly mention the role and significant association of IL-1 family cytokine in Osteoarthritis. Therefore, further study is needed to find out the exact role of IL-1 family cytokines in OA. It may be an unpredictable task to know the overall effect that will be positive.

REFERENCES

Aoki Y, Ohtori S, Ino H, Douya H, Ozawa T, Saito T, Moriya H, Takahashi K. Disc inflammation potentially promotes axonal regeneration of dorsal root ganglion neurons innervating lumber intervertebral disc in rats. Spline, 2004, vol.29:30, 2621-2626.

Benito MJ, Veale DJ, FitzGerald O, Van Den Berg WB, Bresnihan B. Synovial tissue inflammation in early and late osteoarthritis. Annals of the Rheumatic Diseases, 2005.64: 1263-1267

Brenn D, Richter F, Schaible H. Sensitization of unmyelinated sensory fibres of the joint nerve to mechanical stimuli by interleukin-6 in the rat: An inflammatory mechanism of joints pain. Arthritis Rheum. 2007 doi: 10. 1002/art.22282.

Chenoufi HL, Diamant M, Rieneck K, Lund B, Stein GS, Lain JB. Increased mRNA expression and protein secretion of interleukin-6 in primary human osteoblasts differentiated *in vitro* from rheumatoid and osteoarthritic bone. Journal of Cellular Biochemistry, 2001. 84:666-678.

Choi YS, Park JK, Kang EH, et al. Cytokine signalling-1 suppressor is inducible by IL-1 beta and inhibits the catabolic effects of IL-1 beta in chondrocytes: its implication in the paradoxical joint-protective role of IL-1 beta. Arthritis Research & Therapy, 2013.15:R191.

Dominick J. Blasioli, MS, and David L. Kaplan. The Roles of Catabolic Factors in the Development of Osteoarthritis. Tissue Engineering part B, 2013. dio:10.1089/ten.teb.0377.

Farahat MN, Yanni G, Poston R, Panayi GS. Cytokine expression in synovial membrane of patients with rheumatoid arthritis and osteoarthritis. Annals of the Rheumatic Diseases, 1993. Vol. 52. 12: 870-875.

Goldring MB, Birkhead J, Sandell LJ, Kimura T, Krane SM. Interleukin 1 suppresses expression of cartilage-specific types II and IX collagens and increases types I and III collagens in human chondrocytes. J. Clin. Invest, 1988.82:2026-2037.

Grell M, Douni E, Wajant H. The transmembrane form of tumor necrosis factor is the prime activating ligand of the 80 kDa tumor necrosis factor receptor. Cell, 1995, vol. 82: 793-802.

Guerne P A, Carson D A, and Lotz M. IL-6 production by human articular chondrocytes. Modulation of its synthesis by cytokines, growth factors, and hormones *in vitro*. Journal of Immunology, 1990, vol.144, 2: 499-505.

Hammacher A, Ward LD, Weinstock J, Treutlein H, Yasukawa K, Simpson RJ. Structure-function analysis of human IL-6: identification of two distinct regions that are important for receptor binding. Protein Science, 1994, vol. 3: 12, 2280-22933.

Laporte SL, Juo ZS, Vaclavikovaet at. Molecular and structural basis of cytokine receptor pleiotropy in the interleukin-4/13 system. Cell, 2008.132:257-272.

Lutzner J, Kasten P, Günther KP, Kirschner S. Surgical options for patients with osteoarthritis of the knee. Nat Rev Rheumatol, 2009.5:309-16.

Malemud CJ. Cytokines as therapeutic targets for osteoarthritis. BioDrugs, 2004.18:23-35

Mueller TD, Zhang JL, Sebald W, Duschl. Structure, binding, and antagonistic in the IL-4/IL-13 receptor system, Biochemiaet Biophysica Acta- Molecular Cell Research, 2002.1592:237-250.

Nakata K, Ono K, Miyazaki J, Olsen BR, Muragaki Y, Adachi E, Yamamura K, et al. Osteoarthritis associated with mild chondrodysplasia in transgenic mice expressing alpha 1 (IX) collagen chains with a central deletion. Proceedings of the National Academy of Sceinces of the United States of America. 1993; 90(7):2870-2874. Dio10.1073/pnas.90.7.2870.

Obreja O, Biasio W, Andatsch M, Lips KS, Rathee PK, Ludwig A, Rose-John S, Kress M. Fast modulation of heat-activated ionic current by pro inflammatory interleukin 6 in rat sensory neurons. Brain.2005, 128, 1634-1641.

Pelletier JP, Martel-Pelletier J, Abramson SB. Osteoarthritis, an inflammatory disease: potential implication for the selection of mew therapeutic targets. Arthritis Rheum, 2001.44:1237-1247.

Sandell L, Aigner T. Articular cartilage and changes in arthritis. An introduction: cell biology of osteoarthritis. Arthritis Res, 2005 doi: 1186/ar148.

Schafers M, Marziniak M, Sorkin L, Yaksh T,.Cyclooxygenase inhibition in nerve-injury and TNF induced hyperalgesia in the rat. Exp Neurology. 2004, doi: 10.1016.

Sommer C. Recent findings on how pro inflammatory cytokines cause pain: peripheral mechanisms in inflammatory and neuropathic hyeralgesia. Neuroscience letters, 2004 dio: 10.1016.

Steeve KT, Marc P, Sandrine T, Dominique H, Yannick F. IL-6, RANKL, TNF-aplha/IL-1: interrelations in bone resorption pathophysiology. Cytokines and Growth Factor Reviews, 2004. 15:49-60.

Tsang YT, Neelamegham S, Hu Y, et al. Synergy between L-selectin signaling and chemotactic activation during neutrophil adhesion and transmigration. J Immunol, 1997.159:4566-77.

❐

Chapter-19

Hepatitis C Virus and its Genotype

Amar Deep[1,2], Ajay Kumar[2] and Suchit Swaroop[1*]

[1]*Experimental and Public Health Laboratory, Department of Zoology, University of Lucknow, Lucknow*

[2]*Department of Medical Gastroenterology, King George's Medical University, Lucknow*

**Email of Corresponding Author: drsswaroop@gmail.com*; jsa.amardeep@gmail.com; drajaymd12345@gmail.com*

ABSTRACT

Hepatitis C virus (HCV) is a hepatotropic virus of family Flaviviridae and genus Hepacivirus having single stranded RNA of positive polarity as genomic material. A large number of genotypes have been identified among hepatitis C virus isolates from all over the world. Hepatitis C virus (HCV), an important cause of chronic hepatitis, cirrhosis and hepatocellular carcinoma (HCC), shows a considerable genetic heterogeneity among hepatitis C virus isolates from all over the world. At least six main groups of sequence variants are recognized. Hepatitis C virus (HCV) infection is a global public health burden, causing an increasing level of liver-related morbidity and mortality due to the disease progression. Unfortunately, in many countries, there is a lack of robust epidemiological data, especially HCV genotypes distribution, upon which to base country specific prevention, diagnosis and treatment strategies in order to reduce the disease burden represented by HCV. The natural history of disease and response to treatment may be related to the genotype of HCV in a particular patient.

Keywords: Hepatitis C virus, chronic hepatitis.

1. INTRODUCTION

Chronic hepatitis C is the most common cause of chronic liver disease and cirrhosis which further develop into HCC. HCV was initially isolated from the serum of a person with non-A, non-B hepatitis in 1989 by Choo et al. [Choo et al 1989] Shortly after the cloning of HCV, this newfound virus was discovered to be the cause of approximately 90% of non-A, non-B hepatitis in the U.S. According to National Institutes of Health Consensus Development Conference Statement, approximately 170 million people are affected with HCV worldwide, comprising ~3% of the global population. HCV is also known as silent killer because most of the times discovered only by routine serological, biochemical and radiological testing. Many attempts to identify the natural history and progression of hepatitis C infection have been made and several aspects remain to be elucidated. [Lauer et al 2001]

HCV replicates in the cytoplasm of hepatocytes but is not directly cytopathogenic. Persistent infection appears to rely on rapid yield of virus and continuous cell-to-cell spread, along with a lack of vigorous T-cell immune response to HCV antigens. The HCV turnover rate can be rather high with replication ranging between 10^{10} to 10^{12} virions per day, and a predicted viral half-life of 2 to 3 hours. [Neumann, et al 1998] The rapid viral replication and lack of error proofreading by the viral RNA polymerase are reasons why the HCV RNA genome mutates frequently. [Bukh et al 1995] In understanding the long-term outcomes of HCV infection, clinicians may identify the patients at risk for HCV-related complications and offer treatments to prevent further morbidity and mortality.

2. MODES OF TRANSMISSION

The transmission of HCV is mainly through exposure to infected blood. Risks for transmission include blood transfusion, intravenous drug abuse, high-risk sexual activity, organ transplantation from an infected donor, occupational exposure, hemodialysis, household exposure, birth to an infected mother, and intranasal cocaine abuse. According to the U.S. Centers for Disease Control and Prevention (CDC), the most common risk factors for acute HCV infection in the U.S. from 1991- 1995 were high-risk drug (60%) and sexual behaviors (20%). Other modes of transmission (occupational, hemodialysis, household, and perinatal) accounted for approximately 10% of infections. A potential risk factor can be identified in approximately 90% of persons with HCV infection. In the remaining 10%, no recognized source of infection can be identified, although majority of the persons in this category are associated with low socioeconomic status. [Alter M.J. 1997].

3. HEPATITIS C VIRUS GENOMIC RNA

The HCV genomic RNA contains three distinct regions [Scheel & Rice 2013, Penin et al 2004]: (1) a 5′UTR or noncoding region; (2) a long open reading frame (ORF) of more than 9000 nucleotides (nt); and (3) a short 3′ UTR. The HCV 5'UTR contains 341 nucleotides located upstream of the ORF translation initiation codon. The 5'UTR contains the internal ribosomal entry site which forms a stable pre-initiation complex by direct binding with the 40S ribosomal subunit for the HCV polyprotein translation. The long ORF encodes a polyprotein of approximately 3000 amino acids, which will be further processed by host and viral proteases. The 3'UTR is principally involved in minus-strand priming during HCV replication. The nucleotide sequence variability is distributed throughout the entire viral genome. The 5'UTR is the most conserved region in the genome while the regions encoding envelope proteins (E1, E2) are the most variable ones. Thus, the highly conserved 5'UTR region is usually the target of choice for HCV genome detection across different genotypes.

4. VIRAL PROTEINS

The long ORF encodes a polyprotein which is processed co- and post translationally by host and viral proteases into at least 10 different proteins, which are arranged in the order of NH2-C-E1- E2-p7-NS2-NS3-NS4A-NS4B-NS5A-NS5B-COOH. Host signal peptidase is required for the cleavages at C-E1, E1-E2, E2-p7, and p7-NS2 junctions.

Host signal peptide peptidase is also needed for the further maturation of HCV core proteins. Two viral proteases are involved in the processing of HCV nonstructural proteins: NS2, a zinc-dependent metalloprotease that cleaves between NS2 and NS3; and NS3/4A, a serine protease that cleaves between the NS3-NS4A, NS4A-NS4B, NS4BNS5A, and NS5A-NS5B junctions (Figure 1). These viral proteins remain associated with intracellular membranes after processing.

5. ACUTE HEPATITIS C

Acute hepatitis C infection is infrequently diagnosed because the majority of acutely infected individuals are asymptomatic. In the transfusion setting, where acute onset of HCV infection has been best documented, 70% to 80% of cases were asymptomatic. [McCaughan et al 1992] About 20% to 30% of adults with acute HCV infection may develop clinical symptoms. The symptomatic onset ranges from 3 to 12 weeks after exposure. [Alter et al 2000, Thimme et al 2001] Symptoms may include malaise, weakness, anorexia, and jaundice. Increased serum alanine aminotransferase (ALT) levels, signifying hepatocyte necrosis, begin rising 2 to 8 weeks after exposure, and often reach levels of greater than 10 times the upper limits of normal. HCV RNA can be detected in the serum within 1 to 2 weeks after exposure. The antibody to HCV, as detected by enzyme immunoassay, becomes positive near the onset of symptoms, approximately 1 to 3 months after exposure. Up to 30% of patients will test negative for anti-HCV at onset of their symptoms, making anti-HCV testing unreliable in diagnosis of acute infection. [Farci et al 1991, 1996; Thimme et al., 2001].

6. CHRONIC HEPATITIS C

Chronic hepatitis C is marked by the persistence of HCV RNA in the blood for at least 6 months after onset of acute infection. HCV is self-limiting in only 15%-25% of patients in whom HCV RNA in the serum becomes undetectable and ALT levels come back to normal. 1`About 75%-85% of infected patients do not clear the virus by 6 months, and chronic hepatitis develops. The rate of chronic HCV infection is affected by many factors which include, the age at time of infection, gender, ethnicity, and the development of jaundice during the acute infection.

7. MECHANISM OF HCV GENOTYPE EVOLUTION

During replication of the hepatitis C virus, RNA-dependent RNA polymerase frequently introduces random nucleotide errors. This results in a relatively high rate of spontaneous nucleotide substitutions, between 10^{-2} and 10^{-3}/site/year. These replication errors yield 10 to 100 viable replacements per site per year. Whether a given nucleotide substitution results in a viable mutant is site dependent; different regions of the hepatitis C viral genome vary in their level of genetic conservation. Highly conserved areas include those coding for crucial functions, such as polymerase and other on structural proteins, core protein, the 59 noncoding region, and the internal ribosome entry site. Nucleotide substitutions in these regions are less likely to be viable, because changes may result in altered structure that precludes function. Other regions of the genome are more tolerant of nucleotide substitutions, including the hypervariable regions. Such areas would be optimal for identifying subtle sequence variations, as in quasispecies analyses. The high rate of random nucleotide substitutions introduced by the viral RNA polymerase and the constant introduction of replication errors results in slow genetic evolution. As a consequence, HCV is a highly heterogeneous virus, with only approximately 70% homology between all isolates. The basis for classification into genotypes is genetic, or molecular, relatedness.

8. HCV GENOTYPE

Due to the error prone viral RNA-dependent RNA polymerase (HCV NS5B protein), HCV behaves in infected patients as a complex mixture of genetically distinct but closely related variants referred to as a quasispecies, and is produced within HCV infected patients [Gómez et al 1999]. There is 1%-5% variant in HCV nucleotide sequence from a single infected patient. Collection of nucleotide substitutions in the virus has resulted in diversification into distinct subtypes and even genotypes. Therefore, the HCV RNA genome sequences are extremely heterogeneous. At present, HCV is classified into eleven genotypes (designated as 1 to 11) differing in their nucleotide sequence by 30%-50%, six of them are the major ones (genotypes 1 to 6) [Chayama et al 2011, Wang 2013]. Within HCV genotype, several subtypes (designated as a, b, c, etc.) can be defined that differ in their nucleotide sequence by 15%-30% [Bukh 1995, Simmonds 2004]. The prevalence of HCV genotypes and subtypes is geographically different [Gower et al 2014, Messina et al 2015]. At present, genotype 1 is the most prevalent (46%) globally, followed by genotype 3, genotype 2 and genotype 4. Various genotypes have different infectivity and pathogenicity, thereby influencing the rate of progression to cirrhosis and the risk of HCC. HCV heterogeneity would also result in different responses to anti-viral treatments, e.g., genotypes 1 and 4 are more resistant to interferon-based therapies than genotypes 2 and 3.

9. DISTRIBUTION AND PREVALENCE OF HCV GENOTYPE

The most common genotype is G3 (66.7%), followed by G1 (15.5%) and G4 (3.7%). Low percentage of G2, G5 and G6 are reported (Table 2), in both the studied countries like India and Pakistan, G3 counts over the half of all the genotypes described (54.4% and 79.0%), and there is No genotypes distribution data are available from Afghanistan and Bangladesh.

Continents	G1 (%)	G2 (%)	G3 (%)	G4 (%)	G5 (%)	G6 (%)	Mixed
Africa	26.3	23.7	6.3	28.1	12.2	-	3.4
North Africa/Middle East	27.3	0.8	6.3	65.3	0.3	-	-
America	74.5	10.2	10.6	1.7	0.1	0.3	2.6
Asia	46.6	18.6	22.4	1.0	0.1	7.0	4.3
Australasia	55.0	6.5	36.0	1.2	-	1.3	-
Europe	64.4	5.5	25.5	3.7	0.1	0.1	0.7
Total (excludes Oceania)	49.1	11.0	17.9	16.8	2.0	1.4	1.8

Table 1: Global prevalence of hepatitis C virus genotypes [Petruzziello et. al., 2016]

Regions	G1 (%)	G2 (%)	G3 (%)	G4 (%)	G5 (%)	G6 (%)	Mixed
Central Sub-Saharan Africa	12.3	4.0	0.8	82.9	-	-	-
East Sub-Saharan Africa	36.2	26.8	7.4	16.6	13.0	-	-
Southern Sub-Saharan Africa	31.4	1.2	12.6	12.4	35.7	-	6.7
West Sub-Saharan Africa	25.5	62.9	4.4	0.6	-	-	6.6
North Africa and Middle East	27.3	0.8	6.3	65.3	0.3	-	-
North America	66.3	13.1	15.7	4.3	-	0.6	-
Caribbean	83.0	7.2	2.1	0.6	-	0.1	7.0
Andean Latin America	86.0	2.0	10.0	-	-	-	2.0
Central Latin America	74.6	21.6	3.3	0.1	0.1	-	0.3
Southern Latin America	72.0	13.3	13.5	0.9	0.1	0.1	0.1
Tropical Latin America	64.8	4.6	30.2	0.2	0.1	-	0.1
Central Asia	70.4	8.6	19.6	-	-	-	1.4

East Asia	53.5	31.7	5.4	0.1	-	3.3	6.0
Pacific Asia, High-Income	58.7	39.7	0.4	0.1	-	0.5	0.6
South Asia	15.5	1.9	66.7	3.7	0.1	0.5	11.6
Southeast Asia	35.2	11.1	19.9	0.9	0.4	30.8	1.7
Australasia	55.0	6.5	36.0	1.2	-	1.3	-
Central Europe	70.0	3.2	21.0	4.9	-	0.1	0.8
Eastern Europe	68.1	4.3	26.6	0.5	-	-	0.5
Western Europe	55.1	8.9	29.0	5.8	0.2	0.1	0.8

Table 2: Regional estimates of hepatitis C virus genotypes [Petruzziello et al., 2016]

According to global prevalence and genotype distribution of hepatitis C virus infection in 2015: a modelling study for Asia region is as follows:

	prevalence 2015*	(1000s) 2015* Genotype†	1a	1b	1c	1 (other)	2	3	4	5	6	Mixed or other
Asia Pacific high income												
Japan	0·7% (0·3–0·8)	857 (364–1024)	0	64·8%	0	0	34·2%	0	0	0	0	1·0%
South Korea	0·5% (0·3–0·5)	231 (148–261)	3·0%	45·4%	0	4·3%	45·3%	0·8%	0·2%	0	1·0%	0
Asia, Central												
Armenia	..	..	5·2%	36·2%	0	1·7%	18·9%	38·0%	0	0	0	0
Azerbaijan	1·9% (1·3–2·1)	190 (128–212)	2·9%	64·2%	0	0	6·7%	26·0%	0·2%	0	0	0
Georgia	4·2% (3·0–4·2)	165 (120–169)	1·9%	37·6%	0	0	24·5%	34·3%	0	0	0	1·7%
Kazakhsta	2·8 (1·9–3·2)	508 (334–572)	2·5%	52·5%	0	0	10·0%	35·0%	0	0	0	0
Mongolia	6·4% (4·3–7·9)	194 (134–237)	0	98·8%	0	0	1·2%	0	0	0	0	0
Tajikistan	..	..	0	82·7%	0	0	5·8%	7·7%	0	0	0	3·8%
Uzbekistan	4·3% (3·6–5·0)	1292 (902–1524)	2·9%	64·2%	0	0	6·7%	26·0%	0·2%	0	0	0
Asia, East												
China	0·7% (0·5–0·8)	9795 (6675–10 832)	1·4%	56·8%	0	0	15·4%	8·7%	0	0	6·3%	11·4%
Hong Kong	0·2% (0·1–0·3)	15 (6–22)	4·3%	62·4%	0	0	3·2%	2·8%	0	0	27·4%	0
Taiwan	2·1% (1·3–3·7)	489 (310–877)	2·6%	45·5%	0	0·7%	39·5%	1·0%	0·2%	0	0·5%	10·1%
Asia, South												
Afghanistan	0·5% (0·3–0·8)	181 (85–258)	35·2%	2·8%	0	0	0	62·0%	0	0	0	0
India	0·5% (0·4–0·8)	6245 (4748–10 957)	9·0%	16·1%	0	3·3%	0	64·1%	7·3%	0·3%	0	0
Nepal	..	..	11·3%	6·6%	0	21·4%	0	58·4%	0	0	0	2·4%
Pakistan	3·8% (2·8–3·9)	7172 (5363–7487	4·8%	1·2%	0	1·0%	3·8%	79·0%	1·6%	0·1%	0·1%	8·3%
Asia, southeast												
Cambodia	1·6% (0·9–1·7)	257 (147–272		24·0%	0	0	0	20·0%	0	0	56·0%	0
Indonesia 2046)	0·5% (0·2–0·8) 25·6%	1289 (443–		39·0%	0	3·0%	9·3%	9·4%	3·6%	0	0	10·0%
Laos	..	..	0	4·4%	0	0	0	0	0	0	95·6%	0
Malaysia 405)	1·2% (0·8–1·3) 0	382 (240–		0	0	35·8%	0·7%	62·3%	0·7%	0	0·4%	0
Myanmar	..	..	4·1%	6·9%	0	0	0·7%	39·3%	0	0	49·0%	0
Philippines 651)	0·6% (0·3–0·6) 70·7%	614 (353–		2·5%	0	0	26·4%	0	0·2%	0	0·2%	0
SriLanka	..	..		46·9%	0	0	37·5%	0	0	0	0	15·6%
Thailand	0·7% (0·4–0·7)	463 (255–487)	4·4%	13·0%	0	0	0	47·8%	0	0	34·8%	0
Vietnam	1·1% (0·6–1·2)	1066 (580–1116)	30·0%	17·1%	0	0	1·1%	1·1%	0	0	50·8%	0

Table 3: Modelled 2015 hepatitis C viraemic prevalence and chronically infected population (all ages) and genotype distribution in Asia

*2015 year-end estimate is a model output projection based on historic data. †Genotype distribution data are either taken from the literature or based on regional averages in the absence of country-specific data.

10. CONCLUSION

Hepatitis C virus (HCV) infection is one of the chief global health problems, and its paradigm varies region wise depending on its historical and present risk factors. A country-specific policy of prevention, diagnosis and treatment management could reduce this disease burden, but unfortunately, in many countries, there is a lack of epidemiological data upon which to base these strategies, being HCV infection rates typically focused on quantifying only the anti-HCV prevalence with no attention to HCV genotypes distribution. Globally, G1 accounts for 49.1% of all anti-HCV infections among adults making it the most common, followed by G3 (17.9%), G4 (16.8%), G2 (11.0%), G5 (2.0%) and G6 (1.4%). [Hanafa et al 2013; Seivert et al 2011, Hope et al 2014] Undefined or mixed genotypes accounts for 1.8% of the total HCV infections. In Asia the two most common genotypes are G1 (46.6%) and G3 (22.4%), G3 the latter largely driven by its high prevalence in South Asia (66.7%), where in India and Pakistan G3 followed by genotype 3a and 3b, shows the highest percentages of the whole continent (54.4% and 79.0%, respectively).

REFERENCES

Alter H.J and Seeff L.B. Recovery, persistence, and sequelae in hepatitis C virus infection: a perspective on long-term outcome. Semin Liver Dis, 2000. 20(1): 17-35.

Alter M.J. The epidemiology of acute and chronic hepatitis C. Clin Liver Dis, 1997. 1(3): 559-68.

Bukh J, Miller R.H., and Purcell R.H. Genetic heterogeneity of hepatitis C virus: quasispecies and genotypes. Semin Liver Dis, 1995. 15(1): 41-63.

Chayama K, Hayes CN. Hepatitis C virus: How genetic variability affects pathobiology of disease. J Gastroenterol Hepatol 2011; 26 Suppl 1: 83-95.

Choo Q.L, et al. Isolation of a cDNA clone derived from a blood-borne non-A, non-B viral hepatitis genome. Science, 1989. 244(4902): 359-62.

Farci P, et al. A long-term study of hepatitis C virus replication in non-A, non-B hepatitis. N Engl J Med, 1991. 325(2): 98-104.

Farci P, et al. Hepatitis C virus-associated fulminant hepatic failure. N Engl J Med, 1996. 335(9): 631-4.

Global prevalence and genotype distribution of hepatitis C virus infection in 2015: a modelling study. Polaris Observatory HCV Collaborators. Lancet Gastroenterol Hepatol. 2017 Mar; 2(3): 161–176. Published online 2016 Dec 16. doi: 10.1016/S2468-1253(16)30181-9.

Gómez J, Martell M, Quer J, Cabot B, Esteban JI. Hepatitis C viral quasispecies. J Viral Hepat 1999; 6: 3-16.

Gower E, Estes C, Blach S, Razavi-Shearer K, Razavi H. Global epidemiology and genotype distribution of the hepatitis C virus infection. J Hepatol 2014; 61: S45-S57.

Hope VD, Eramova I, Capurro D, Donoghoe MC. Prevalence and estimation of hepatitis B and C infections in the WHO European Region: a review of data focusing on the countries outside the European Union and the European Free Trade Association. Epidemiol Infect 2014; 142: 270-286.

Lauer GM, Walker BD. Hepatitis C virus infection. N Engl J Med. 2001; 345:41–52.

McCaughan G.W, et al. Clinical assessment and incidence of hepatitis C RNA in 50 consecutive RIBA-positive volunteer blood donors. Med J Aust, 1992. 157(4): 231-3.

Messina JP, Humphreys I, Flaxman A, Brown A, Cooke GS, Pybus OG, Barnes E. Global distribution and prevalence of hepatitis C virus genotypes. Hepatology 2015; 61:

77-87.

Mohd Hanafah K, Groeger J, Flaxman AD, Wiersma ST. Global epidemiology of hepatitis C virus infection: new estimates of age specifc antibody to HCV sero-prevalence. Hepatology 2013; 57: 1333-1342.

National Institutes of Health Consensus Development Conference Statement: Management of hepatitis C 2002 (June 10-12, 2002). Gastroenterology, 2002. 123(6): 2082-99.

Neumann AU, et al. Hepatitis C viral dynamics in vivo and the antiviral efficacy of interferon-alpha therapy. Science, 1998. 282(5386): 103-7.

Penin F, Dubuisson J, Rey FA, Moradpour D, Pawlotsky JM. Structural biology of hepatitis C virus. Hepatology 2004; 39: 5-19.

Petruzziello A, Marigliano S, Loquercio G, Cozzolino A, Cacciapuoti C. Global epidemiology of hepatitis C virus infection: An up-date of the distribution and circulation of hepatitis C virus genotypes. World J Gastroenterol 2016; 22(34): 7824-7840.

Scheel TK, Rice CM. Understanding the hepatitis C virus life cycle paves the way for highly effective therapies. Nat Med 2013; 19: 837-849

Sievert W, Altraif I, Razavi HA, Abdo A, Ahmed EA, Alomair A, Amarapurkar D, Chen CH, Dou X, El Khayat H, Elshazly M, Esmat G, Guan R, Han KH, Koike K, Largen A, McCaughan G, Mogawer S, Monis A, Nawaz A, Piratvisuth T, Sanai FM, Sharara AI, Sibbel S, Sood A, Suh DJ, Wallace C, Young K, Negro F. A systematic review of hepatitis C virus epidemiology in Asia, Australia and Egypt. Liver Int 2011; 31 Suppl 2: 61-80.

Simmonds P. Genetic diversity and evolution of hepatitis C virus--15 years on. J Gen Virol 2004; 85: 3173-3188.

Thimme R, et al. Determinants of viral clearance and persistence during acute hepatitis C virus infection. J Exp Med, 2001. 194(10): 1395-406.

Wang Y. Scotomas in molecular virology and epidemiology of hepatitis C virus. World J Gastroenterol 2013; 19: 7910-7921.

❒

Chapter-20

ANTIOXIDANTS AS A THERAPEUTIC TARGETS OF COPD

Saurabh Kumar[1], Akancha Sahu[2], Monisha Banerjee[2] and Suchit Swaroop[1]*

[1]Experimental and Public Health Laboratory, Department of Zoology, University of Lucknow, Lucknow-226007
[2]Molecular and Human Genetics Lab, Department of Zoology, University of Lucknow, Lucknow-226007
**Email of Corresponding Author: drsswaroop@gmail.com*; saurabhdhuriya110@gmail.com*

ABSTRACT

Chronic Obstructive Pulmonary Disease (COPD) is categorised by the development of airflow limitation that is usually progressive. In a recent update by WHO, COPD has become the 3rd leading cause of death and is estimated to be second by 2020. It can include some combination of emphysema, chronic bronchitis (CB), and small airway obstruction. The mechanistic basis underlying COPD is complex and can involve recurrent inflammation, oxidative stress (oxidant/antioxidant imbalance), protease/antiprotease imbalance, environmental insult, and host genetics. Many studies show an imbalance between proteases and endogenous antiproteases plays an important role in the pathogenesis of COPD. Among the factors that contribute to this imbalance are oxidative stress, genetic polymorphism and an abnormal inflammatory response to the respiratory tract injury caused by smoking. Cigarette smoking serves as the major risk factor for developing COPD and is also the major source of oxidants/reactive oxygen species (ROS) to the lungs and the body in exposed individuals. Oxidative stress results from an imbalance between exogenous, harmful, oxidants and endogenous, protective, antioxidants. Antioxidant enzymes present in the airway include glutathione-S-transferase, superoxide dismutase and catalase, amongst others. Antioxidants may play a significant role as therapeutic agents to prevent the progression of the disease like COPD.

Keywords: Chronic Obstructive Pulmonary Disease (COPD), Oxidative stress, Antioxidants.

1. INTRODUCTION

The Global Initiative for Chronic Obstructive Lung Disease (GOLD) recently defined chronic obstructive pulmonary disease (COPD) as ‘a common preventable and treatable disease’ as is characterized by persistent airflow limitation that is usually progressive and associated with enhanced chronic inflammatory response in the airways and the lung to noxious particles or gases. Exacerbations and comorbidities contribute to the overall severity in individual patient (GOLD, 2015). COPD is the fourth leading cause of death in the world (WHO, 2013). It results in an economic and social burden that is both substantial and increasing (Lopez *et al.* 2006, Pauwels *et al.* 2001, Lawels *et al.* 1990, Mathers *et al.* 2006). COPD prevalence, morbidity, and mortality vary across countries and across different groups within countries. COPD is the result of cumulative exposures to health catastrophes over decades. The prevalence of COPD is related to the prevalence of tobacco smoking, although in many countries, outdoor, occupational and indoor air pollution the latter resulting from the burning of wood and other biomass fuels are major COPD risk factors (Salvi and Barnes, 2009). Inhaled particles (smoke or other sources) cause lung inflammation, a normal response that appears to be modified in individuals who develop COPD. This chronic inflammatory response may induce parenchymal tissue destruction (resulting in emphysema) and disrupt normal repair and defence mechanisms (resulting in small airway fibrosis), which in turn lead to air trapping and progressive airflow limitation.

Chronic inflammatory changes with increased numbers of inflammatory cell types, and structural changes resulting from repeated injury and repair are found in the airways, lung parenchyma, and pulmonary vasculature of patients with COPD (Hogg, 2004). Few studies have shown that a decreased exacerbation rate is associated with improved quality of life. Thus, lowering the exacerbation rate would be expected to decrease hospitalisations and have important health economic benefits. Several classes of drugs with potential for reducing exacerbations have been investigated, with variable evidence for their use: vaccines, bacterial extracts, inhaled steroids and long acting bronchodilators, phosphodiesterase inhibitors, and mucolytic agents (Burge *et al.* 2000, Calverley *et al.* 2003, Vincken *et al.* 2002).

2. EPIDEMIOLOGY

The Global Burden of Disease Study reports, 251 million people have moderate to severe COPD in 2016. Globally, more than 3.17 million deaths were caused by the disease in 2015, which corresponds 5% of all deaths globally in that year (WHO, 2013, Lopez *et al.* 2006).

In 2002, COPD was the fifth leading cause of death. Total deaths from COPD are projected to increase by more than 30% in the next 10 years unless urgent action is taken to reduce the underlying risk factors, especially tobacco use. Estimates show that COPD becomes in 2030 the third leading cause of death worldwide.

3. SYMPTOMS

COPD symptoms are less prominent until the last stage is achieved and the condition of the patient gets worst as the time increases. Some etiological signs of the disease are:

- Shortness of breath, especially during physical activities
- Wheezing
- Chest tightness

- A chronic cough that may produce mucus (sputum) that may be clear, white, yellow or greenish
- Blueness of the lips or fingernail beds (cyanosis)
- Frequent respiratory infections
- Swelling in ankles, feet or legs

Exacerbations are among one of the common symptoms of COPD, during which their symptoms become worse than usual day-to-day variation and persist for at least several days.

4. ROLE OF ANTIOXIDANTS

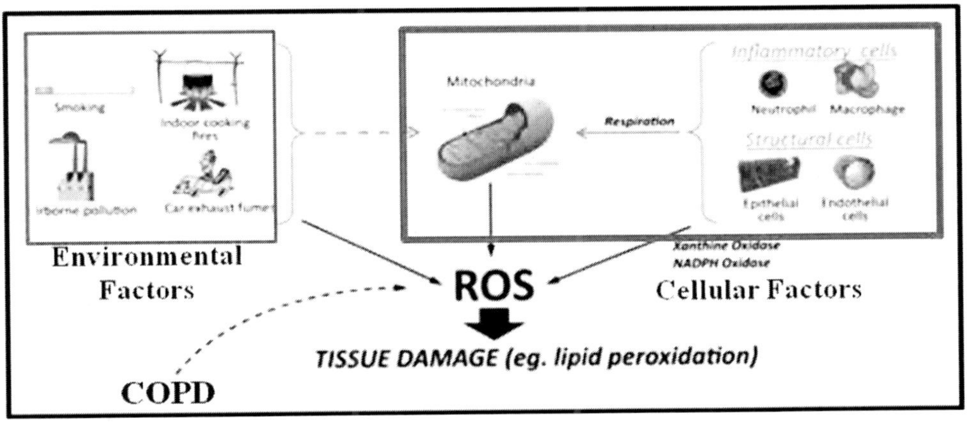

The pathogenic triad of COPD consists of oxidative stress, protease-antiprotease imbalance, and inflammation, of which oxidative stress forms prime component. Oxidative stress affects airways by myriad mechanisms including mucus hypersecretion, damage to airway epithelium, neutrophil influx, airway inflammation, and increased apoptosis. Several processes lead to oxidative stress-related tissue damage, primary among them is lipid peroxidation which leads to the formation of various lipid hydroperoxides and aldehydic products. Many markers of oxidative stress have been shown to have direct correlation among themselves and severity of airway obstruction represented by forced expiratory volume 1s (FEV). Chronic smoking exposes the respiratory tree and lungs to reactive oxygen species (ROS), resulting in oxidative stress and injury. This triggers production of other ROS and lipid peroxidation and subsequent pulmonary inflammation (Rahman and Adcock, 2006). The glutathione-S- transferase (GST) genes code for a family of enzymes that detoxify some of the harmful contents of tobacco smoke (Mannervik, 1985). Microsomal epoxide hydrolase (EPHX1) is expressed in bronchial epithelial cells and metabolises highly reactive epoxide intermediates in cigarette smoke (Bartsch et al. 1992, Tingle et al. 1993,). Heme oxygenase-1 (HMOX1) is an enzyme important in heme metabolism, which catalyses the oxidative cleavage of heme, resulting in the release of carbon monoxide, bilverdin and iron (Tenhunen et al. 1968). Bilverdin breakdown into bilirubin, which scavenges local ROS, thus HMOX1 contributes to the generation of antioxidants. It is present at higher concentrations in the lungs of smokers than nonsmokers, suggesting up-regulation in these circumstances, presumably because of a response to increased ROS (Maestrelli et al. 2001).

Oxidative stress results from an imbalance between exogenous, harmful, oxidants and endogenous, protective, antioxidants (MacNee, 2000). Antioxidant enzymes present in the airway include glutathione-S-transferase, superoxide dismutase and catalase (Young et al. 2006).

5. SUMMARY

In the race of being big and rich, in present scenario our life has become fast and full of rush. To cope with that stress, we became addicted to many of the drugs and habits like smoking and drinking. These addictions may momentarily work by elevating the levels of dopaminergic neurotransmitters but on long term usage may lead to the hazardous conditions like COPD. To maintain the homeostasis of inner self body applies many of the pathways like activation of immune responses, antioxidants and apoptotic pathways. But somehow, after certain stage it becomes difficult for body too to maintain pace with the diseases. By evolving newer therapeutic targets, we may help in controlling the adverse effects of such incurable disease. Antioxidants may play a significant role in providing potent targets for the cure of disease like COPD. Hence the above written chapter describes the role of antioxidants and some target molecules for the cure of COPD.

REFERENCES

Bartsch H, Petruzzelli S, De Flora S, et al. Carcinogen metabolism in human lung tissues and the effect of tobacco smoking: results from a case--control multicenter study on lung cancer patients. Environ Health Perspect 1992, 98:119-124.

Burge PS, Calverley PMA, Jones PW, et al. Randomised, double blind, placebo-controlled study of fluticasone propionate in patients with moderate to sever chronic obstructive pulmonary disease: the ISOLDE trial.BMJ 2000; 320: 1297–303.

Calverley P, Pauwels R, Vestbo J, et al. Combined salmeterol and fluticasone in the treatment of chronic obstructive pulmonary disease: a randomised controlled trial. Lancet 2003: 361: 449–56.

Global Initiative for Chronic Obstructive Lung Disease (GOLD), 2015. Global strategy for the diagnosis, management, and prevention of COPD. Available from http://www.goldcopd.org/uploads/users/files/GOLDReport_2015.pdf

Hogg JC. Pathophysiology of airflow limitation in chronic obstructive pulmonary disease. Lancet 2004; 364:709-21.

Lopez AD, Shibuya K, Rao C, et al. Chronic obstructive pulmonary disese: current burden and future projections. Eur Respir J 2006;27: 397-412.

MacNee W. Oxidants/antioxidants and COPD. Chest 2000, 117(5 Suppl 1):303S-17S.

Maestrelli P, El Messlemani AH, De Fina O, et al. Increased expression of heme oxygenase (HO)-1 in alveolar spaces and HO-2 in alveolar walls of smokers. Am J Respir Crit Care Med 2001, 164(1):1508-1513.

Mannervik B: The isoenzymes of glutathione transferase. Adv Enzymol Relat Areas Mol Biol 1985, 57:357-417.

Mathers CD, Loncar D. Projections of global mortality and burden of disease from 2002 to 2030. PLoS Med 2006; 3: e 442.

Mahal A, Karan A, Engelgau M. Washington DC: The economic implications of non-communicable disease for India. The International Bank for Reconstruction and Development/The World Bank 2009; 14.

Pauwels RA, Buist AS, Calverley PM, Jenkins CR, Hurd SS. Global strategy for the diagnosis, management, and prevention of Chronic obstructive pulmonary disease.

Rahman I, Adcock IM. Oxidative stress and redox regulation of lung inflammation in COPD. Eur Respir J 2006; 28:219–242.

Salvi SS, Barnes PJ. Chronic obstructive pulmonary disease in non smokers. Lancet 2009;374: 733-43.

Tenhunen R, Marver HS, Schmid R. The enzymatic conversion of heme to bilirubin by microsomal heme oxygenase. Proc Natl Acad Sci U S A 1968, 61(2):748-755.

Tingle MD, Pirmohamed M, Templeton E, et al. An investigation of the formation of cytotoxic, genotoxic, protein-reactive and stable metabolites from naphthalene by human liver microsomes. Biochem Pharmacol 1993, 46(9):1529-1538.

Vincken W, van Noord JA, Greefhorst APM. Improved health outcomes in patients with COPD during 1 yr's treatment with tiotropium, Eur Respir J 2002; 19: 209–16.

World Health Organization. Burden of COPD. [Last accessed on 2013 Jun 5]. Available from: http://www.who.int/respiratory/copd/burden/en/

Young RP, Hopkins R, Black PN, et al. Functional variants of antioxidant genes in smokers with COPD and in those with normal lung function. Thorax 2006, 61(5):394-399.

❒

Chapter-21

COMPARISON OF AGE ESTIMATES FROM SAGITTAE, LAPILLI AND ASTERISCI IN SPOTTED SNAKEHEAD, *CHANNA PUNCTATA* (BLOCH, 1793)

M. Afzal Khan*, Salman Khan, Ankita,
Irfan Ahmed and Masood Nadeem

Section of Fishery Science & Aquaculture, Department of Zoology,
Aligarh Muslim University, Aligarh-202002 (U.P.), India
Email of Corresponding Author: khanmafzal@yahoo.com

ABSTRACT

The spotted murrel, Channa punctata (Bloch, 1793), is a warm-water teleost widely available in lakes, ponds and rivers as well as in confined water bodies of southeast Asia. Different type of otoliths (sagittae, lapilli and asterisci) were collected from 210 specimens of C. punctata from the Ganga River. Standard procedures were followed to prepare and study the otoliths for age reading. Among the three types of otoliths, sagittae showed highest (88.5%) values of agreement between readers' age estimates followed by the lapillus (65.4%) and asteriscus (47.2%). Mean values of age estimates from different structures, when compared using ANOVA followed by DMRT, showed that maximum age estimates obtained from sagittae were significantly (p<0.05) higher than asteriscus but comparable (p>0.05) to the values obtained from lapillus. The age estimates from sagittae and lapillus were insignificantly different (p<0.05) from each other. Due to the highest values of percent agreement and lowest average percent error and coefficient of variation values between two readers, sagittae were considered to be the most suitable otolith for ageing C. punctata.

Keywords: Ageing precision, otoliths, *Channa punctata*, River Ganga.

1. INTRODUCTION

Fish age determination is an important tool in fishery biology especially for the assessment of life history, growth rate, age at first maturity, and population dynamics. Knowledge on these aspects has important and useful applications in fisheries resource management. Age of fishes can be estimated using a variety of hardparts, including scales, otoliths, opercular bones, fin rays, cleithra, vertebrae, and urohyal bone. Some hard parts provide more accurate and precise estimates of age than others. On account of its characteristics of being metabolically inert and continuously growing throughout the fish life, otoliths are very commonly used for estimating fish age and growth. Age estimation is often accompanied by several sources of error that can have significant effects on many population parameters estimates (Campana 2001). Understanding the precision of structures used for age estimation can reduce variation and increase the utility of age and growth information (DeVries and Frie1996) that may provide useful information to managers. Comparison of age estimates between structures is an alterative technique to validation that may provide useful information on the accuracy and bias of age estimating structures (Sylvester & Berry, 2006). The most reliable method may vary among species and it is necessary to evaluate precision and accuracy of bony structures used for age estimation (Polat et al., 2001). Several studies have focused on comparing ages estimated from different bony structures in an attempt to quantify the most suitable age estimate and to identify possible bias associated with each structure, e.g. in black crappies, *Cyprinus carpio* (Phelps et al. 2007), *Labeo rohita*, Catla catla, and *Channa marulius* (Khan and Khan, 2009), *Salvelinus confluentus* (Zymonas and McMahon, 2009), *Cirrhinus mrigala* (Khan et. al. 2011a), *Clarias gariepinus* (Khan et. al., 2011b), *Clarias batrachus, Heteropneustes fossilis* and *Wallago attu* (Khan et. al. 2013a), *Hypophthalmichthys molitrix* (Seibert and Phelps, 2013), *Channa argus argus* (Pang-hua Gu et. al. 2013), *Channa punctata* (Khan et. al. 2013b), *Schizopyge curvifrons, Schizopyge niger* and *Schizothorax esocinus* (Sabah and Khan, 2014), *Hoplias malabaricus* (Lonzano et. al. 2014) and *Labeo bata* (Khan et. al. 2015), *Sperata aor* (Khan et al. 2016) and *Channa striata* (Khan et al. 2017) etc. To date, no published information is available on the precision of age estimates from different type of otoliths in *C. punctata*. Therefore, the present study was undertaken with the objective to evaluate and compare different type of otoliths (i.e., sagittae, lapilli and asterisci) in order to identify the most suitable ageing structure that can provide precise age estimate in *C. punctata.*

2. MATERIALS AND METHODS

A total of 210 samples of *C. punctate* were collected monthly from January 2017 to June 2017 from the River Ganga at Narora (27°30' N, 78°25' E), U.P., India (Fig. 1). Samples were identified according to Talwar and Jhingran, 1991. Total length (TL) of the fish was measured to the nearest cm while body weight was recorded to the nearest gram as total weight (TW) including gut and gonads. Otoliths are situated in the membranous labyrinth in the inner ear of fishes. The utricle, saccule and lagena contains lapillus, sagitta and asteriscus, respectively. Otoliths were removed from otic capsules by opening the otic bulla. Otoliths were washed, cleaned and read whole by immersion in alcohol (Fig. 2). All otoliths were examined under a stereozoom microscope with 10× and 20× magnification under reflected light (Fig. 3). An opaque and a translucent zone together were regarded as a year's growth; age was considered to be the total number of translucent zones.

3. CALCULATIONS AND STATISTICAL ANALYSIS

Age estimates were compared by calculating the average percent error (APE), coefficient of variation (CV), and percent agreement (PA) between the readers and between the pairs of ageing structures. To calculate APE, we used the formula presented by Beamish & Fournier (1981):

$$APE_j = 100\% \times \frac{1}{R}\sum_{i=1}^{R}\frac{|x_{ij} - x_j|}{X_j}$$

Where, xij is the ith age determination of the jth fish, xj is the average age calculated for the jth fish, and R is the number of times each fish was aged.

The coefficient of variation (Camapana, 2001) was calculated as the ratio of standard deviation over the mean, and can be written as:

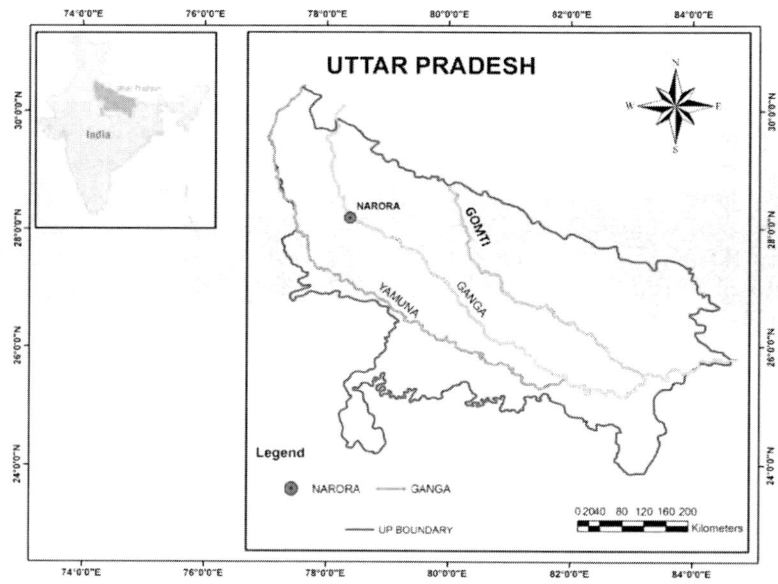

Fig. 1. Showing sampling site.

$$CV_j = 100\% \times \frac{\sqrt{\sum_{i=1}^{R} \frac{(x_{ij} - x_j)^2}{R-1}}}{X_j}$$

Where, CV_j is the coefficient of variation for the jth fish.

Both APE and CV have been widely used as statistically sound measures of ageing precision in fishes (Campana, 2001). But PA, although used as an index of ageing precision in fish by many researchers (Hoxmeier et al., 2001; Sabah and Khan, 2014) is not considered as a suitable measure of precision by several authors (Beamish and Fournier, 1981; Chang, 1982; Campana et al., 1995). Percent agreement may be expressed as the percentage of the number of observations showing similar age estimates to the total number of observations on age estimates. Percent agreement was calculated using the templates for calculating ageing precision by Sutherland (2006). Age bias graphs (Campana et al., 1995) were constructed to examine potential biases between readers and between pairs of ageing structures.

Mean age readings obtained from various bony structures were subjected to one-way analysis of variance (ANOVA) followed by Duncan's multiple range test (DMRT) (Gomez and Gomez, 1984) in order to explain whether the readings from different bony structures of the same species showed significant differences among themselves (Khan and Khan,

2009). Although the mean age estimate is not an indicator for the reliability of ageing structure, it may provide useful information regarding over- or under- estimation of age by a structure irrespective of the fish size-class. This may prove useful in selecting the structure(s) that may be used for age estimation in the absence of the most suitable ageing structure.

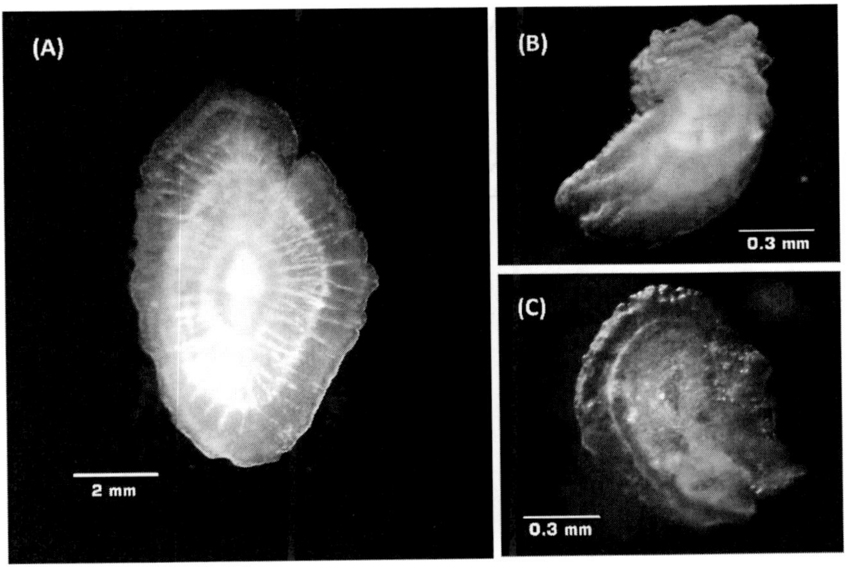

Fig.2. The otoliths in *Channa punctata*, (A) sagitta, (B) lapillus and (C) asteriscus.

4. RESULTS

Percent agreement between the age readings of two readers was highest for sagitta (88.5%) followed by lapillus (6.54%) and asteriscus (4.72%). Also, sagitta exhibited least values for CV (3.97%) and APE (1.24%) (Table 1). Age bias graphs between two readers for sagitta, lapillus and asteriscus are presented in Figure 3. In case of sagitta, no age bias was found between readers, while small error was present in lapillus. Disagreement between readers increased with age for asteriscus, as indicated by larger standard error bars for older fishes. Mean values of age estimates from different otoliths, when compared using ANOVA followed by DMRT, showed that maximum age estimates obtained from sagitta were significantly ($p<0.05$) higher from asteriscus, but comparable ($p>0.05$) to the values obtained from lapillus (Table 2).

Otoliths	APE	CV	PA
Sagittae	1.24	3.97	88.5
Lapilli	13.1	16.8	6.54
Asterisci	17.8	22.3	4.72

Table 1. Comparison of PA, APE, and CV of the age readings between different otoliths in *C. punctata*.

Otoliths	Mean values of age estimates
Sagittae	2.68[a]
Lapilli	2.51[ab]
Asterisci	2.30[b]

Table 2. Comparison of mean values of age estimates from different otoliths in *C. punctata*. Values having similar superscripts are insignificantly different (p>0.05) from each other.

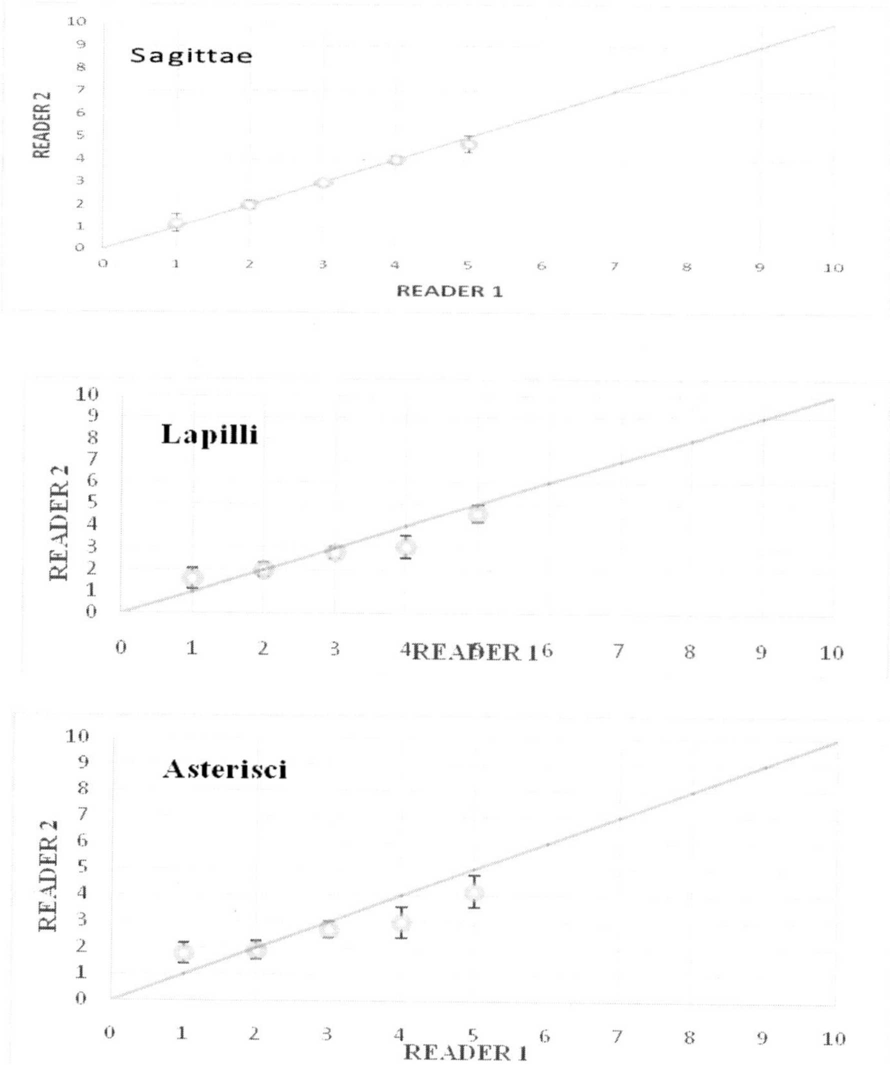

Fig. 3. Age bias graphs between two independent readers for sagitta, lapillus and asteriscus age estimates. Each error bar represents the 95% confidence interval, and the solid line indicates the theoretical 1:1 agreement line of age estimates between readers. Points above the line indicate ages that were overestimated, where as points below the line indicate ages that were underestimated

5. DISCUSSION

Otoliths have several advantages for estimating age because otoliths are not subjected to resorption and theirgrowth is acellular rather than by ossification (Secor et al. 1995). Higher precision of age estimates from otoliths has led several authors to recommended otoliths over other structures for estimating age (Khan et al. 2017; Hoxmeier et al. 2001). In present study, agreement between age estimates from sagittae and from other otoliths varied between readers (Table 1 and 2). Comparison of age estimates from the three ageing structures within the current study revealed that sagittae provided the most suitable age estimates in *C. punctata*. The suitability of otoliths for age estimation is also supported bythe fact that otoliths do not show reabsorption and their growth is acellular rather than by calcification (Secor et al. 1995) and also because otoliths are reported to be metabolically inert and thus do not reflect physiological changes that may occur throughout the life of fish (Phelps et al. 2007). Otoliths continue to grow and form annuli even as body growth slows, and asymptotic length is reached, and annuli reasbsorption does not appear to occur during periods of food limitation or stress (DeVries & Frie, 1996). Sagittae were reported to be the most suitable ageing structure in a number of fish species such as *Pylodictis olivaris* (Nash & Irwin, 1999), *Ictalurus punctatus* (Buckmeier, Irwin, Betsill, & Prentice, 2002), *Clarias gariepinus* (Khan, Khan, & Miyan, 2011), *Channa punctata* (Khan et al. 2013a) and *Channa striata* (Khan et al. 2017). Khan et al. (2013b) reported that otoliths give the precise way for age estimation in *Clarias batrachus* compared to vertebrae and pectoral spines. Ages estimated from the lapillus were less precise than sagittae because of poor clarity of rings near the primordia and because of the formation and fusion of numerous accessory primordia. On the other hand, asteriscus was very difficult to interpret. The processing and preparation were really difficult because it was very delicate. The asteriscus is also considered inappropriate because of its ambiguous core (Fig. 2C). The asteriscus, in addition, is reported to be absent at hatching and formed later than the sagitta and lapillus in several fish species (Campana et al. 1987; Oxenford et al. 1994; Hoff et al. 1997). This also suggests the asteriscus being unsuitable for otolith increment reading. In contrast, David et al. (1994) suggested asteriscus as the best ageing structure for *Sciaenaps acellatus*.

6. CONCLUSION

It may be concluded from the study that sagittae are the most suitable otolith for ageing C. punctata. The information generated will be useful to fisheries managers and researchers to select the most appropriate structure for age estimation in the selected fish species.

ACKNOWLEDGEMENT

The authors are thankful to the Chairman, Department of Zoology, Aligarh Muslim University, Aligarh, India, for providing the necessary facilities for the study.

REFERENCES

Beamish, R. J.; Fournier, D. A., 1981: A method for comparing the precision of a set of age determinations. Can. J. Fish. Aquat. Sci. 38, 982-983.

Buckmeier, D. L.; Irwin, E. R.; Betsill, R. K.; Prentice, J. A., 2002: Validity of otoliths and pectoral spines for estimating ages of channel catfish. N. Am. J. Fish. Manage, 22, 934–942.

Campana, S. E.; Annand, M. C.; McMillan, J. I., 1995: Graphical and statistical methods for determining the consistency of age determinations. T. Am. Fish. Soc. 124, 131-138.

Campana, S.E., 2001: Accuracy, precision and quality control in age determination, including a review of the use and abuse of age validation methods. J. Fish Biol. 59 (2), 197-242.

Campana, S.E.; Gagné, J.A.; Munro, J; 1987: Otolith microstructure of larval herring *(Ciupea harengus)*: image or reality? Can. J. Fish. Aquat. Sci., 44, 1922- 1929.

Chang, W. Y. B., 1982: A statistical method for evaluating the reproducibility of age determination. Can. J. Fish. Aquat. Sci. 39, 1208–1210.

DeVries, D. R.; Frie, R. V., 1996: Determination of age and growth. Pages 483–512 in B. R. Murphy and D. W. Willis, editors. Fisheries techniques, 2nd edition. American Fisheries Society, Bethesda, Maryland.

David, A.W.; Isely, J.J.; Grimes, C.B., 1994: Differences between the sagitta, lapillus, and asteriscus in estimating age and growth in juvenile red drum, *Sciaenaps acellatus*. Fish. Bull.,92, 509-515.

Gomez, K. A.; Gomez, A. A., 1984: Statistical procedures for agricultural research. Singapore: John Wiley and Sons. p. 680.

Gu, P.; Xiang, J.; Chen, Y.; Li, Y.; Tang, Jun.; Xie, S.; Chen, Y., 2013: A Comparison of Different Age Estimation Methods for the Northern Snakehead. N. Am. J. Fish. Manage.33, 994-999.

Hoff, G.R.; Logan, D.J.; Markle, D.F.; 1997: Otolith morphology and increment validation in young Lost Liver and shortnose suckers. Transact. Amer. Fish. Soc. 126, 488-494.

Hoxmeier, R.J.H.; Aday, D.D.; Wahl, D.H., 2001: Factors influencing precision of age estimation from scales and otoliths of bluegills in Illinois reservoirs. N. Am. J. Fish. Manage. 21, 374-380.

Khan, M. A.; Khan, S., 2009: Comparison of age estimates from scale, opercular bone, otolith, vertebrae and dorsal fin ray in *Labeo rohita* (Hamilton), *Catla catla* (Hamilton) and *Channa marulius* (Hamilton). Fish. Res. 100, 255-259.

Khan, M. A.; Khan, S.; Khan, S., 2017: Precision of age estimates in striped snakehead, Channa striata (Bloch, 1793) from the Ganga River and its tributaries (rivers Gomti and Yamuna). J. Appl. Ichthyol.33, 230-235.

Khan, M. A.; Khan, S.; Miyan, K., 2011a: Precision of aging structures for Indian major carp, *Cirrhinus mrigala*, from the River Ganga. J. Freshwat. Ecol. 26 (2), 231-239.

Khan, M. A.; Nazir, A.; Khan, S., 2016: Assessment of growth zones on whole and thin-sectioned otoliths in *Sperata aor* (Bagridae) inhabiting the river Ganga, J. Ichthyol. 56, 242-246.

Khan, S.; Khan, M.A.; Miyan, K., 2011b. Comparison of age estimates from otoliths, vertebrae, and pectoral spines in African sharptooth catfish, *Clarias gariepinus* (Burchell).

Estonian J. Ecol. 60, 183-193.

Khan, S.; Khan, M. A.; Miyan, K., 2013a: Evaluation of ageing precision from different structures of three threatened freshwater fish species, *Clarias batrachus, Heteropneustes fossilis* and *Wallago attu.* Folia Zoologica, 62, 103–109.

Khan, S.; Khan, M. A.; Miyan, K., 2013b: Precision of age determination from otoliths, opercular bones, scales and vertebrae in the threatened freshwater snakehead, *Channa punctata* (Bloch, 1793). J. Appl. Ichthyol. 29, 757–761.

Khan, S.; Khan, M. A.; Miyan, K., 2015: Precision of age estimates from different ageing structures of *Labeo bata* (Hamilton) collected from river Ganga, India. Indian J. Fisheries, 62, 110–114.

Nash, M. K.; Irwin, E. R. 1999: Use of otoliths versus pectoral spines for ageing adult flathead catfish. In E. R. Irwin, W. A. Hubert, C. F. Rabeni, L. Schramm Jr & T. Coon (Eds.), Proceedings of the international ictalurid symposium (pp. 309–316). Bethesda, Maryland: American Fisheries Society, Symposium 24.

Oxenford, H.A.; Hunte, W.; Deane, R.; Campana, S.E.; 1994: Otolith age validation and growth-rate variation in *flyingfish (Hirundichthys affinis)* from the eastern Caribbean. Mar. Biol., 118, 585-592.

Phelps, Q. E.; Edwards K. R.; Willis, D. W., 2007: Precision of five structures for estimating age of common carp. N. Am. J. Fish. Manage. 27, 103–105.

Polat, N.; Bostanci, D.; Yilmaz, S., 2001: Comparable age determination in different bony structures of *Pleuronectes flesus luscus* Pallas, 1811 inhabiting the Balck sea. Turk. J. Zool. 25, 441-446.

Secor, D. H.; Trice, T. M.; Hornick, H. T., 1995: Validation of otolith-based ageing and comparison of otolith and scale-based ageing in mark-recaptured Chesapeake Bay striped bass, Morone saxatilis. Fishery Bulletin, 93, 186–190.

Sabah; Khan, M. A., 2014: Precise age estimation and growth of three Schizothoracinae fishes from Kashmir valley. Zool. and Ecol. 24, 16-25.

Seibert, J.R.; Phelps, Q.E., 2013. Evaluation of aging structures for silver carp from Midwestern U.S. Rivers. N. Am. J. Fish. Manage. 33, 839-844.

Sutherland, S. J., 2006: Templates for calculating ageing precision. http: //www. nefsc.noaa.gov/fbi/age-prec/) (accessed 2018-01-15).

Sylvester, R. M.; Berry, C. R., 2006: Comparison of white sucker age estimates from scales, pectoral fin rays, and otoliths. N. Am. J. Fish. Manage. 26, 24-31.

Talwar, P. K.; Jhingaran, A. G., 1991: Inland Fishes of India andAdjacent Countries. Vol. 1. Oxford & IBH Publishing Co., NewDelhi.

Zymonas, N.D.; McMahon, T.E., 2009: Comparison of pelvic fin rays, scales and otoliths for estimating age and growth of bull trout, *Salvelinus confluentus*. Fish. Manage. Ecol. 16, 155-164.

Chapter-22

STOCK DELINEATION IN FISH USING OTOLITH SHAPE AND CHEMISTRY

Farah Bano, Tarique Mahmood and M. Serajuddin*
Fish Biology Research Lab, Department of Zoology, University of Lucknow, Lucknow-226007, Uttar Pradesh, India
**Email of Corresponding Author: lu.fisheries@gmail.com*

ABSTRACT

Otolith has become a key tool to identify and delineate fish stock units. Its successful function as a natural tag in stock discrimination studies is based on the ability to distinguish among different groups due to variation in shape and chemical composition. Here, we intend to provide a short overview of the continued development of holistic approaches combining otolith shape and chemistry used in stock delineation. Ultimately, this is carried out in the course of success to promote the interconnection between management actions and biological processes, and unambiguously resolve stock structure and scientific fisheries management. Here, we briefly summarize the general concept of otolith, its shape and chemistry and their application in stock delineation studies which is an example of relatively simple work that could quickly lead to tangible benefits in the future.

Keywords: Stock Delineation, Otolith shape, Otolith chemistry, Stock

1. THE STOCK CONCEPT AND METHODOLOGIES FOR STOCK IDENTIFICATION STUDIES

The identification of fish stocks is must to study the dynamics and structure of the fishery management units. According to Ihseen et al. (1981) a fish stock is an intra-specific group of randomly mating individuals with temporal or spatial integrity. Stock structure is basic to the sustainable fisheries management, because it is the unit at which population assessments are undertaken and management measures are applied (Begg et al., 1999; Abaunza et al., 2008a). Hence, with this respect the efficiency of the management of a given fishery depends on the correct delineation of stocks (Cadrin et al., 2013).

A range of methodologies have been used including direct and indirect methods to demonstrate the stock structure (Cadrin et al., 2005). An indirect method (evidence of spatial structuring) includes differences in life history traits (i.e. growth, reproduction and distribution) and seasonal variation among fisheries. Direct methods used for stock delineation are considered to be artificial, genetic, phenotypic and parasitic markers (Baldwin et al., 2012; Catalano et al., 2014).

2. OTOLITHS AND THEIR CHEMICAL COMPOSITION

Otoliths are paired calcified structures of all teleost (Osteichthyes) fishes and intended for hearing and balancing (Popper et al., 2005). Fish otoliths are composed of calcium carbonate crystals suspended in a protein matrix and deposited as aragonite. They are located in the inner ear contains a left and right membranous labyrinth enclosed in bony otic capsules at the rear of the neurocranium. The labyrinth includes three semicircular canals oriented in different planes and three compartments: the sacculus, lagena and utriculus.

Each compartment contains one otoliths; the sagittae, asteriscus, lapillus respectively (Berra and Aday, 2004). These three bones differ in their shape and size in different groups of fishes and out of these three, sagittae are the largest in most of the bony fishes except in Cypriniformes and Siluriformes fishes (Harvey et al., 2000; Assis 2003).

3. APPLICATION OF OTOLITH SHAPE IN STOCK DELINEATION

Over the past few years, otoliths have been the hub of interest of various ecologists and fisheries biologists (Martinho et al., 2013; Black et al., 2014; Disspain et al., 2015). As per the fisheries management point of view, populations delineated on the basis of phenotypic otolith traits can be considered as separate management units (Cadrin and Friedland 1999; Swain and Foote, 1999). The study of morphological features of fish otoliths are important to identify fishes in paleoichthyology (Nolf, 1995), the trophic ecology of fish predators (Santos et al., 2014a) and in functional morphology (Parmentier et al., 2001). Otolith-based approaches particularly its morphology is informative and used for analyzing the morphological characteristics and species identification (Sadighzadeh et al., 2014) and stock assessment (Adams et al., 2004; Turan 2006).

According to Campana (2004) otolith morphology ranges from ellipsoidal to more complex shapes, with protrusions and invaginations. In generalized otoliths models, the otoliths are described as medially convex and distally concave bodies while the sulcus acusticus is an important part of the otolith as it is among the most important features for fish species identification and does not show much variation among the fish samples (Parmentier et al., 2002; Parmentier et al., 2007. The otolith shape can be coded for genetically (Ali et al., 2013) and many studies believe that it is environmental factors (which included depth, temperature, food and mineral availability, habitat category) that play a greater role in determining otolith morphology (Lombarte, 1992; Lombarte and Lleonart, 1993; Arellano et al., 1995; Aguirre and Lombarte, 1999; Paxton, 2000; Volpedo et al., 2003; Reichenbacher et al., 2007. Particularly, the sulcus morphology is consistent among the species of a single genus and this feature is likely to be controlled genetically (Nolf, 1985; Gauldie, 1988) some characteristics (e.g. sulcus area, depth of the sulcus, rostrum size and the sulcus area: otolith area ratio) vary according to ecological factors.

These environmental factors are temperature, depth, food and mineral availability, as well as habitat type (Aguirre and Lombarte, 1999; Arellano et al., 1995; Lombarte, 1992; Lombarte and Lleonart, 1993; Paxton, 2000; Reichenbacher et al., 2007; Volpedo et al., 2003).

4. APPLICATION OF OTOLITH CHEMISTRY IN STOCK DELINEATION

Otolith chemistry have an increasingly prominent role in stock assessment, identification, delineation and its management practices (Ferguson et al., 2011; Khan et al., 2012; Miyan et al., 2014). The dissimilarity in the chemistry of otolith is considered to be highly imperative for the reason that there is a continuous deposition of chemicals throughout the life of the fish which remain immune to deformation (Green et al., 2009). The trace elements are deposited into growing surface of the otoliths, derived from the ambient water in which fish lives and act as recorder for environmental chemistry (Campana, 1999; Silva et al., 2011) this is a which is a complex process involving transport of dissolved ions over several membranes (Payan et al., 2004). However, the incorporation of elements is influenced by environmental conditions and physiological or genetic too (Elsdon et al., 2008; Sturrock et al., 2014). Elemental concentrations in the habitats were suggested to be influence by many factors including temperature, salinity, water chemistry, growth rate, diet, ontogeny, stress and physiology (Kalish, 1992; Bath et al., 2000; Elsdon et al., 2008; Buckel et al., 2004; Walther et al., 2010).

5. CONCLUSION

Otolith functions as good phenotypic markers and suggested to be more effective tools, found to compensate the lack of genetic data (Begg et al., 1999; Castonguay et al., 1991; Cardinale et al., 2004; Stransky et al., 2005). In some cases, molecular markers and genetic techniques fails to differentiate the stocks as it is time consuming, costlier and requiring infrastructure and equipment (Hartl & Clark, 1989; Bektas & Belduz, 2009). Here, rather than an exhaustive enumeration of studies addressing stock structure, have not been discussed but we intended to provide a short overview of the major achievements in stock delineation using otolith shape and otolith chemistry.

REFERENCES

Abaunza P, Murta AG, Campbell N, Cimmaruta R, Comesan˜a AS, Dahle G, Gallo E, Garc´ıa Santamar´ıa MT, Gordo LS, Iversen SA, MacKenzie K, Magoulas A, Mattiucci S, Molloy J, Nascetti G, Pinto AL, Quinta R, Ramos P, Ruggi A, Sanjuan A, Santamar´ıa MT, Santos AT, Stransky C, Zimmermann C (2008a) Considerations on sampling strategies for an holistic approach to stock identification: the example of the HOMSIR project. Fish Res 89:104–113.

Adams, DC, Rohlf FJ, Slice DE (2004) Geometric morphometrics: Ten years of progress following the 'revolution'. Ital J Zool 71:5-16.

Aguirre, H. & Lombarte, A., (1999). Ecomorphological comparisons of sagittae in *Mullus barbatus* and *M. surmuletus*. Journal of Fish Biology 55(1): 105-114.

Ali A, Khaled S, Bettina R (2013) Inter-population differences in otolith morphology are genetically encoded in the killifish *Aphanius fasciatus* (Cyprinodontiformes). Scientia Marina 77(2): 269-279.

Arellano RV, Hamerlynck O, Vinex M, Mees J, Hostens K, Gijselinck W (1995). Changes in the ratio of the sulcus acusticus area to the sagitta area of *Pomatoschistus minutus* and *P. Iozanoi* (Pisces: Gobiidae). Marine Biology 122: 355-360.

Assis CA (2003) The lagenar otoliths of teleosts: Their morphology and its application in species identification, phylogeny and systematic. Journal of Fish Biology 62, 126-895.

Baldwin RE, Banks MA, Jacobson KC (2012) Integrating fish and parasite data as a holistic solution for identifying the elusive stock structure of Pacific sardines *(Sardinops sagax)*. Rev Fish Biol Fisher 22:137–156.

Bath, GE, Thorrold SR, Jones, CM, Campana SE, McLaren, JW, Lam JW (2000). Strontium and barium uptake in aragonitic otoliths of marine fish. Geochim. Cosmochim. Ac. 64(10): 1705-1714.

Begg GA, Friedland KD, Pearce JB (1999) Stock identification and its role in stock assessment and fisheries management: an overview. Fish Res 43:1–8.

Bektas Y, Belduz AO (2009). Morphological variation among Atlantic horse mackerel, *Trachurus trachurus* populations from Turkish coastal waters. Journal of Animal and Veterinary Advances 8: 511–517.

Berra TM, Aday DD (2004) Otolith description and age-and-growth of *Kurtus gulliveri* from northern Australia. Journal of Fish Biology 65(2):354–362.

Black, B.A., Sydeman, W.J., Frank, D.C., Griffin, D., Stahle, D.W., Garcia-Reyes,

M., Rykaczewski, R.R., Bograd, S.J., Peterson, W.T., 2014. Six centuries of variability and extremes in a coupled marine-terrestrial ecosystem. Science 345, 1498–1502.

Buckel, J.A., Sharack, B. L., Zdanowicz, V.S. 2004. Effect of diet on otolith composition in *Pomatomus saltatrix,* an estuarine piscivore. Journal of Fish Biology 64, 1469-1484.

Cadrin SX, Friedland KD (1999) The utility of image processing techniques for morphometric analysis and stock identifcation. Fish Res 43:129-139 Camacho J (1995) Análisis.

Cadrin SX, Friedland KD, Waldman JR (2005) Stock identification methods. Applications in fishery science. Elsevier, Amsterdam

Cadrin, S.X., Karr, L.A., Mariani, S., 2013. Stock identification methods: an overview, in: 276 Cadrin, S.X., Kerr, L.A., Mariani, S. (Eds.), Stock Identification Methods. Applications in 277 Fishery Science. pp. 3–6.

Campana SE (2004). 'Photographic Atlas of Fish Otoliths of the Northwest Atlantic Ocean.'(NRC Research Press: Ottawa.)

Campana SE (1999). Chemistry and composition of fish otoliths: pathways, mechanisms and applications. Marine Ecology Progress Series 188, 263-297.

Cardinale M, Doering-Arjes P, Kastowsky M, Mosegaard H (2004). Effects of sex, stock, and environment on the shape of known-age Atlantic cod *(Gadus morhua)* otoliths. Canadian Journal of Fisheries and Aquatic Sciences 61, 158–167. doi:10.1139/ F03-151.

Castonguay M, Simard P, Gagnon P (1991). Usefulness of Fourier analysis of otolith shape for Atlantic mackerel *(Scomber scombrus)* stock discrimination. Can. J. Fish. Aquat. Sci. 48, 296–302.

Catalano SR, Whittington ID, Donnellan SC, Gillanders BM (2014) Parasites as biological tags to assess host population structure: guidelines, recent genetic advances and comments on a holistic approach. Int J Parasitol Parasites Wildl 3:220–226.

Disspain MCF, Ulm S, Gillanders BM (2015). Otoliths in archaeology: methods, applications and future prospects. J. Archaeol. Sci. Rep., http://dx.doi.org/10. 1016/j. jasrep.2015.05.012.

Elsdon TS, Wells BK, Campana SE, Gillanders BM, Jones CM, Limburg KE, Secor DH, Thorrold, SR, Walther BD (2008). Otolith chemistry to describe movements and life-history parameters of fishes: hypotheses, assumptions, limitations and inferences. Oceanography and Marine Biology: An Annual Review 46, 297-330.

Ferguson GJ, Ward TM, Gillanders BM (2011). Otolith shape and elemental composition: complementary tools for stock discrimination of mulloway *(Argyrosomus japonicus)* in southern Australia. Fisheries Research 110, 75-83.

Gauldie RW (1988). Function, form and time keeping properties of fish otoliths. Comparative Biochemistry and Physiology Part A: Physiology A 91, 395-402.

Green BS, Mapstone BD, Carlos G, Begg GA (2009). Tropical otoliths- Where to next? In: Green B.S., Mapstone B.D., Carlos G., Begg G.A. (eds) Tropical Fish Otoliths: Information for Assessment, Management and Ecology. Reviews: Methods and Technologies in Fish Biology and Fisheries, vol 11. Springer, Dordrecht.

Hartl DL, Clark AG (1989). Principles of population genetics. Sinauer Associates Inc, Sunderland, MA, pp. 682.

Harvey JT, Loughlin TR, Perez MA, Oxman DS (2000). Relationship between Fish Size and Otolith Length for 63 Species of Fishes from the Eastern North Pacific Ocean. NOAA Technical Report NMFS, 150, 8 pp.

Ihssen PE, Booke HE, Casselman JM, McGlade JM, Payne Utter FM (1981). Stock identification: materials and methods. Canadian Journal of Fisheries Aquatic Sciences 38, 1838-1855

Kalish JM (1992). Formation of a stress-induced chemical check in fish otoliths. Journal of Experimental Marine Biology and Ecology 162, 265-277.

Khan MA, Miyan K, Khan S (2012). Morphometric variation of snakehead fish, *Channa punctatus* population from three Indian rivers. Journal of Applied Ichthyology 29, 637-642.

Lombarte, A,. Palmer, M., Matallanas, J., Gómez-Zurita, J., and Morales-Nin, B. 2010. 762 Ecomorphological trends and phylogenetic inertia of otolith sagittae in Nototheniidae. Environ. Biol. Fish. 89(3-4): 607-618.

Lombarte A, Lleonart J (1993). Otolith size changes related with body growth, habitat depth and temperature. Environmental Biology of Fishes 37(3): 297-306.

Lombarte A (1992). Changes in otolith area: Sensory area ratio with body size and depth. Environmental Biology of Fishes 33(4): 405-410.

Martinho F, van der Veer HW, Cabral HN, Pardal MA (2013). Juvenile nursery colonization patterns for the European flounder *(Platichthys flesus):* a latitudinal approach. J. Sea Res. 84, 61–69.

Miyan K, Khan MA, Khan S (2014). Stock structure delineation using variation in otolith chemistry of snakehead, Channa punctata (Bloch, 1793), from three Indian rivers. Journal of Applied Ichthyology 30, 881-886.

Nolf D (1985). Handbook of Paleoichthyology. Otolithi piscium. Stuttgart, Springer Verlag. 10, pp. 141.

Parmentier E, Cloots R, Warin R, Henrist C (2007). Otolith crystals (in Carapidae): growth and habit. Journal of Structural Biology 159, 462-473.

Parmentier E, Lagardere F, Vandewalle P (2002). Relationships between inner ear and sagitta growth during ontogenesis of three Carapini species and consequences of life-history events on the otolith microstructure. Marine Biology 141, 491-501.

Parmentier E, Vandewalle P, Lagardère F (2001). Morphoanatomy of the otic region in carapid fishes: eco-morphological study of their otoliths. Journal of Fish Biology 58, 1046–1061. doi:10.1006/JFBI.2000.1511.

Paxton JR Fish otoliths: do size correlate with taxonomic group, habitat and Luminescence. Philosophical Transactions of the Royal Society of London B 365: 1299-1303 (2000).

Payan P, de Pontual H, Boeuf G, Mayer-Gostan N (2004). Endolymph chemistry and otolith growth in fish. Comptes Rendus Palevol. 3, 535–547.

Popper AN, Ramcharitar J, Campana SE (2005). Why otoliths? Insights from inner ear physiology and fisheries biology. Mar. Freshw. Res. 56(5): 497-504.

Reichenbacher, B., Sienknecht, U., Küchenhoff, H. & Fenske, N. 2007. Combined otolith morphology and morphometry for assessing taxonomy and diversity in fossil and extant killifish (Aphanius prolebias). Journal of Morphology 268(10): 898-915.

Sadighzadeh Z, Valinassab T, Vosugi G, Motallebi AA, Fatemi MR, Lombarte, A., Tuset, VM (2014). Use of otolith shape for stock identification of John's snapper, Lutjanus johnii (Pisces: Lutjanidae), from the Persian Gulf and the Oman Sea. Fisheries Research 155, 59-63.

Santos MB, Saavedra C, Pierce GJ (2014a). Quantifying the predation on sardine and hake by cetaceans in the Atlantic waters of the Iberian peninsula. Deep Sea Research Part

II: Topical Studies in Oceanography 106, 232–244. doi: 10.1016/j.dsr2.2013.09.040

Silva DM, Santos P, Correia, AT (2011). Discrimination of *Trisopterus luscus* stocks in northern Portugal using otolith elemental fingerprints. Aquatic Living Resources 24, 85-91.

Stransky C, Murta AG, Schlickeisen J, Zimmermann C (2008). Otolith shape analysis as a tool for stock separation of horse mackerel *(Trachurus trachurus)* in the Northeast Atlantic and Mediterranean. Fish. Res. 89(2): 159-166.

Sturrock AM, Trueman CN, Milton JA, Waring CP, Cooper MJ, Hunter E (2014). Physiological influences can outweigh environmental signals in otolith microchemistry research. Mar. Ecol. Prog. Ser. 500, 245–264.

Swain DP, Foote CJ (1999) Stocks and chameleons: the use of phenotypic variation in stock identificaton. Fish Res 43:113–128.

Turan C (2006) The use of otolith shape and chemistry to determine stock structure of Mediterranean horse mackerel *Trachurus mediterraneus* (Steindachner). Journal of Fish Biology 69, 165-180.

Volpedo A, Echeverrı, DD (2003) Ecomorphological patterns of the sagitta in fish on the continental shelf off Argentine. Fisheries Research 60(2–3): 551-560

Walther BD, Kingsford MJ, Callaghan, MDO, McCulloch MT (2010) Interactive effects of ontogeny, food ration and temperature on elemental incorporation in otoliths of a coral reef fish. Environmental Biology of Fishes 89: 441-451.

❐

Chapter-23

COMMERCIAL FISHERIES IN PILI RESERVOIR OF UTTAR PRADESH, INDIA

Nagma and M. Afzal Khan*

Section of Fishery Science and Aquaculture, Department of Zoology, Aligarh Muslim University, Aligarh 202 002 (UP), India.
**Email of Corresponding Author: khanmafzal@yahoo.com*

ABSTRACT

The present investigation was undertaken to study fish yield and catch composition based on the use of various fishing craft and gears adopted by fishermen for fishing in Pili reservoir at Rehar village in district Bijnor, Uttar Pradesh. The study was conducted monthly for one year from February 2013 to January 2014. Documentation of crafts and gears in freshwater sector of India is scanty. During the present investigation one craft and four different kinds of gears were observed in the reservoir. The craft was a wooden boat while the gears were gill net, cast net, scoop net, and hook and line. These fishing methods have been practiced by the local fishing communities for long time. Average fish production was 1.63 kg/ha/yr.

Keywords: Fishing craft, fishing gears, Pili reservoir, Bijnor.

1. INTRODUCTION

Reservoirs in India play an important role in fish production and contribute significantly to the livelihood of local communities. Reservoirs contribute the single largest inland fishery resource both in terms of size and production potential (Kamble and Mudkhede, 2013). Reservoirs are an important source of fish production in Uttar Pradesh. Presently, the area under reservoir fisheries in the state of Uttar Pradesh has been estimated to be more than 1.57 lac hectares and with the constant addition of new reservoirs and impoundments, this area is likely to further increase in the coming years (Sultan, 2013). In spite of the large area available under reservoirs, the utilization of these water bodies from fisheries point of view is poor. The average productivity from all categories of reservoirs is estimated at about 30 kg/ha/year (Yadava and Sugunan, 2009), while the potential exists for manifold increase as demonstrated in some small, medium and large reservoirs in the country. The productivity from reservoir can be increased through a number of approaches like better management measures, higher value for fish catch through improvement in processing and marketing and through more equitable distribution of benefits.

Pili reservoir is a manmade impoundment constructed in the year 1966. Pili reservoir is an earthen dam, 15.4 km long with maximum depth of 19.0 meter and top width of 6.0 meter. The dam is located near village Rehar and about 2.5 km North of Afzalgarh - Kashipur road on river Pili in Afzalgarh Block, Bijnor district of Uttar Pradesh. The reservoir is located at latitude 29° 21' N - 29° 24' N and longitude 78°46' E - 78°49' E. The river Pili is the main river to impound the reservoir mainly for the purpose of irrigation of the agriculture fields as well as for the pisciculture. The Pili reservoir harbors a variety of fish species especially Indian major carps, minor carps, catfishes, murrels and few exotic fishes. It contributes significantly to the total fish production in Bijnor district. The craft and gears were mostly old-fashioned, indigenous and locally made. The crafts, gears and the fishing methods play a vital role in the scientific and rational exploitation of any water body and its management (Kolekar, 2009).Fishing craft and gears adopted in the Pili reservoir were simple tools used by local fishers for commercial catches of fish production. The use of fishing crafts and gears have been documented by several researchers in freshwater systems (Kingdom and Kwen, 2009; Mohammed, 2012; Baruah et al., 2013; Islam et al., 2013 and Nagma, 2015). Also, a number of studies were undertaken on the fishing methods of Indian reservoirs (Mathai and George, 1972; Varghese et al., 1982; Sugunan, 1995 and Sakhare, 2007). However, the fishing methods used in the Pili reservoir have so far not been documented. Therefore, the present study was undertaken to analyse the use of fishing craft and gears and fish production in Pili reservoir.

2. MATERIALS AND METHODS

The present investigation on the use of fishing craft and gears in Pili reservoir was conducted monthly for a period of one year from February 2013 to January 2014. During the field work the information regarding the various fishing craft and gears and their operational procedure were collected by the personal interviews of local fishermen community and local fish contractor as well as by personal observation during the fishing operations.

Data related to carp seed stocking and annual fish yield were obtained from Fisheries Department of Bijnor, Uttar Pradesh, India. Materials for the study comprised of daily catch record maintained by the Fisheries department during the period 2007 to 2011. Data was used to study fish production and catch composition. The categories formulated by the Fisheries department to represent different fish species in catch were category A- major carps of more than 1.5 kg, category B- major carps 0.5 to 1.5 kg, category C- major carps less than 0.5 kg, and category D- weed fishes, cat fishes, prawns etc.

3. RESULTS AND DISCUSSION

Pili reservoir is one of the important sources of freshwater in Bijnor. It also serves as the main water body for commercial fish production which is managed by Fisheries department of Bijnor. The ichthyofauna of the reservoir comprised of *Catla catla*, *Cirrhinus mrigala, C. reba*, *Cyprinus carpio*, *Labeo rohita, L. calbasu, L. gonius, L. bata, Systomus sarana, Puntius ticto, P. sophore*, *Chela bacaila*, *Ctenopharyngodon idella*, *Hypophthalmichthys molitrix*, *Esomus danricus*, *Amblypharyngodon mola*, *Oxygaster bacaila, O. gora, Wallago attu*, *Sperata seenghala*, *Mystus cavasius, M. vittatus*, *Bagarius bagarius*, *Heteropneustes fossilis*, *Clarias batrachus, C. gariepinus*, *Channa striatus, C. punctatus*, *Mastacembelus armatus*, *Notopterus notopterus* and *Chitala chitala*. Nagma and Khan (2013) reported a total of 36 fish species belonging to six orders, 11 families and 23 genera from various water resources of Bijnor district. Moreover, in Pili reservoir of district Bijnor, a total of 31 species belonging to five orders, nine families and 21 genera were reported (Nagma and Khan, 2015). One craft and four gears were observed to be utilized for fishing in Pili reservoir; a wooden, non-mechanized boat and gears such as gill net, cast net, scoop net and hook and lines. The different fishing gears along with the fish species caught by them are listed in Table1.

Gears	Mesh size	Species caught
Gill net	5 -10 cm	Mostly Indian major carps and large catfishes
	10-20 cm	Indian major carps
Cast net	1-4 cm	Fingerlings of major carps, minnows, small carps and small catfishes
Scoop net	5-7 cm	Major carps
	1-3 cm	Minor carps and occasionally catfishes
Hook and Line	-	Murrels and catfishes

Table 1. Fishing gears used in Pili reservoir.

Fishing Craft:

The one and only fishing craft used in Pili reservoir was a Plank boat (wooden and non-mechanized). The wooden boat is locally manufactured. The length of boat usually ranges between 5-7 meters. The boats were painted inside by synthetic paints while outside they are coated with dammar for protection from water. It also gives long life to the craft besides making the boat leak proof. The average age of plank boat is nearly 8-10 years. Fishing gear such as cast net is also operated from the boat.

Fishing Gears:

(a) ***Gill net :*** Gill net is most common fishing gear used by the fishermen. Maximum fish harvesting in the reservoir was performed by gillnet. For fish harvesting, gillnet of different mesh size was used by the fishers. The mesh size of net shows variations and depend upon the species and size of different targeted groups. Gill net were either weaved by the local fishermen themselves or procured from the local market.

(b) ***Cast net:*** Cast net is another important fishing gear used in the Pili reservoir. The cast net is used throughout the reservoir. The advantage of cast net is it can be used single handedly. The fishermen can operate it mainly from the boat or from the bank of reservoir. However, it takes some time to learn the skill of successfully operating the cast net. Generally small fishes such as minor carps, catfishes and other small species were caught through cast net.

(c) ***Scoop net:*** The Scoop net is very useful fishing equipment in shallow areas of reservoir. This is a circular net having a long handle. This is mainly used from the boat or from bank of reservoir especially near outgoing canal of reservoir to catch mainly carps and catfishes.

(d) ***Hook and line:*** This might be one of the oldest and famous fishing methods all over the world. A metal hook is tied with one end of nylon thread and the other end of nylon thread is tied to a bamboo stick. Earthworms, grasshoppers, or small fishes placed into the hook as bait. The nylon line with baited hook is placed into reservoir water for 2-6 hours. However, the fishermen place the baited hook in water overnight and collect the fish in the morning. This is not a commercial fishing method and mostly the fish caught is consumed by the fishermen family. Mostly, the murrels, catfishes and other shallow water species are caught by this gear.

In the present study an assessment of fish population and production was made on the basis of the data on the annual stocking of carp fry and commercial fish landing of the reservoir. Table 2 shows the fish production and yield in the Pili reservoir from the year 2007-08 to 2011-12. Average annual production for all the five years of Pili reservoir was 1.63 kg/ha/year. Initially for the first two years, the catch was high (25100 and 51900 kg) but in the last three years the yield drastically reduced. Fish yield in impoundments depend not only on geo-morpho and physico-chemical character, but much more on manner and level of exploitation (Sugunan, 2002). Although, one craft and four gears were used for fishing in Pili reservoir, productivity of this reservoir is much below the national average of 30 kg/ha (Paul et.al, 2017).

Year	Stocking of fry/ ha/year	Annual production (kg)	Annual yield kg/ ha/year
2007-2008	3333	25100	1.67
2008-2009	3333	51900	3.46
2009-2010	3333	23400	1.56
2010-2011	3333	10800	0.72
2011-2012	3333	11100	0.74
Total		122300	8.15

Table 2. Fish production during the period 2007-08 to 2011-12 in Pili reservoir.

It reveals a poor carp production at an average of 1.63 kg/ha/year against an average stocking rate of 3,333 fry/ha /year. Since, fry size at the time of stocking is 20mm-25mm there could have been increased mortality that led to low production. Thus, to improve the production, stocking of advanced fingerlings is strongly recommended. The fish production of the Pili reservoir remained in the range of 0.72 to 3.46 kg/ha/year, which is very low as compared to national average catch from large reservoir at 11.43 kg/ha/yr (Ayyappan, 2011). There are published reports on the fish production from various reservoirs, viz, 4.1 to 25.08 kg/ha/yr in Pong dam , 3.7 to 14.24 kg/ha/yr in Rihand, 0.53 to 1.471 kg/ha/yr in Tenughat, 0.55 to 1.10 kg/ha/yr in Kangsabati, 6 kg/ha/yr in Kodana, 0.52 to 13.3 kg/ha/yr in Gandhisagar, 10.5 kg/ha/yr in Hirakud, 3.5 to 11 kg/ha/yr in Santhamur, 5.54 kg/ha/yr in Tungabhadra, 08 to 35.30 kg/ha/yr in Pilit and 42 to 56 kg/ha/yr in Shardarsagar reservoir (Srivastava and Tamot, 2002). The present fish production was much less than the average reservoir fish production of the country or other states.

The behavior of this reservoir is like a large seasonal pond due to its less productivity. The reservoir is full of water at maximum flood level or get flooded during monsoon and the juveniles are being stocked in September/October. In November/December, the water from the reservoir is released for Rabi crops and by the end of February the reservoir is at its minimum water level while in pre-monsoon summers either it dries up completely or else the water level gets drastically reduced. This affects the fish production in two ways-

1. Auto stocking fails during monsoon and post monsoon months because of the absence of broodstocks, as the water body is seasonal and not perennial.
2. The growth rate of the stocked fishes in post-monsoon months is adversely affected because of the harsh winter when water temperature drops up to 5^0C.

In post-winter months when fish may show accelerated growth rate, the water from the reservoir is released for irrigation purpose, thus reducing the total water volume required for good growth rate of the fish.

Category A fishes constituted nearly 33%, category B fishes constituted 1.6%, category C fishes constituted 13% and category D fishes constituted nearly 53% of Pili reservoir in five years (Table 3). In addition to this, stocking of fry size is 20-25 mm. Positive correlation between seed stocking, increased carp population and better productivity proved efficiency of proper seed stocking measures as a definite tool for production enhancement in reservoirs (Sugunan, 2002). Thus, stocking density, survival rate and water quality should be monitored regularly to enhance the productivity in the selected reservoir.

Category	Period (2007-08 to 2011-12) Average catch (kg)	Percentage
A	40300	32.95
B	1900	1.55
C	15900	13.00
D	64200	52.49
Total yield (kg)	122300	

Table 3. Category wise catch composition during in period 2007-08 to 2011-12 in Pili reservoir.

4. CONCLUSION

Indigenous as well as exotic fish species were observed in Pili reservoir. In terms of commercial fisheries, it is one of the important water bodies of the district Bijnor. However, the reservoir is having a lot of problems regarding fish harvesting. Most of the fishing methods adopted in the reservoir were traditional and unscientific. Fishing of juveniles should be banned by implementing the mesh size restriction for fishing. Unauthorized poaching of fish species should be effectively prevented. Efforts should be made to stock sufficient number of healthy fish seeds of carps especially Indian major carps, and simultaneously the species other than the carp that are not compatible for polyculture should be removed. Boats should be regularly inspected and repaired, and over-age boats should be replaced with new ones. Proper training of modern fishing methods should be provided to the fishers by reservoir authorities for commercial yield of fish production.

Considering the higher rate of culture fishery production in other states, there is a need for further improvement and development of the culture fishery in the Pili reservoir. It is therefore, suggested that attention should be paid to provide organic manure to increase the primary productivity of the reservoir and also, to raise composite fish culture with stocking density manipulation of various carp species and with regular supplementary feed.

ACKNOWLEDGMENT

We are thankful to the Chairman, Department of Zoology, Aligarh Muslim University, Aligarh, India for providing necessary facilities for the study. First author acknowledges the financial support from UGC, New Delhi (No. F. 15-189/12 (SA-II). Authors are grateful to Mr. N. S. Rehmani, Deputy Director, Uttar Pradesh Government Fisheries Department, Moradabad for permission to conduct the investigations.

REFERENCES

Ayyappan, S. (2011). Handbook of Fisheries and Aquaculture. ICAR publication, 238-248.

Baruah D., Dutta A. and Pravin P. (2013). Traditional Fish Trapping Devices and Methods in the Brahmaputra Valley of Assam, Indian J Traditional Knowledge, 12(1),123-129.

Islam M.R., Das B., Baruah D., Biswas S.P. and Gupta A. (2013). Fish Diversity and Fishing Gears used in the Kulsi River of Assam, India, Annals of Biological Research, 4(1),289-293.

Kamble, A.T. and Mudkhede, L. M. (2013). Study of fish fauna and productivity of Loni reservoir, Tq. Kinwat (Maharashtra). International Journal of Biomedical and Advance Research, 04 (03), 155-159.

Kingdom T. and Kwen K., (2009). Survey of Fishing Gears and Methods in Lower Taylor Creek Area, Bayelsa state, Nigeria, World Journal of Fish & Marine Sciences, 1(4),313-319.

Kolekar V, (2009). An account of observations on fishing gear and crafts of Mandovi estuary of Goa. Journal of Inland Fisheries Society India, 41 (1): 41-46.

Mathai T.J. and George N.A. (1972). A Note on the comparative catch efficiency of Nylon over Cotton Gillnets in Reservoir Fishing, Fish Technology, 9(1), 81-82.

Mohammed O.M. (2012). A Short review on: Fishing boats used in Sudan Freshwater fisheries, Bull. Environ. Pharmacol. Life Sci., 1(6),93-99.

Nagma, Khan M.A. (2013). Studies on freshwater fish fauna of district Bijnor in Western Uttar Pradesh, India. International Journal of Life Sciences Biotechnology and Pharma Research, 2(3):410-417.

Nagma M. (2015). Inland Fishery Resources of District Muzaffarnagar in North India. LAP Lambert Academic Publishing, Germany. ISBN: 978-3-659-68152-3pp. 178.

Nagma, and Khan, M.A., (2015). Observations on ichthyofauna and water quality of Pili reservoir in district Bijnor of Uttar Pradesh.International Journal of Fisheries and Aquatic Studies,2(6):106-112.

Paul, T.T, Palaniswamy R, Manoharan S, Unnithan U, Sarkar U.K. (2017) Management Strategies for Reservoirs Fisheries. Journal of Aquaculture Research and Development,8: 492. doi: 10.4172/2155-9546.1000492

Sakhare V.B. (2007). Reservoir Fisheries and Limnology, Narendra Publishing House, New Delhi.

Srivastava, A. and Tamot, P. (2002). Liminified status of Kernian Reservoir with special reference to Socio-economic survey of threatened fish Mahseer. AE–9, The Sixth Indian Fisheries forum, CIFE, Mumbai, India, p.90.

Sugunan V.V. (1995). Reservoir Fisheries of India, Fisheries Aquaculture Department, FAO Corporate Documentary Repository, FAO, Rome.

Sugunan, V.V. (2002). Enhancement an effective tool for increased fish population. Fishing Chimes. 21 (10-11): 21-26.

Sultan, S. (2013). Fisheries in Uttar Pradesh. National fisheries Development Board, Hyderabad. India.

Varghese M.D., George V.C., Khan A.A. and Iyer H. (1982). Shore seines for Reservoirs, Part-II, Studies on Optimal Mesh Size, Fish Technology, 20(2),97-100.

Yadava, Y. S. and Sugunan, V. V. (2009). Guidelines for Culture based fisheries management in small reservoirs in India. Department of Animal Husbandry, Dairying and Fisheries, Ministry of Agriculture, Government of India.

❐

Chapter-24

DNA BARCODING OF FISH: AN OVERVIEW

Graish Kumar, Priyanka Agarwal and M. Serajuddin*

Fish Biology Research Lab
Department of Zoology, University of Lucknow, Lucknow-226007
**Email of Corresponding Author: lu.fisheries@gmail.com*; bodhgayagk@gmail.com*

ABSTRACT

Fishes are good source for protein, vitamin D, vitamin B12, iodine, selenium and omega-3 fatty acids and has a very significant contribution in food security and economic welfare. Therefore, Proper identification of fish species is an imperative for management of fisheries and authentication of food products. In this context, DNA barcoding allows rapid and proficient means of fish identification. DNA barcoding is an advanced taxonomic method for identification of different organism's belonging to a specific species based on short genetic marker. The concept of DNA barcoding uses the cytochrome c oxidase subunit 1 mitochondrial gene (cox1 or COI) as a genetic marker which is suggested as unique barcode region for animals and plants. The advantages of barcoding fishes involve species identification for all potential users, including taxonomists and most notably, enabling identifications where traditional methods are not valid. It provides an idea that we need to construct a digital barcode hologram eventually and this digital barcode hologram needs to expand for all fishes, so that, we can easily identify the fish species as soon as possible with the help of the barcode reader. Two main global barcoding initiatives for fish are FISH-BOL and SHARKBOL.

Keywords: Fish identification, DNA barcode, Freshwater fishes, Mitochondrial cytochrome c oxidase.

1. INTRODUCTION

DNA barcoding is an advanced taxonomic method for identification of different organism's belonging to a specific species on the basis of short genetic marker (Hebert et al., 2003). The main purpose of DNA barcoding is to identify an unknown sample of pre-existing classification that's why it has some variations from molecular phylogeny (Kress et al., 2005). DNA barcoding promises to resolve all the queries and controversies related to biodiversity assessment and taxon identification which are spreading due to various leading scientific approaches like morph taxonomy analysis, sibling species identification, etc. (Blaxter, 2003). In other words, crusaders for DNA barcoding are proposing that a short fragment of DNA may be helpful to describe and discriminate between all forms of life which exist on this Earth (Frey & Frey, 2004; Kraytsberg et al., 2004). For example, Segment of 600 base pairs of mitochondrial gene cytochrome c oxidase I (COI) are mostly used barcode regions which provide as the base of a global bio identification system for animals, protists and plants. In spite of the fact that barcodes are sometimes used to try hard to identify unknown species or assess even if species should be combined or separated (Koch, 2010). Part of Internal Transcribed Spacer 2(ITS2) between rRNA genes is used in case of fungi whereas multiple regions are used mostly in case of plants. One of the most acceptable and commendable approach is the use of molecular in spite of morphological data for identifying taxa, which has long been a fundamental idea of many biologists (Busse et al., 1996; Blaxter 2004). The main purpose is to enhance an alliance between molecular and morphological taxonomists for fast species identification (Summerbell et al., 2005). DNA barcodes can also provide some useful information about taxa that are moderately well studied, inspite of those that require more research (Krishnamurthy & Francis, 2012). A squabble is made over that, DNA barcoding can offer useful taxonomic data but should be used with proper attention to preclude it from being used out of circumstance (Pilgrim et al., 2011). To enlighten the biodiversity assessments, a new concept of DNA barcoding has been used; however, this perspective should be used in combination with other methods for more influenced conservation efforts. The applications of DNA barcoding involve, for example, knowing about the plant leaves even when flowers or fruits are not sprouting, identifying insect larvae, identifying the diet of an animal, depending on its stomach contents or faeces (Valentini et al., 2009b, Soininen et al., 2009). Hence, DNA barcoding is very good tool to be used largely for discrimination and identification purposes.

2. MITOCHONDRIAL COI GENE

Cytochrome c oxidase I (COX1) is a protein in humans that is encoded by MT-COI gene which is recognized as mitochondrially prearranged cytochrome c oxidase I (MT-CO1). All eukaryotic cells carries mitochondria and animal mitochondrial DNA (mtDNA) has a relatively quick mutation rate than nuclear genetic markers (Boursot and Bonhomme, 1986; Delsuc et al., 2003; Hassanin et al., 2013), consequentially in the causation of diversity within and between populations above relatively brief evolutionary time scales (thousands of generations). In eukaryotes, the gene is called COX1, CO1, or COI. Cytochrome c oxidase I is the key subunit of the cytochrome c oxidase complex. DNA barcoding used the cytochrome c oxidase subunit 1 mitochondrial gene (cox1 or COI) as a genetic marker which is very highly effective for discriminating between vertebrate and invertebrate species (Rodrigues et al., 2017). For population genetics and phylogeographic studies, the mitochondrial cytochrome oxidase c subunit 1 (COI) gene is one of the most adequate markers away from the animal kingdom (Avise, 1994).

3. ADVANTAGES OF DNA BARCODING IN FRESHWATER PERSPECTIVE

Freshwater fishes are important source of protein, vitamin D, vitamin B12, iodine, selenium and omega-3 fatty acids. Proper identification of fish species is important for management of fisheries and authentication of food products. DNA barcoding allows fast and efficient means of fish identification (Trivedi et al, 2016). About 94% of all freshwater fisheries occur in developing countries (FAO, 2007). They present food and source of revenue for millions of the world's poorest people and also supply to the overall economic well being by means of export products trade, tourism and recreation (Worldfish center, 2002). It is estimated that freshwater fish's makeup more than 6% of the world's annual animal protein supplies for human (FAO, 2007). This leads to pressure on the fragile biodiversity of fresh water environment. There are several advantages of DNA barcoding including confirmation and well being assessment of seafood, wildlife forensics, preservation genetics and revealing of insidious alien species (IAS). Universal DNA barcoding hard work in the marine and freshwater habitat include MarBOL, CeDAMar, CMarZ, SHARK-BOL, FISH-BOL etc. In conjugation with latest and quicker techniques like high-throughput sequencing, DNA barcoding can serve as an effective modern tool in marine biodiversity assessment and conservation (Trivedi et al., 2016).

4. IDENTIFICATION OF FISH ON THE BASIS OF DNA BARCODING

As we know that DNA barcoding are the backbone for taxonomic identification, characterization and discovery of unknown species to facilitate the studies of biodiversity; It is more interesting approach to all researchers to appreciate genetic and evolutionary associations by collection of molecular, morphological and distributional data. An increase in the population of human's day-by-day causes an enormous threat to biodiversity, therefore species recognition has significant role in biodiversity preservation (Frankham, 2002; Frankham, 2003; Frankham et al., 2005). On the other hand, recognition of species is very complicated task; so many controversies have been put forward to the present day, related from the definition of the species concept to the definition of the doorstep for speciation (Carstens et al., 2013). Fish DNA barcoding depends on the sequencing of a constant area of Cytochrome C Oxidase type I (COI) gene, that plays a significant role and a precise tool for species identification, authentication, and phylogenetic analysis.The Fish Barcode of Life Initiative (FISH-BOL) (Ward et al., 2009) is a universal effort to organize a congregation of a standardised DNA barcode library for all fish species, one that is available from voucher specimens with reliable taxonomic identifications (Steinke & Hanner, 2011).The advantages of barcoding fishes involve species identification for all potential users, including taxonomists; highlighting specimens that signify a range expansion of known species; declining previously unrecognized species; and possibly most notably, enabling identifications where traditional methods are not valid. An example is the likely identification of groupers causing Ciguatera fish poisoning from meal remnants (Schoelinck et al., 2014).Since its beginning in 2005 FISH-BOL, it has been created an important and some valuable public resource in the shape of an electronic database containing DNA barcodes for nearly 10000 species, images, and geospatial coordinates of examined specimens (Becker et al., 2011). FISH-BOL thus complements and raising existing informative resources, such as Catalog of Fishes, FishBase and diverse genomics databases.

5. UNIVERSAL STATUS AND FUTURE PROSPECTS OF DNA BARCODING IN FISHES

DNA barcoding has gained extensive validation as a main tool for species identification and discovery. The arrival and too much misuse of polymerase chain reaction (PCR) followed by the reducing cost of DNA sequencing has driven molecular-based genetics next to ubiquity, by the primary field of biological sciences (Hebert et al., 2003). The growing market for fish and seafood products has been accompanied by an increasing number of cases of mislabelling and replacement (Clark, 2015).However various DNA techniques have been implemented to identify intended cases of market substitution and misbranding (Bossier, 1999; Martinez et al., 2005; Wong and Hanner, 2008; Rasmussen and Morrissey, 2008; Cohen et al., 2009; Rehbein, 2009b; Miller and Mariani, 2010; Handy et al., 2011). The repercussion of DNA barcoding is steadily increasing in a variety of fields of natural sciences. For ecologists, it is a tool with several possibilities (Valentini et al., 2009a), most of them are related to biodiversity inventories. Some scientific studies have been demonstrated its influence in identifying both marine and freshwater fishes (Alvarez & Nicieza, 2003; Ballard & Whitlock, 2004; Besansky et al., 2003). At present, records are obtainable globally for over 41771 fishes, representing 6566 fish species on the Barcode of Life Data System (BOLD) (Guiliano & Blaxter, 2006). Indian balance sheet for about 7.8% of the global finfish diversity, out of which 2553 are native fin fishes and

1563 are marine finfishes. In an attempt to barcode and classify the fishes in India, about 2334 fish species have been barcoded, till date (Nagpure, 2012). Therefore, the results propose that the species boundaries require to be examined immediately for the heavily exploited populations, targeted by the fish industry, to collect the information regarding accurate conservation strategies, scheduling and planning (Ward et al., 2005; Zemlak et al., 2009). Approximately 31,000 currently known fish species, 25% have been processed experimentally, with at least one species from each of all 89% fish families barcoded. DNA barcoding of fish have accomplished more than 10 years and still fish DNA barcoding is in early life period. It provides an idea that we need to construct a digital barcode hologram eventually and this digital barcode hologram needs to expand for all fishes. We can easily identify the fish species as soon as possible with the help of the barcode reader. Though, it is not an easy task. To understand the barcode sequences of fishes, the next-generation digital information storage system may also be urbanized. Rather than the COI barcode marker, various marker gene (like ND gene) should be put forward from fish mitochondrial DNA for the variegation of marker gene of fishes. It may also be helpful to develop fish DNA barcoding more simply as well as rapidly.

6. CONCLUSION

In the present situation, association and optimization of global efforts to create and share taxonomic knowledge of aquatic organisms are obligatory. DNA barcoding is enlightening itself as an enormously valuable tool in this way. DNA barcoding has become gradually more common since it was proposed in 2003. At this time, more than one million records are accessible in the BOLD system, which is the official depository of DNA barcode data. The latest large-scale project, iBOL, will go faster the creation of reference barcode libraries and will make possible the application of this simple identification method. In future, DNA barcoding will become a regular identification protocol for a variety of organisms. DNA barcoding in confirmation and traceability systems for seafood/fish products is hailed by stakeholders like scientists and eNGOs as an essential improvement that should be adopted to fight the health, safety and economic concerns nearby the mislabelling of fish products, however questions related to the applicability and suitability of the technology remains. The victorious adoption of DNA barcoding is also dependent upon its approval by leading international organizations like the FAO/WHO as a methodological standard. If DNA barcoding's profit can be translated into economic benefits at the state level by reducing and identifying cases of fraud, then trading partners are more probably to judge it as a yardstick methodology for species identification for fish species, for other commercially important food products as well as endangered flora and fauna.

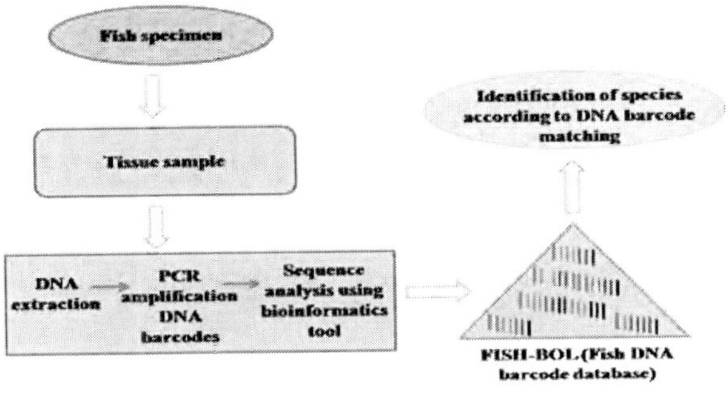

Fig 1. Graphical representation of Fish DNA barcoding.

REFERENCES

Alvarez, D. & Nicieza, A. G. (2003). Predator avoidance behaviour in wild and hatchery-reared brown trout: the role of experience and domestication. Journal of Fish Biology, 63(6), 1565-1577.

Avise, J. C. (1994). Molecular Markers, Natural History and Evolution: Natural History and Evolution.Springer Science & Business Media.

Ballard, J. W. O. & Whitlock, M. C. (2004). The incomplete natural history of mitochondria. Molecular ecology, 13(4), 729-744.

Becker, S., Hanner, R. & Steinke, D. (2011). Five years of FISH-BOL: brief status report. Mitochondrial DNA, 22(sup1), 3-9.

Besansky, N. J., Severson, D. W. & Ferdig, M. T. (2003). DNA barcoding of parasites and invertebrate disease vectors: what you don't know can hurt you. TRENDS in Parasitology, 19(12), 545-546.

Blaxter, M. (2003). Molecular systematics: counting angels with DNA. Nature, 421(6919), 122.

Blaxter, M. L. (2004). The promise of a DNA taxonomy. Philosophical Transactions of the Royal Society of London B: Biological Sciences, 359(1444), 669-679.

Bossier, P. (1999). Authentication of seafood products by DNA patterns. Journal of food science, 64(2), 189-193.

Boursot, P. & Bonhomme, F. (1986). Génétique et évolution du génome mitochondrial des Métazoaires. Genetique, selection, evolution, 18(1), 73.

Busse, H. J., Denner, E. B. & Lubitz, W. (1996). Classification and identification of bacteria: current approaches to an old problem. Overview of methods used in bacterial systematics. Journal of biotechnology, 47(1), 3-38.

Carstens, B. C., Pelletier, T. A., Reid, N. M. & Satler, J. D. (2013). How to fail at species delimitation. Molecular ecology, 22(17), 4369-4383.

Clark, L. F. (2015). The current status of DNA barcoding technology for species identification in fish value chains. Food Policy, 54, 85-94.

Cohen, N. J., Deeds, J. R., Wong, E. S., Hanner, R. H., Yancy, H. F., White, K. D., ... & Gerber, S.I. (2009). Public health response to puffer fish (tetrodotoxin) poisoning from mislabeled product. Journal of Food Protection, 72(4), 810-817.

Delsuc, F., Stanhope, M. J. & Douzery, E. J. (2003). Molecular systematics of armadillos (Xenarthra, Dasypodidae): contribution of maximum likelihood and Bayesian analyses of mitochondrial and nuclear genes. Molecular phylogenetics and evolution, 28(2), 261-275.

Frankham, R., Briscoe, D. A. & Ballou, J. D. (2002). Introduction to conservation genetics. Cambridge university press.

Frankham, R. (2003). Genetics and conservation biology. Comptes Rendus Biologies, 326, 22

Frankham, R. (2005). Genetics and extinction. Biological conservation, 126(2), 131-140.

FAO, 2007. The state of world aquaculture and fisheries 2006, FAO Fisheries and Aquaculture Department, Food and Agriculture organization of the United nations, Rome, Italy.

Frey, J. E. & Frey, B. (2004). Origin of intra-individual variation in PCR-amplified mitochondrial cytochrome oxidase I of Thrips tabaci (Thysanoptera: Thripidae): mitochondrial heteroplasmy or nuclear integration? Hereditas, 140(2), 92-98.

Guiliano, D. B. & Blaxter, M. L. (2006). Operon conservation and the evolution of trans-splicing in the phylum Nematoda. PLoS genetics, 2(11), e198.

Handy, S. M., Deeds, J. R., Ivanova, N. V., Hebert, P. D., Hanner, R. H., Ormos, A., ... & Yancy, H. F. (2011). A single-laboratory validated method for the generation of DNA barcodes for the identification of fish for regulatory compliance. Journal of AOAC International, 94(1), 201-210.

Hanner, R., Becker, S., Ivanova, N. V., & Steinke, D. (2011). FISH-BOL and seafood identification: Geographically dispersed case studies reveal systemic market substitution across Canada. Mitochondrial DNA, 22(sup1), 106-122.

Hassanin A., An J., Ropiquet A., Nguyen T.T.& Couloux A. Combining multiple autosomal introns for studying shallow phylogeny and taxonomy of Laurasiatherian mammals: Application to the tribe Bovini (Cetartiodactyla, Bovidae). Mol Phylogenet Evol. 2013 Mar;66(3):766-75. doi: 10.1016/j.ympev.2012.11.003. Epub 2012 Nov 15. PubMed PMID: 23159894

Hebert, P. D., Cywinska, A. & Ball, S. L. (2003). Biological identifications through DNA barcodes. Proceedings of the Royal Society of London B: Biological Sciences, 270(1512), 313-321.

Koch, H. (2010). Combining morphology and DNA barcoding resolves the taxonomy of western Malagasy Liotrigona Moure, 1961 (Hymenoptera: Apidae: Meliponini). African Invertebrates, 51(2), 413-421.

Kraytsberg, Y., Schwartz, M., Brown, T. A., Ebralidse, K., Kunz, W. S., Clayton, D. A.& Khrapko, K. (2004). Recombination of human mitochondrial DNA. Science, 304(5673), 981-981.

Kress, W. J., Wurdack, K. J., Zimmer, E. A., Weigt, L. A. & Janzen, D. H. (2005). Use of DNA barcodes to identify flowering plants. Proceedings of the National Academy of Sciences, 102(23), 8369-8374.

Krishnamurthy, P. K. & Francis, R. A. (2012). A critical review on the utility of DNA barcoding in biodiversity conservation. Biodiversity and Conservation, 21(8), 1901-1919.

Martinez, I., James, D. & Loréal, H. (2005). Application of modern analytical techniques to ensure seafood safety and authenticity. FAO, Roma (Italia).

Miller, D. D. & Mariani, S. (2010). Smoke, mirrors, and mislabeled cod: poor transparency in the European seafood industry. Frontiers in Ecology and the Environment, 8(10), 517-521.

Nagpure, N. S., Rashid, I., Pathak, A. K., Singh, M., Singh, S. P. & Sarkar, U. K. (2012). FBIS: A regional DNA barcode archival & analysis system for Indian fishes. Bioinformation, 8(10), 483.

Pilgrim, E. M., Jackson, S. A., Swenson, S., Turcsanyi, I., Friedman, E., Weigt, L. & Bagley, M. J. (2011). Incorporation of DNA barcoding into a large-scale biomonitoring program: opportunities and pitfalls. Journal of the North American Benthological Society, 30(1), 217-231.

Rasmussen, R. S. & Morrissey, M. T. (2008). DNA-based methods for the identification of commercial fish and seafood species. Comprehensive reviews in food science and food safety, 7(3), 280-295.

Rehbein, H. & Oehlenschlager, J. (Eds.). (2009). Fishery products: quality, safety and authenticity. John Wiley & Sons.

Rodrigues, M. S., Morelli, K. A. & Jansen, A. M. (2017). Cytochrome c oxidase subunit 1 gene as a DNA barcode for discriminating Trypanosoma cruzi DTUs and closely related species. Parasites & vectors, 10(1), 488.

Schoelinck, C., Hinsinger, D. D., Dettaï, A., Cruaud, C. & Justine, J. L. (2014). A phylogenetic re-analysis of groupers with applications for ciguatera fish poisoning. PLoS One, 9(8), e98198.

Soininen, E. M., Valentini, A., Coissac, E., Miquel, C., Gielly, L., Brochmann, C., ... & Taberlet, P. (2009). Analysing diet of small herbivores: the efficiency of DNA barcoding coupled with high-throughput pyrosequencing for deciphering the composition of complex plant mixtures. Frontiers in Zoology, 6(1), 16.

Steinke D.& Hanner R. (2011). The FISH-BOL collaborators' protocol. Mitochondrial DNA, 22(S1), 10–14.

Summerbell, R. C., Lévesque, C. A., Seifert, K. A., Bovers, M., Fell, J. W., Diaz, M. R. and Crous, P. W. (2005). Microcoding: the second step in DNA barcoding. Philosophical Transactions of the Royal Society of London B: Biological Sciences, 360(1462), 1897-1903.

Trivedi S., Aloufi A.A., Ansari A.A. & Ghosh S.K. (2016). Role of DNA barcoding in marine biodiversity assessment and conservation: An update. Saudi Journal of Biological Sciences, 23, 161–171.

Valentini, A., Pompanon, F. & Taberlet, P. (2009a). DNA barcoding for ecologists. Trends in Ecology & Evolution, 24(2), 110-117.

Valentini, A., Miquel C., Nawaz M.A., Bellemain E., Coissac E., Pompanon F., Gielly L., ... & Taberlet P. (2009b). New perspectives in diet analysis based on DNA barcoding and parallel pyrosequencing: the trnL approach. Mol. Ecol. Res.,9, 51–60.

Ward, R. D., Hanner, R. & Hebert, P. D. (2009). The campaign to DNA barcode all fishes, FISH-BOL. Journal of fish biology, 74(2), 329-356.

Ward, R. D., Zemlak, T. S., Innes, B. H., Last, P. R. & Hebert, P. D. (2005). DNA barcoding Australia's fish species. Philosophical Transactions of the Royal Society of London B: Biological Sciences, 360(1462), 1847-1857.

Wong, E. H. K. & Hanner, R. H. (2008). DNA barcoding detects market substitution in North American seafood. Food Research International, 41(8), 828-837.

WorldFish Centre, 2002. Fish: An issue for everyone. A concept paper for fish for all.

Zemlak, T. S., Ward, R. D., Connell, A. D., Holmes, B. H. & Hebert, P. D. (2009). DNA barcoding reveals overlooked marine fishes. Molecular Ecology Resources, 9, 237-242.

❐

Chapter-25

LENGTH-WEIGHT RELATIONSHIP, LENGTH-LENGTH RELATIONSHIP AND CONDITION FACTOR OF *NOTOPTERUS CHITALA* AND *NOTOPTERUS NOTOPTERUS* COLLECTED FROM THE RIVER GANGA

Mohd. Gufran Khan and M. Afzal Khan*

Section of Fishery Science and Aquaculture, Department of Zoology, Aligarh Muslim University, Aligarh-202002, Uttar Pradesh, India
Email of Corresponding Author: khanmafzal@yahoo.com

ABSTRACT

Length-weight relationship, length-length relationship and condition factor were estimated for Notopterus chitala and Notopterus notopterus collected from Narora site of the river Ganga. The parameters a and b of the length-weight relationships were estimated after the logarithmic transformation of the equation: W = aLb. The length-length relationships were estimated using linear regressions. Condition factor (K) was estimated by the equation: $K = W/L^3 \times 100$. Length-weight relationship indicated isometric growth pattern for N. chitala (b = 3.01) and negative allometric growth pattern for N. notopterus (b = 2.49). Length-weight and length-length relationships for both the target fish species were highly significant (R2 > 0.9, P < 0.001). The condition factor values for N. chitala and N. notopterus were 0.72 and 1.02 suggesting that the condition is less suitable for N. chitala as compared to N. notopterus.

Keywords: Length-weight relationship, Condition factor, Ganga River, Notopterus chitala, Notopterus notopterus.

1. INTRODUCTION

Notopterus chitala (Hamilton-Buchanan, 1822) and *Notopterus notopterus* (Pallas, 1769) are commonly known as humped featherback and grey featherback. They are distributed in deep and clear waters of rivers, beels, reservoirs, ponds and lakes located in India, China, Sri Lanka, Nepal, Bangladesh, Pakistan, Thailand, Malaya, Indonesia, and Cambodia (Talwar and Jhingran, 1991). In India these fish species are reported from rivers; Indus, Ganges, Brahmaputra and Mahanadi (Froese and Pauly, 2017). Both fish species are having food, sport and ornamental value (Sarkar et al., 2006).

Length-weight relationships (LWR) provide information on basic biological parameters that are useful in fisheries management of the target fish species. These relationships are often used to estimate biomass from length observation, to calculate condition factor and to convert growth-in-weight equation to growth-in-length equation (Khan et al., 2012; Nazir and Khan, 2017).

Condition factor is a quantitative parameter of the state of well-being of the fish that will determine present and future population success by its influence on growth, reproduction and survival. The condition of a fish reflects recent physical and biological circumstances, and fluctuates by interaction among feeding conditions, parasitic infections and physiological factors (Richter, 2007).

The present study provides information on length-weight relationship, length-length relationship and condition factor in the target fish species collected from river Ganga.

2. MATERIALS AND METHODS

Sampling:

Notopterus chitala (N=130) and *Notopterus notopterus* (N=70) were collected during the period September 2014 to September 2015 from the river Ganga at Narora (27°30'N, 78°25' E), Uttar Pradesh, India. Fishes were collected using different types of fishing gears such as cast net, drag net and gill net.

The length-weight relationship of each fish species was calculated by using logarithmic transformation of linear regression equation following Garcia (2010),

$$\text{Log } W = \text{Log } a + b \text{ Log } TL$$

where, W is the total weight (gm), TL is the total length (cm), a and b are the intercept and slope of the regression, respectively. Total length (TL) and standard length (SL) were also measured to nearest cm to calculate the TL vs. SL relationship. The condition factor (K) was measured using equation $K = 100\ W/L^3$ where W= weight of fish in gm, L= total length in cm.

All the statistical analyses were performed using MS-excel and SPSS (version 22).

3. RESULTS

The estimated parameters of length-weight relationship of *Notopterus chitala* and *Notopterus notopterus* are presented in Table 1. Moreover, the length-length relationship parameters are shown in Table 2.

Species	Range of total length (cm)	a	b	R^2
Notopterus chitala	22-94	-2.158	3.01	0.992
Notopterus notopterus	13-30	-1.332	2.49	0.999

Table 1. Estimated parameters of length-weight relationship (LWR) of *Notopterus chitala* and *Notopterus notopterus*. a = intercept; b = slope or regression coefficient; R^2 = coefficient of determination.

Species	TL(cm)	Equation	a	B	R^2
Notopterus chitala	22-94	TL= a+b SL	-2.088	0.998	0.998
Notopterus notopterus	13-30	TL= a+b SL	0.556	0.902	0.994

Table 2. Relationship between total length (TL) and standard length (SL) of *Notopterus chitala* and *Notopterus notopterus*. a = intercept; TL= total length; SL= standard length; b = slope; R^2= coefficient of determination.

The condition factor value was 0.72 of *Notopterus chitala* and 1.02 for *Notopterus notopterus* suggesting that the condition of *N. notopterus* was better as compared to *N. chitala* collected from the river Ganga.

4. DISCUSSION

In this study, *N. chitala* exhibited isometric growth pattern (b = 3) inferring that the fish length increased in equal proportions to its body weight at constant specific gravity whereas *N. notopterus* showed negative allometric growth pattern (b = 2.49) suggesting that the fish length increased more than its weight or fish became slenderer as it increased in weight. Length-length relationship of *N. chitala* and *N. notopterus* showed significant linear relationships among the length parameters such as total length (TL) and standard length (SL). The value of the condition factor revealed that the environmental conditions were relatively less suitable for *N. chitala* growth as compared to *N. notopterus*; which showed better condition in the selected river.

ACKNOWLEDGEMENT

The authors are thankful to the chairman of the Department of Zoology, Aligarh Muslim University, Aligarh, India for providing the necessary facilities for the study. The first author is grateful to the University Grants Commission for providing Non-NET fellowship.

REFERENCES

Froese, R.; Pauly, D., Editors. 2017: FishBase. World Wide Web electronic publication. www.fishbase.org, (06/2017).

Garcia, L. M. B., 2010: Species composition and length-weight relationship of fishes in Candaba wetland on Luzon Island, Philippines. J. Appl. Ichthyol. 26, 946-948.

Khan, M. A.; Khan, S.; Miyan, K., 2012: Studies on Length-weight and Length-length relationships of four freshwater fishes. J. Fish. Aquat. Sci.7, 481-484.

Nazir, A.; Khan, M. A., 2017: Length-weight and Length-length Relationships of *Cirrhinus mrigala* (Cyprinidae) and *Xenentodon cancila* (Belonidae) from the River Ganga. J. Ichthyol. 57, 787-790.

Richter, T. J., 2007: Development and evaluation of standard weight equations for bridge lip sucker and large-scale suckers. N. Am. J. Fish. Manage. 27, 936-939.

Sarkar, U. K.; Deepak, P. K.; Negi, R. S.; Singh, S. P.; Kapoor, D., 2006: Captive Breeding of endangered fish *Chitala chitala* (Hamilton-Buchanan) for species conservation and sustainable utilization. Biodivers. Conserv. 15, 3579-3589.

Talwar, P. K.; Jhingaran, A. G., 1991: Inland fishes of India and Adjacent Countries. Vol.1.Oxford & IBH Publishing Co., New Delhi.

❐

Chapter-26

HABITAT ,DISTRIBUTION AND FEEDING SPECIALIZATIONS OF TWO SPECIES OF SPINY EELS *MACROGNATHUS ACULEATUS* (BLOCH) AND *MACROGNATHUS PANCALUS*(HAM.)

Aysha Aziz Faridi [1*], M. Serajuddin[2], M. Moshahid A. Rizvi[1] and Rustam Ali [3]

[1]*Department of Biosciences, Jamia Millia Islamia, New Delhi-110025, India*
[2]*Department of Zoology, University of Lucknow, Lucknow -226007, U.P., India*
[3]*Department of Zoology, S.P. Mukherjee Degree College, Bhadohi, U.P., India*
**Email of Corresponding Author: faridiaysha@gmail.com*; rizvijmi@gmail.com*

ABSTRACT

Spiny eels Macrognathus aculeatus and Macrognathus pancalus are quite palatable and nutritious too. The demand of these fishes always exceeds their supply in the market. The Spiny eels are restricted to Asia and Africa only and occur in a variety of freshwater habitats in the plains as well as in the hills of India, common in rivers ,ponds, stream and estuaries. For this study the samples of both sexes of different age and size groups of these species were collected on monthly basis from Delhi and NCR by using drag and cast nets. Measurements of different body parameters such as total length, standard length, body depth, body width and length of alimentary canal were taken with the help of fish measuring board and fine pointed compass. Close examination of the organs of alimentation and their correlation with feeding habits were carried out by applying the methods given by Pillay (1952) and Lagler (1956). A strong dentition at the angle of preopercle, followed by three to five more along its lower border ,well developed stomach, short intestine besides other characteristic features showed the carnivorous orientation of the fish. Juveniles and adults both are predacious in nature preferring live aquatic annelids ,crustaceans and insects larvae. The fish patrols a body of water throughout its length and breadth but spends major part of its life at the bottom of water bodies.

1. INTRODUCTION

Spiny eels, *Macrognathus pancalus* and *Macrognathus aculeatus* are quite palatable as a table fish and nutritive too they are economically important inland water teleostean fish. Commonly known as Patya and Malga but locally called as Goingee. *M. pancalus* is the most beautiful among the spiny eels. The smaller size of this fish has often been placed into aquaria and kept by hobbyists.

A survey of literature showed that very scanty information is available on the biology of spiny eels despite palatability and consumer appeal of this group of teleosts Karim and Hossain (1972) studied the sexual maturity and fecundity of *M. pancalus*. Srivastava (1975) studied the unusual development of the caudal fin of this species. Sikder and Das (1980) carried out works on skin structure and Mittal et al. (1994) on histochemical analysis of glycoproteins. Qaayum and Qasim (1961) studied the spawning frequencies and breeding season of Rhynchobdella aculeata (*Macrognathus aculeatus.*)

Rahman (1989) reported the availability of spiny eels in rivers, canals, beels, ponds and inundated fields throughout Bangladesh. Despite of its palatability and consumer appeal a little information is available on the biology of this fish. The studies carried out on this fish are related to the seasonal variation in fat and water contents (Prasad et al., 1983) and functional anatomy of digestive organs of *M. aculeatus* (Srivastava, 1970). Dutta (1989), Serajuddin and Mustafa (1994) and Serajuudin et al. (1998) investigated the food and feeding habits of a closely related species, *Mastacembelus armatus*. In 2004 Serajuddin studied Intraspecific diversity of riverine populations of spiny eel, *M. armatus*.

Demand for the spiny eels almost always exceeds supply. The fish sold in the market are bought almost entirely from their natural population which, in turn, showing the signs of dwindling due to their reckless exploitation. This situation demands their scientific cultivation, but to develop a method for their culture in inconceivable without a sound knowledge of their biology. Keeping in mind the paucity of information on the biology of this species, the present study focusing on the habitat, distribution and feeding specialization of these two species of spiny eels.

2. MATERIALS AND METHODS

Monthly samples were collected by using drag and cast nets from rivers and ponds of Delhi and NCR. The specimens were preserved in 10% formalin and brought to laboratory for their investigation. Measurement of different body parameters such as Total length (TL), Standard Length (SL), Body Depth (BD), Body Width (BW) and Length of Alimentary canal (LAC) were taken with the help of fish measuring board and compass. Specimens were examined for their dentitions and salient features of their morpho-biology. Close examination of organs of alimentation and their correlation with feeding habits was carried out.

Relative Gut Index was computed by using the formula

RGI= Length of Gut/ Length of Fish X 100

Gut contents were removed and qualitative as well as quantitative analysis were carried out by using the methods as given by Lagler (1956).

3. OBSERVATIONS AND DISCUSSION

Habitats and Distribution:

The Spiny eels, M. pancalus and M. aculeatus are restricted to Asia.They have been reported from India as well as Pakistan, Sri Lanka , Bangladesh, Myanmar, Thailand , Malyasia and southern China. In India they are found in Assam, Punjab, Orissa, Andhra Pradesh, Tamil Nadu , Kerala and Mahrashtra. They inhabit a variety of fresh water habitats in India and are common to rivers, lakes, ponds, streams and estuaries. Fish patrols all along the length of the water bodies but spends major parts of its life at the bottom.

Feeding Specializations:

Body is slender. Mouth is equipped with fine but firm jaws. The upper jaw is longer and projects over the lower one. The pointed mouth facilitates the probing of food items which may be concealed under submerged objects and bottom deposits.The teeth are numerous, small, sharp and strong on both jaws and are villiform and do not show any enlargement into canine or incisor.The nature of dentitions indicates that it helps in grasping and holding the active prey and preventing its escape.The gap of mouth is broad enough to support the intake of small to medium sized shrimps.The alimentary canal is broadly distinguished into foregut (comprising esophagus and stomach), midgut (intestine) and hindgut (rectum).the stomach is well developed, elongated and thick, the muscular walls and internal lining is wavy. In the absence of mastication being done in mouth cavity, the stomach's ability to deal with the whole prey assumes importance.

Intestine is short, thick walled and almost straight and constitutes about half of the total length of alimentary canal. Intestine opens into rectum, which is short and contains fecal matter. The diameter of rectum is less as compared to its intestine. These fishes feeds on variety of food items such as shrimps, annelids, crustaceans , insects and its larvae, submerged plants parts and rarely mud.(table-3&4) A wide range of food item spectrum indicates that these fishes are euryphagous. The alimentary canals were found to be vacant during spawning and breeding season.

Relative Gut Index (RGI):

Relative gut indices of these fishes are given in Table 1 and 2.

Absence of any appreciable difference in this index is juvenile and adult in both species of spiny eels indicated that the growth does not involve any major shift in their basically carnivorous orientation of the dietary habit.

RGI value in specimens of different sizes of M. aculeatus varied from 0.53 to 0.63 with an average value of 0.58. RGI value to specimens of different sizes of *M. pancalus* varied from 0.50 to 0.60 with an average value of 0.50.

Total Length of Fish (cm)	Number of observations	Relative Gut Index
5-10	12	0.55
11-15	10	0.56
16-20	22	0.61

Table 1. Relative gut index of different sizes of *M. aculeatus*

Total Length of Fish (cm)	Number of observations	Relative Gut Index (MEAN)
5-10	20	0.51
11-15	28	0.59
16-20	26	0.58
21-25	08	0.66

Table 2. Relative gut index of different sizes of *M. pancalus*

Food items	Numerical count	Frequency of occurrence (%)	Gravimetric Index (%)
Teleostomi			
Puntius	15	44	83.2
esomus	08	12	

Aquatic insects (larval stage)			
Ephemeroptera	Numerous	19.1	7.20
Diptera	Numerous	33.5	19.81
Hemiptera	04	17.5	0.55
Unidentified	35	8.4	4.32
Crustacea			
Prawn	79	26.6	67.4
Eubranchipus	77	21.1	0.92
Daphnia	16	11.2	
Annelids			
Earthworms	Numerous	23.2	18.6
Mollusca			
Gastropods	05	3.9	0.22
Digested matter		35.9	36.5
Miscellaneous		11	7.41

Table 3. Gut contents of the spiny eel, *M. aculeatus*

Food items	Numerical count	Frequency of occurrence (%)	Gravimetric Index (%)
Annelids	Numerous	24.5	12
Earthworms			
Crustacea	59	44.2	76.02
Prawn	22	16.0	03.76
Caridina	18	11.0	00.68
Daphnia	05	06.8	22.39
Branchinella			
Aquatic insects			
Chaoborus	66	54.3	96.33
Tipula	13	21.3	04.56
Unidentified	Numerous	21.5	03.43
Teleostomi			
Esomus Unidentified	08	08.6	11.67
scales	31	59.6	66.21
Mollusca	36	31.6	01.89
Gastropods	05	04.6	01.32
Digested matter		28.5	25.67
Aquatic vegetation	Numerous	08,8	0.425

Table 4. Gut contents of the spiny eel, *M. pancalus*

ACKNOWLEDGEMENT

The authors thank the Head, Department of Biosciences, Jamia Millia Islamia University for the facility and administrative support. The study was funded by UGC, New Delhi.

REFERENCES

Dutta, S.P.S. (1989,1990) food and feeding ecology of *M. armatus* from Gadhigarh stream, Jammu, Matsya 1005 & 16:66.

Karim, M.A., and Hossain, A. (1972) Studies on the biology of *Mastacembelus pancalus* (spiny eel, Hamilton) in artificial ponds. Part II, sexual maturity and fecundity. Bang. J. Biol. Agric. Sci., 1(2), pp. 15-18

Lagler, K.F. (1956) Freshwater fishery biology 2nd ed Iowa WMC Brown Co.U.S.A.pp-421

Mittal, A. K., T. Ueda, D. Fujimori and K.Yamda 1994. Histochemical analysis of glycoprotein in the unicellular glands in the epidermis of an Indian freshwater fish, *Mastacembelus pancalus*. Histochem. J., 26: 666-677

Nikolsky, G.V (1963)The Ecology of Fishes-Academic Press. London and New York

Qasim, S.Z., and Qayyum, A. (1961) Spawning frequencies and breeding season of some freshwater fishes with special reference to those occurring in the plains of northern India. Indian J. Fish., 8, pp. 24-43.

Prasad, S., Pandey B.N. and Sinha D.P. (1983) Seasonal variation in fat and water content of the Indian freshwater mud eel *Macrognathus aculeatus* Z.Tierphysiol.Tierer narhr Futtermitot lkd.49;117-126

Rahman A.K.A.(1989) Freshwater fishes of Bangladesh Univ. Dhaka p-364 Srivastava, V.M.S.(1970) Functional anatomy of digestive organs of a freshwater fish *Rhynchobdella aculeata* (Ham.) PartIB Histology ves. cst.spol.zool.34:136-142

Serajuddin, M. and Mustafa S.(1994) Feeding specializations in adult spiny eel *Mastacembelus armatus* Asi.Fish.Sci.7:63-65

Serajuddin, M., Khan A.A. and Mustafa S.(1998) Food and feeding habits of the spiny eel Mastacembelus armatus Asi.Fish.Sci.11:271-278

Serajuddin, M.(2004) Intraspecific diversity of riverine populations of spiny eel, Mastacembelus armatus.App. Fish.and aqua. 4.pp.25-29

Venkatachalam, U., Sreedevi, P.R., Senthil kumar, D., Munirasu, S., Kiruba A. and Ramasubramanian,V.(2013) Impact of seasonal variation and feeding on reproductive behavior of freshwater spiny eel *Mastacembelus armatus* from Cauvery river Asi.Paci. Jour. Of Repr.2(3)189-195

❐

Chapter-27

BIRD COMMUNITY STRUCTURE IN URBAN HABITAT

Farah Akram and Orus Ilyas*

Department of Wildlife Sciences, Aligarh Muslim University, Aligarh. U.P. India-202002
**Email of Corresponding Author: orus16@gmail.com*; farahakram2013@gmail.com*

ABSTRACT

Urban areas spread very rapidly, and it has a strong influence on wildlife especially birds. Present study reveals the bird community structure along a high to low congested urban areas. We examine relationships between bird communities and habitat structure. Three different sites were selected on the basis of human population and building structure viz, high medium and low congested areas. A total of 41 bird species were recorded during the six-month study (Jan 2013 to June 2013). Bird density was found maximum in high congested area followed by medium low congested area whereas bird diversity and richness were found maximum in low congested area followed by medium and high congested areas. Our result conclude that highly congested and disturbed areas decline the bird diversity and richness but favors the density of few dominating species while low congested areas supports the high diversity and richness as it includes green parks and low disturbed area as far as it is emerging as a new habitat for birds.

Keywords: Urbanization, Bird community Structure, Habitat variables.

1. INTRODUCTION

Today human settlements or urbanization have a strong influence on bird communities. It produces series of fundamental changes on the landscape, these changes include vegetation removal, construction of infrastructure like (roads, channels etc), introduction exotic flora and fauna species, increase in feral pets etc. So, with the global high rate of urbanization and the rapid loss of wild habitat, cities are now viewed as challenging ecosystems for sustaining biotic communities (Shochat, 2004) The growth of urban centers has profound effects on natural ecosystems. During the 2000s research on urban bird populations and communities focused on global patterns, as well as processes and mechanisms that lead to the two globally recognized patterns: increased overall population densities and decrease in species diversity compared with wild lands (Shochat, 2004).

Although the study of urban birds has a fairly long history, urban ecosystems have been largely ignored throughout many decades of ecological research (Miller and Hobbs, 2002; Collins et. al., 2000). Since the early 1990s, a different view emerged, accepting urban settings as ecosystems that are structured and function like other natural ecosystems (Rebele, 1994; Grimm et. al., 2000; McKinney, 2002; Miller and Hobbs, 2002). This theoretical view represents an emerging realization that by now, most of the world's land is managed and dominated by humans (approximately 50% of the human population lives in cities). Wildlands are continuously converted to agricultural fields and urban areas. Consequently, urban environments can no longer be viewed as lost habitat for wildlife, but rather as new habitat that, with proper management, has the potential to support diverse bird communities.

Birds are highly visible and sensitive to alterations in habitat structure and function; consequently, they serve as excellent indicators of changes and stresses in urban ecosystems (Clergeau et. al., 1998; Blair, 1999; Savard et. al., 2000).

From the above views it conclude that urban areas emerging as a new ecosystem so there is a need to know what are the parameters that helps to adapt this new ecosystem and which area in urban that is most preferred by birds so the aim of the present study to investigate the bird community structure at different level of urbanization in urban areas from high to low congested areas.

2. MATERIALS AND METHODS

Study Area:

The survey was conducted in Aligarh city of U.P India (Fig. 1). The city, Aligarh is an important city of Uttar Pradesh and comes in the metropolitan cities in India that are known worldwide for Aligarh Muslim University and a lock factory.

It is situated in the northern region of UP. It is bounded by the river Ganga and Yamuna. The extreme parallels of latitude are 27^0 29' and 28^0 11' North and 77^0 38' and $78^0$38, East longitude. It lies in the upper Gangetic plains, areas of remarkable fertility. It has uniform surface and large constituted of the alluvial filling. It experiences tropical monsoon type of climate.

Intensive Sites for Study:

An urban area is built up with roads, houses, shops, offices, entertainment centers, public buildings, etc. It is a proper planned settlement built up according to the process called urbanization. Urban areas have a higher population density (more people in a smaller space), more congested and extreme overcrowding as compare to rural areas and also have more pollution and traffic.

As related to my research, I was selected 3 sites in the study area according to the density of human population and created building structures in a particular area.

High Congested

High congested areas are defined those areas in which houses / shops or any type of building situated very close and transportation also high.

Medium Congested

Medium congested areas are those areas where buildings are not closely distributed

Low Congested

Low congested areas are those where houses or shops etc are sparsely distributed. It includes open areas like parks and small water body. Out of these three habitats 30 points were located and 10 plots in each habitat were laid.

Data Collection:

Habitat Structure and Variables

The present study was conducted during January 2013 through June 2013 for a base line understanding on bird community structure at different level of urbanization. It is necessary to sample vegetation and building infrastructure in order to correlate the birds with the habitat type. Vegetation sampling was conducted along with the bird sampling in the selected sampling sites. The number of individuals of tree species was recorded in 10 m radius circular plot for species density, diversity and richness estimation. All shrubs were recorded within the radius of 3 m. For herbs and grasses, quadrate method was used wherein quadrates of 0.5 × 0.5 m were laid.

Bird Survey

Point count method (Sutherland, 2006) was used to document the population of different bird species in the area. During the survey period bird species were recorded in different habitat types for 20 min at each point within close width of 30-meter radius circular plot. The points were randomly established with a minimum interval of 200-300 m in different habitat, so as to avoid repetition and biasness. Repetitive counts were taken periodically at each point.

Data Analysis

The number and individuals of bird species recorded were considered and subsequently density, Shanon indices diversity and Margleef richness were estimated for different habitats. Species diversity and richness were calculated by using modified version of "SPECDIVER BAS" (Ludwig and Reynolds, 1988).

One-way ANOVA performed to examine the mean differences in bird density, diversity and richness parameters in different habitat types.

Pearson product moment correlation was performed to establish the correlation between bird density, diversity and richness with habitat parameters. The tests were performed by SPSS 17.0 software.

3. RESULTS

Vegetation Structure in Different Categories in Urban Habitat:

Table 1 shows that vegetative structure is prominent in the low congested area where in density, diversity and richness of all the parameters viz a viz trees, shrubs, herbs and grasses is maximum and minimum in high congested habitat. However, results were found to be significant for all parameters ($p<0.05$).

Bird Abundance:

Over All Bird Species Densities in Urban Habitat

Table 2 shows that the result of mean densities of different bird species in urban habitat. Total 41 species were recorded in the urban habitat in which density of rock pigeon *(Streptopelia senegalensis)* was found to be maximum (15.844/ha) followed by rose ringed parakeet *(Psittacula krameri)* (6.566/ha) and density of common hoopoe *(Upupa epops)*, Egyptian vulture *(Neophron percnopterus)*, Indian pond heron *(Ardeola grayii)*, Oriental magpie robin *(Copsychus saularis)*, Hume's warbler *(Phylloscopus humei)*, Yellow wagtail *(Motacilla flava)* and Grey wagtail *(Motacilla cinerea)* were found to be minimum (0.0393/ha) in urban habitat.

Bird Species Density, Diversity and Richness In Different Categories of Urban Habitat

Table 3 shows that bird density was found to be maximum in highly congested area (68.09±7.84) whereas minimum in low congested area (48.95±5.32). However the result was not found to be significant ($F_{3\ 86}$ =2.055, P>>0.05). Bird diversity was found to be maximum in low congested area (1.37±0.06) whereas minimum in high congested area (1.22±0.08).however the result were not found to be significant ($F_{3\,86}$=1.846, P>>0.05) Bird richness was found to be maximum in low congested area (1.58±0.09) whereas minimum in high congested area(1.46±0.11) however the result were not found to be significant ($F_{3\,86}$=0.961, P>>0.05).

Further Post hoc Scheffe's test shows that categorization of urban habitat i.e. high, medium and low congested areas were not significantly different in relation to bird density, diversity and richness.

Relationship of Birds with Urban Habitat Parameters:

Table 4 shows that in urban habitat bird density was found to negatively correlated with the habitat parameters except grass diversity and richness. However result were found to be significant only for tree diversity (r= -.483, p<.01) and tree richness (r=-.468, p<.01). But bird diversity and richness was found to be positively correlated with most of the habitat parameters except grass diversity and richness, however result were not found to be significant.

4. DISCUSSION AND CONCLUSION

Present study is the preliminary assessment of bird community structure in different habitats along a high to low congested urban areas. Though it was conducted very short duration, but the results were impressive, and much diverse avifauna was found. Total 41 bird species were recorded at the study sites.

We categorized urban area in three different habitats that was high, medium and low congested areas. The structure and composition of birds in urban areas varied with the high to low congested areas. High congested area has high density but low diversity and richness and vice versa in low congested areas. Since green parks and low disturbed areas included in low congested areas that have more vegetation and ground cover that provide the food, shelter and low predation risk due to which diversity and richness is maximum in this category. As per other authors (Emlen, 1974; Mills et. al., 1989; Chace and Walsh, 2006; Sandstrom et. al., 2005) vegetation cover increases, species diversity. Hence among these categories of urban those have more vegetation had the greater diversity but lower density or we can say there is a linear positive relationship of bird diversity and richness with more vegetative structure across the high to low urbanized gradient.

Urban bird densities are normally extremely high (Emlen, 1974; Marzluff, 2001; Chace and Walsh, 2006). However, the present study also reveals the density of birds is higher in high urbanized area. This is due to the some synanthropic species such as rock pigeon, dove etc. which are dominating in urban areas. The first major direct contribution towards the bird community structure of built environment may be an increased supply of holes, crevices, and ledges for nesting. In the more intensely developed areas these would be nearly exclusive alternatives restricting urban niches to ledge- and cavity-nesting species (Lancaster & Rees 1979). Second contribution is the availability of food because food is considered as very important resource for survival and in urban habitat it is easily available and also in huge quantity. Food abundance is the most common mechanism described in the literature for the increase in bird densities (Emlen, 1974; Bolger, 2001; Marzluff, 2001; Mennechez and Cleurgeau, 2001). So, birds prefer urban habitat and also change their feeding habit according to the availability of food. Many studies also show that cities have high abundance of birds despite the high densities of domestic and feral predators. In most cases, diversity in urban habitats decreases or remains similar to wildlands (Marzluff, 2001; Chace and Walsh, 2006).

Inspite of man-made features of urban environment, many ecological niches for birds are affected but few niches like food and nesting place provided by man has a major effect on density and distribution of birds in urban areas. Hence from our study we conclude that Bird species diversity and richness is inversely related to the apparent increase of built up structure and it has a negative impact on the bird diversity and richness but positive impact on bird density only due to few species those adapted themselves according to disturbances /activities creating by human because availability of resources such as food and shelter, in the forms of ornamental vegetation created in lawns, backyards and parks nest sites in building. Only a few species can exist and thrive in the most built parts of the city where vegetation is almost absent, such as business districts and industrial zones. Therefore, despite declining natural productivity and species diversity, total bird density does not decline across the urban gradient. Overall, available habitat and food in highly urbanized environments strongly favor the few cavity-nesting, ground-feeding graminivorous or omnivorous bird species that have already come to dominate. From this survey we conclude that open areas are the alternatives of natural habitat in urban areas as they provide the natural niche to the birds in man-made environment.

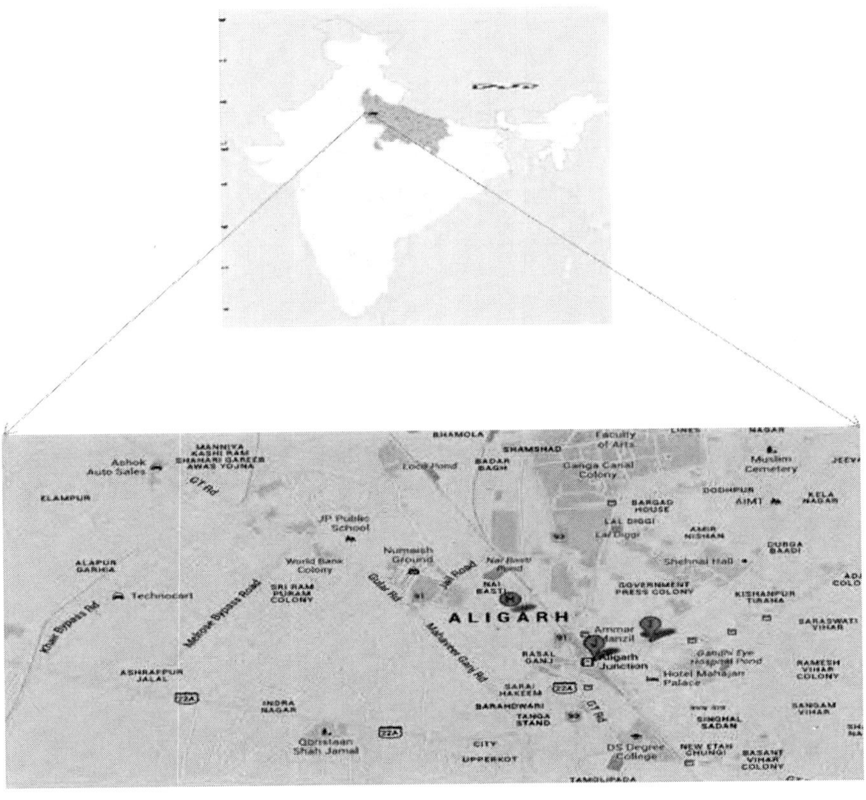

Fig. 1. Map of Study Area

	High congested	Medium congested	Low congested
Tree density	57.85±11.20	81.31±13.11	155.92±24.39
Tree diversity	.48±.15	.60±.13	.91±.14
Tree richness	.84±.25	.95±.21	1.24±.17
Shrub density	64.34±64.34	117.95±83.40	743.10±193.45
Shrub diversity	0±0	0±0	.19±.10
Shrub richness	0±0	0±0	.26±.14
Herb density	7272.73±7272.72	40000±21081.85	132000±335700
Herb diversity	0	.13±.09	.46±.16
Herb richness	0	.18±.12	.47±.16
Grass density	43636.36±22533.72	160000±54569.02	1250000±84852.81
Grass diversity	0	.13±.09	.18±.09
Grass richness	0	.11±.08	.14±.07

Table 1. Vegetative structure (Mean values± SE) in different habitat of urban.

Order	Family	Name of species	Density in urban
Galliformes	Phasiniadae	Indian peafowl	0.2752
Piciformes	Megalaimidae	Brown headed barbet	0.2359
		Coppersmith barbet	0.2359
Bucerotiformes	Bucerotidae	Indian grey hornbill	0.3538
Upupiformes	Upupidae	Common hoopoe	0.0393
Coraciiformes	Halcyonidae	White throated kingfisher	0.314
	Meropidae	Green bee-eater	0.0786
Cuculiformes	Cuculidae	Asian koel	0.5897
Psittaciformes	Psittacidae	Rose ringed parakeet	6.566
Columbiformes	Columbidae	Rock pigeon	15.844
		Laughing dove	4.0496
		Spotted dove	0.1965
		Eurasian collared dove	0.9829
		Yellow footed green pigeon	0.0786
Gruiformes	Rallidae	white brested waterhen	0.1572
		Common moorhen	1.0222
Falconiformes	Accipitridae	Black kite	1.4547

		Egyptian vulture	0.0393
Ciconiiformes	Ardeidae	Cattle egret	2.8701
		Indian pond heron	0.0393
Passeriformes	Corvidae	Rufous treepie	0.0786
		House crow	6.1728
		Large billed crow	0.0786
	Dicruridae	Black drongo	0.1965
	Muscicapidae	Oriental magpie robin	0.039
		Brown rockchat	1.2188
	Sturnidae	asian pied starling	0.4718
		Common myna	5.3078
		Bank myna	4.089
	Hirundinidae	Dusky crag martin	0.0786
	Pycnonotidae	Red vented bulbul	1.5726
	Costicolidae	Ashy prinia	0.1572
	Sylviidae	Lesserwhite throat	0.0786
		Common tailorbird	0.0786
		Hume's warbler	0.0393
	Timaliidae	Large grey babler	1.1008
		Jungle babler	0.9043
	Necteriniidae	Purple sunbird	0.4324
	Passeridae	House sparrow	2.0445
	Motacillidae	Yellow wagtail	0.0393
		Grey wagtail	0.0393

Table 2. Species densities/ha in urban habitat.

Areas	**Bird density±SE**	**Bird diversity±SE**	**Birdrichness±SE**
High congested	68.09±7.84	1.22±0.08	1.46±0.11
Medium congested	59.89±6.52	1.31±0.07	1.53±0.09
Low congested	48.95±5.32	1.37±0.06	1.58±0.09

Table 3. Mean bird density/ha, diversity and richness in different areas of urban.

Habitat parameter	Bird density	Bird diversity	Bird richness
Tree density	-.287	.135	.183
Tree diversity	-.483**	.138	.214
Tree richness	.183	.214	.190
Shrub density	-.254	.188	.196
Shrub diversity	-.151	.224	.139
Shrub richness	-.111	.244	.148
Herb density	-.172	.177	.115
Herb diversity	-.085	.060	.016
Herb richness	-.038	.098	.038
Grass density	-.259	.118	.082
Grass diversity	.085	.067	-.014
Grass richness	.106	.057	-.039

Table 4. Relationship of birds with urban habitat parameters,
***= significant at -0.05; **= significant at 0.01; ***= significant at 0.001.**

REFERENCES

Blair, R.B. (1999) Birds and butterflies along an urban gradient: surrogate taxa for assessing biodiversity? *Ecological Applications* 9, pp: 164-170.

Bolger, D. (2001) Urban birds: population, community, and landscape approaches. *Avian ecology and conservation in an urbanizing world* (ed. by J.M. Marzluff, R. Bowman and R. Donelly), pp: 155–177. Kluwer Academic Publishers, Boston, MA, USA.

Chace, J.F. and Walsh, J.J. (2006) Urban effects on native avifauna. *Landscape urban plan*, pp: 46-69.

Clergeau, P., Savard, J.L., Mennechez, G. and Falardeau, G. (1998) Bird abundance and diversity along an urban-rural gradient: A comparative study between two cities on different continents. *Condor* 100, pp: 413-425.

Collins, J.P., Kinzig, A., Grimm, N.B., Fagan, W.F., Hope, D., Wu, J.G. and Borer, E.T. (2000) A new urban ecology. *Am. Sci.* 88, pp: 416–425.

Emlen, J.T. (1974) An urban bird community in Tucson, Arizona: derivation, structure, regulation. *Condor* 76, pp: 184–197.

Grimm, N.B., Grove, J.M., Redman, C.L. and Pickett, S.T.A. (2000) Integrated approaches to long-term studies of urban ecological systems. *Bioscience* 50, pp: 571–584.

Lancasterr, K., and Rees, W. E. (1979). Bird communities and the structure of urban habitats. *Can. J. Zoo*. 57, pp: 2358-2368.

Ludwig, J.A., Reynolds, J.F. (1988) *In Statistical Ecology: A primer on methods and computing*. John Wiley and Sons, New York.

Marzluff, J.M. (2001) Worldwide urbanization and its effects on birds. *Avian ecology and conservation in an urbanizing world.* (ed. by J.M. Marzluff, R. Bowman and R.

Donelly), pp: 19–38. Kluwer Academic Publ., Boston, MA.

McKinney, M.L. (2002) Urbanization, biodiversity, and conservation. *Bioscience* 52, pp: 883–890.

Mennechez, G., and Clergeau, P. (2001) Settlement of breeding European starlings in urban areas: Importance of lawns vs. anthropogenic wastes. *Avian ecology and conservation in an urbanizing world* (ed. by J.M. Marzluff, R. Bowman and R. Donelly), pp: 275–288. Kluwer Academic Publ., Boston, MA.

Miller, J.R., and Hobbs, R.J. (2002) Conservation where people live and work. *Conserv. Biol.* 16, pp: 330–337.

Mills, G.S., Dunning, J.B. and Bates, J.M. (1989) Effects of urbanization on breeding birdcommunity structure in southwestern desert habitats. *Condor* 91, pp: 416–428.

Rebele, F. (1994) Urban ecology and special features of urban ecosystems. *Global Ecol. Biogeogr. Lett.* 4, pp: 173–187.

Sandstrom, U.G., Angelstam, P. and Mikusinski, G. (2005) Ecological diversity of birds in relation to the structure of urban green space. *Landscape Urban Plan.* 77, pp: 39–53

Savard, J.L., Clergeau, P. and Mennechez, G. (2000) Biodiversity concepts and urban ecosystems. *Landscape and Urban Planning* 48, pp 131-142.

Shochat, E. (2004) Credit or debit? Resource input changes population dynamics of city-slicker birds. *Oikos*, 106, pp: 622–626

Sutherland, W.J. (2006). *Ecological census techniques*. A Handbook, 2nd edn. Cambridge University Press, Cambridge, UK.

❒

Chapter-28

ABUNDANCE AND POPULATION STRUCTURE OF PREY PREDATOR SPECIES IN KEOLADEO NATIONAL PARK

Zarreen Syed and Orus Ilyas*

Department of Wildlife Sciences, Aligarh Muslim University, Aligarh-202002
**Email of Corresponding Author:orus16@gmail.com*

ABSTRACT

The Keoladeo National Park is in Rajasthan India covering an area about 29 sq. km. It is primarily a wetland, the wetland area is surrounded with the terrestrial land and which is scrub land and having good population of ungulates like Chital, Sambar, Hog deer and bluebull. Apart from ungulates area is also have some predator species like Jackal. Though the area has been explored in the past. However, no attempt has been made to understand the prey predator abundance and their relationship in KNP. Under this study efforts have been made to document the current status and distribution of prey and predator in Keoladeo National Park. Trail transects were used to assess the abundance of the species. Four forest existing trails were selected and found that the Jackal density was maximum on trail 1 and Chital (spotted deer) and Blue bull were recorded maximum on second trail there were no direct sightings of wild boar as well as Hyaena on any trail. Over all group size was found to be maximum for Chital (2.767±2.5401) followed by Sambar (2.125±1.126) and blue bull (1.85±1.5528), Group size was recorded minimum for Jackal. The sex ratio was female biased.

Keywords: Keoladeo National Park, prey predator abundance, current status, distribution, prey and predator.

1. INTRODUCTION

The KNP in Rajasthan, India covers an area about 29 sq km. it was a primarily a wetland, however significant changes due to frequent drought and scarcity of water in past, have resulted in Large scale reduction in wetland area. And area is being converted into a scrun land. Most of the area have been encroached by the exotic species *Prosopis juliflora*. Changes in landscape leading to alteration in distribution and size of different forest types are often associated with changes in distribution and abundance of animal communities (Gilpin and Soule 1986, Scott 1991). The animal communities in KNP have been explored and documented in past. However no attempt has been made to understand the prey predator densities and their relationship in KNP. Therefore, collecting baseline data on population attributes of animal populations and changes in these attributes overtime is essential and necessary for population studies. This study intends to investigate population status of jackal viz-a-viz ungulate community through both direct and indirect methods i.e by line transects sampling, trails monitoring and pellet group count in Keoladeo National Park.

Study Area:

Keoladeo National Park (27°7.6′ to 27°12.2′N & 77°29.5′ to 77° 33.2′E) is situated on the extreme western edge of the Gangetic basin, forty kilometers South-East of Mathura and 50 km west of Agra and 180 km from Delhi. It is located midway between Delhi & Jaipur (fig. 1). It lies in the Golden triangle of tourism.The keoladeo National park is one of the areas which has got many statuses, apart from the National Park status it has got the status of Ramsar Site, World heritage site. It used to be a bird paradise but due to scarcity of water the population of migratory birds are declining every year and wetlands are converting into terrestrial land. Due to increase in the terrestrial land, the population of ungulates are increasing.

Floristic elements and vegetation series of Rajasthan have been discussed in detail by Meher-Homji (1970) and more recently by Purie et al. (1983). The vegetation of the area is a mixture of xerophytic and semi-xerophytic species consisting predominantly of *Prosopis juliflora, Acacia nilotica, Prosopis cineraria, Salvadora persica, Salvadora oleoides, Capparis separia* and *Capparis decidua*. However, *Acacia catechu* and *Acacia pendula*, typical of Bharatpur area (Meher-Homji 1970), are conspicuous by their absence.

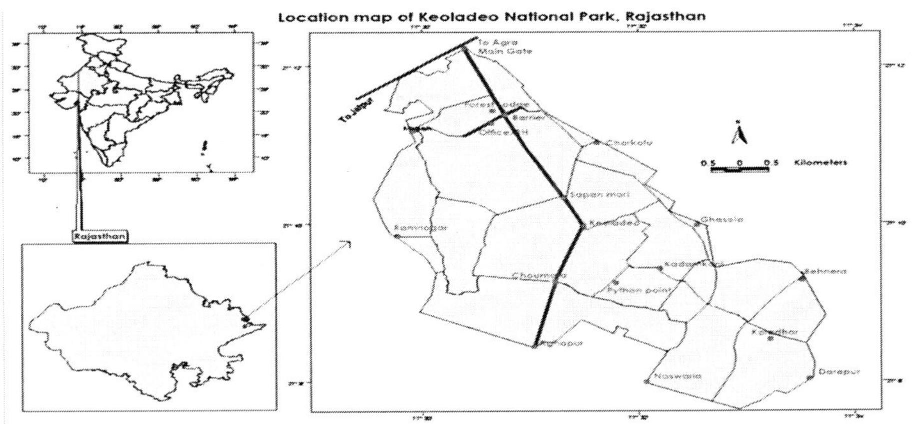

Fig. 1. Location map of Keoladeo National Park, Rajasthan

2. MATERIALS AND METHODS

Data Collection:

The data on population estimate of jackal and ungulate species was collected by quantification of direct evidences at each trail. Four forest existing trails were also monitored at early morning hours to gather the data on population status of jackal and ungulate species and at each sighting of animal the data recorded were i) Trail number ii) Species iii) Group composition and iv) Number of individuals.

GPS was used to record the location and elevation of the trails. Every trail was also monitored three times and the length of the trail was also measured by laying pre-measured rope.

Data Analysis:

The sightings of different ungulate species were used to calculate encounter rates on different trails. The numbers of pellet group of each species in each plot were used to calculate pellet group density (pellet groups/hectare) for each species in each plot. One-way ANOVA was used to test for significant differences in mean pellet group densities of each species among different habitats and blocks following Fowler and Cohen (1986).

Mean group size was calculated for each species on each trail. To calculate the mean group size on each trail, the total number of individuals of a species recorded on each trail was divided by the number of observations.

The percentage of male, female and sub-adults were calculated for each species to know the population structure. In case of Sambar, the sexual dimorphism was not very clear. And Wild boar the number of individuals is very less so no population structure was recorded, besides it, wild boar is extremely shy of human being and lives in more or less dense habitat and also it is very difficult to differentiate the sex of these animals.

3. RESULTS

During the surveys, direct and indirect evidences of Jackal and the ungulate species of the park were collected. They were Jackal, Chital, Nilgai, Wild boar and Sambar. Jackal, Chital and Nilgai were recorded on all the four trails while Sambar were recorded from one trail and the wild boar was not seen directly on any of the trail. Besides, the pellet groups of these animals were also recorded on the trails.

(i) Encounter Rates:

(a)Jackal:

The encounter rate of Jackal (individuals/hour) was maximum on trail 1 (3.106 indviduals/hour) followed by trail 4 (2.478 individuals/hour), then trail 3 (0.8 indvs/hour) and then trail 2 (0.494 individuals/ hour), also the encounter rate of jackal (individuals/ km), follow the same hierarchy i.e. encounter rate of jackal is maximum on trail 1 (1.323 individuals/km), followed by trail 4 (0.959 individuals/km), then trail 3 (0.3144 individuals/ km) and minimum at trail 2 (0.2824 individuals/km) (table 1 & 2).

(b) Chital:

The encounter rate of Chital (individuals/hour) was maximum on trail 2 (61.552 individuals/hour) followed by trail 1 (28.774 individuals/hour), trail 3 (8.946 individuals/ hour) and then trail 4 (3.428 individuals/hour), while the encounter rate of Chital (individuals/km) also follow the same hierarchy i.e. encounter rate of Chital is maximum on trail 2 (33.333 individuals/km), followed by trail 1(12.304 individuals /km), then trail 3 (3.615 individuals/ km) and then trail 4 (1.318 individuals/km) (table 1 & 2).

(c) Bluebull:

The encounter rate of Bluebull (individuals/hour) was maximum on trail 2 (36.616 individuals/hour) followed by trail 3(12.286 individuals /hour), then trail 1 (5.49 individuals /hour) and minimum at trail 4 (2.726 individuals / hour), while the encounter rate of Bluebull in respect of distance also followed the same hierarchy i.e. encounter rate of Bluebull is maximum on trail 2 (20.056 individuals/km), followed by trail 3 (5.031 individuals /km), then trail 1 (2.336 individuals /km) and minimum at trail 4 (0.959 individuals /km) (table 1 & 2).

(d) *Sambar:*

As Sambar was recorded only on trail 1, its encounter rate in respect of time was 3.062 individuals/hour and in respect of distance it was 1.323 individuals/km. while the Wild boar was not recorded on any trail (table 1 & 2).

(ii) *Group Size of Animals on Trails:*

Table 4 shows that the group size of jackal ranged from 1-8 individuals/group. in KNP. 90% of groups with-1 individual and 10% groups with ≤8 individuals. The mean group size (mean group size±SD) of Jackal were maximum on trail 1(1.545±0.6875 individuals/group.) Followed by trail 4 (1.333±0.5164 individuals/group) and the trail 2 and 3 showed the same group size of 1 individual/group having 0 standard deviation. The overall group size of Jackal was 1.38 ±0.58959 individuals/group.

Group size of Chital ranged from 1 to 30 individuals. In case of Chital 70% of groups was found to be with <10 individuals and 10% groups with >20-25 individuals. Although the herd of 70-100 individuals of Chital was also seen several times at the dusk time mostly in the dry wetland area of L block. The mean group sizes of Chital were 3.1176±2.9233 individuals/group, 2.7442±2.4743 individuals/group, 2.5556±1.424 individuals/group, 1.222±0.441 individuals/group on trail 1, 2, 3 and 4 respectively. The overall group size of Chital was 2.767 ±2.5401 individuals/group.

The group size of Nilgai ranged from 2-10 individuals in KNP. 90% of groups with 2-6 individuals and 10% groups with >6 individuals. The mean group size (mean group size±SD) of Nilgai were maximum on trail 2 (2.32±1.9037 individuals/group) followed by trail 3 (1.4545±0.9117 individuals/group) then trail 4 (1.333±0.5164 individuals/group) and then trail1 (1.16±0.4726 individuals/group). The overall group size of Bluebull was 1.85 ±1.5528 individuals/group. The mean group size of Sambar was 2.125±1.126 individuals/group on trail 1, as it was sighted on trail 1 only (table 3).

(iii) ***Population Composition:***

The details of individuals classified in sex categories were provided in the table. A total of 29 individuals of Jackal were classified for the sex composition on different trails. Out of 29 individuals 18 were adult males, 8 were adult females, no young one and sub-adult were recorded while 3 of individuals were unidentified. A total of 429 individuals of Chital were classified for the sex composition. Out of 429 individuals 197 were adult males, 214 were adult females, 3 individuals were sub-adult, 13 were young and 2 of individuals were unidentified. In case of Nilgai, total 211 individuals were classified for the sex composition. Out of which 51 were adult males, 103 were adult females, 33 individuals were sub-adult and 24 were young (table 4). The table 5 shows the male female sex ration of different species. The male and female sex ratio was biased towards the male (2.25:1) while among ungulates it was biased towards the female.

Table. 1. Average Encounter Rate per hour of Prey and Predator on trails.

Trail	Golden Jackal	Spotted deer	Bluebull	Wild boar	Sambar	Hyaena
1	3.106	28.774	5.49	0	3.062	0
2	0.492	61.552	36.616	0	0	0
3	0.8	8.946	12.286	0	0	0

4	2.478	3.428	2.726	0	0	0

Table 2. Average Encounter Rate per km. of Prey and Predator on trails.

Trail	Golden Jackal	Spotted deer	Bluebull	Wild boar	Sambar	Hyaena
1	1.323	12.304	2.336	0	1.323	0
2	0.2824	33.333	20.056	0	0	0
3	0.3144	3.615	5.031	0	0	0
4	0.959	1.318	`.959	0	0	0

Table 3. The Mean Group size± Standard deviation of Jackal and Ungulate species of the parkon different trails.

Trail	Jackal	Chital	Blue bull	Sambar
Trail 1	1.545±0.68755	3.1176±2.9233	1.16±0.4726	2.125±1.126
Trail 2	1±0	2.7442±2.4743	2.32±1.9037	0
Trail 3	1±0	2.5556±1.424	1.4545±0.9117	0
Trail 4	1.333±0.516398	1.222±0.441	1.333±0.5164	0
Overall Gp. Size	1.38±0.58959	2.767±2.5401	1.85±1.5528	2.125±1.126

Table 4. Overall population Composition of Jackal and ungulate species of the park on different trail.

Species	Adult male	Adult female	Sub-adult	Young	Unknown	Total
Jackal	18	8	0	0	29	29
Chital	197	214	3	13	2	429
Nilgai	51	103	33	24	0	211
Sambar	3	9	4	1	0	17

Table 5. Male female sex ratio of prey and predator species.

Species	Male:Female
Jackal	2.25:1
Chital	1:1.09
Blue bull	1:2.02
Sambar	1:3

4. DISCUSSION

The population is an ultimate self-reproducing grouping of conspecific individuals, which occupy different areas over an evolutionary long span of time to form an independent genetic system and evolutionary niche of its own (Haque1990). The most common factor that limits the population of a predator is the prey base number and their availability. It is thus important to record data on population size and population structure of the prey species.

The method of estimating the status, distribution and abundance size of wild population was done directly. The choice of methods depends on a variety of factors such as behavior of species, habitat type, terrain, man power availability and the accuracy required in results. There has been rapid advancement in the field of population estimation using direct method such as line transects or more appropriately distance sampling (Burnham et al. 1980, Buckland et al. 1993). It is the best and robust method provided that the key assumptions are not violated and so we have used this method for population estimation of jackal and ungulate species of the KNP.

The present study has made an attempt to gather a wide spectrum of information on the population of each of the ungulate species i.e. Chital, Nilgai, Sambar and Wild boar excluding the feral cattles, as well as the only predator species of the K.N.P. i.e. Golden Jackal (as the hyaena has a limited population of approx 8-10 individuals only). The knowledge of size or density of a population is often a vital pre-requisite to manage it effectively (Caughly 1977).

The encounter rate of Chital was maximum on trail 2 as this trail passes through the woodland habitat and same was the case with transects as the maximum density of Chital was recorded on transect of block H which is a Woodland. Therefore, this study revealed that Chital prefers the areas or blocks of the park having high tree density. According to Dinerstein (1980), Chital prefers woodland habitat in winter and summer season during the day time. The high density of Chital among all ungulates was probably due to the diversity of plants, which is almost same in all the habitat types and maximum during winter and monsoon (Haque 1990). In winters the maximum area is covered with palatable species like *Vetiveria zizanoides* and *Desmostachya* spp and extensive patches of short grasses and dense thickets attract more chital resulting in high density. The encounter rate of Bluebull was found to be maximum on trail 2 and its density was coming out to be maximum in Grassland habitat as the enough resources was available to them in this habitat.

ACKNOWLEDGEMENT

We thank Dr. P.K. Mathur, Dean WII, Dehradun for providing UNESCO-WII funding to carry out the project in Keoladeo National Park. The Forest department of Keoladeo National Park was as helpful as no other authority could be. We would thank Mr. Anoop K.R, DFO of Keoladeo during the study period for being very cooperative and supportive towards our work. We also thank to the Chairman, Department of Wildlife Sciences, for providing the necessary facilities to carry out the project in field.

REFERENCES

Buckland S.T., Anderson D.R., Burnham K.P. and Laake J.L (1993). Introduction to Distance Sampling: Estimating abundance of biological populations. Oxford University press.

Burnham, et al. (1979). Guidelines for line transect sampling of biological population. J. Wildl. Manage. 43(1).

Burnham, K. P; Anderson D.R. and Laake J.L. (1980). Estimation of density from line transects sampling of biological populations. Wildlife Monograph 44.

Dinerstein E. (1980).An ecological survey of the Royal Karnali-Bardia Wildlife Reserve, Nepal: Part III: Ungulate populations. Biological Conservation. 18 (1) 5-37.

Fowler, J. & Cohen, L. (1986). Statistics for ornithologist, BTO guide no. 22. British Trust of Ornithology. Herrtfordshire. pp175.

Gilpin, N. E. and SouleM. E. (1986). Minimum viable population. Process of species extinction. In:Conservation Biology, the Science of Scarcity and Diversity, Soule, M.E. (Ed.). Sinauer Associates. Inc. Publishers.Pp 19-34.

Haque, N. (1990). Study on the ecology of wild ungulates of Keoladeo National Park, Bharatpur, Rajasthan, Ph.D. dissertation. Dept. of Wildlife Sciences, A.M.U.

Ilyas, O. (2001). Status and conservation of ungulates in the Kumaon Himalayas with special reference to aspects of ecology of Barking deer and Goral. Ph.D. thesis, Dept. of Wildlife Sciences, A.M.U.

Meher-Homji, V.M.(1970). Some phytogeographic aspects of Rajasthan, India. vegetation,12:299-321.

Scott, K. R. (1991). Effect of habitat fragmentation on midwestern raptors. Midwest Raptor Management Symposium and Workshop. Natural Wildlife Foundation. Washington, D.C.

❒

Chapter-29

TAXONOMIC DIVERSITY OF DUNG BEETLES (COLEOPTERA: SCARABAEIDAE: SCARABAEINAE) IN ALIGARH

Prince Tarique Anwar, Syeda Uzma Usman, Shahid Bin Zeya* and Rifhat Aara

Department of Zoology, Aligarh Muslim University, Aligarh, Uttar Pradesh – 202002, India
**Email of Corresponding Author: ta.friday@gmail.com*

ABSTRACT

An attempt is made through faunistic survey to explore dung beetle diversity in Aligarh. These scarab beetles consist of species collectively called true dung beetles, and are heavy-bodied. They vary greatly in their size, colour and habits; exclusively feed on dung and, therefore were collected from dung pats of dogs, cows and human. Their great diversity and ability to adapt, along with being a key component in the working of the biosphere, make them one of the most abundant and successful of all insects. They were known for its bioindicator characteristic. Sensitive towards forest disturbance, dung beetles population and diversity will be less in disturbed and modified area. Species collected in this study represents great pattern of diversity, richness, abundance, distribution of functional groups; and altitudinal and habitat variation in dung beetle assemblage. Using diversity measure agricultural lands found to be more diverse and mixed forests in the sanctuary. The results were based on examination of more than 40 specimens that included eight species within four genera (Digitonthophagus Balthasra, Gymnopleurus Illiger, Onitis Fabricius and Onthophagus Latreille) of the dung beetle tribe Scarabaeinae. The present study tries to give a brief account on the diversity of dung beetles in Aligarh region, so that these results may be utilized in future to access diversity and conservation problems from the region.

Keywords: Survey, Dung beetles, Taxonomy, Diversity

1. INTRODUCTION

Insects are the most diverse of all animal groups. Their numbers are nothing short of remarkable both in terms of diversity and population. This continuous effect of uncovering the faunal diversity living on earth resulted in the description of about 1,552,319 animal species belonging to 40 phyla (Zhang, 2011). Among all the 40 phyla of animals, the phylum Arthropoda which includes spiders, crustaceans, ticks, mites, myriapods and insects alone is represented by about 1,242,040 species and accounts for about 80% of the total in the kingdom Animalia throughout the world (Zhang, 2011). Their success as a group is also by the diversity of habitats in which they live. They are numerous in terrestrial environment such as deserts, forests and grassland. Likewise they are also abundant in freshwater habitats such as ponds, lakes, streams and wetlands. They are scarce in marine habitats, but are more common in brackish waters such as salt marshes and mangrooves. Insects play major role in human life both in primary production and in decomposition, serve as major resource for many vertebrates directly or indirectly, also act as pest to plant and most important they help in plant reproduction by pollinating them. Some insects take man as host thereby causing injuries that even serve for secondary infections.

Insects belong to class Insecta and differ from other allied arthropods by having three segmented body, presence of a pair of antennae on the head, two pair of wings and three pair of legs.

Though the described taxonomic richness of insects is distributed unevenly among the higher taxonomic groups, the order Coleoptera includes about 3,87,100 species and account for about 38% of all species in all the 39 insect orders (Zhang, 2011) throughout the world. It includes beetles which comprise nearly a quarter of all the described species on earth. Coleoptera means "sheath wings", referred to hard forewings which cover the insect's body. Therefore the forewing of Coleopterans is modified into horny or leathery elytra, which is hard and brittle, which almost always meet in a straight line mid- dorsally and cover the hind wing. The hind wing is membranous usually longer than fore wing but they are folded under the forewings. The flight is only powered by the hindwings.

Beetles are found in almost every type of habitat and feed upon all type of plants and animals materials. Most of the beetles are phytophagous, many are predacious or fungivorous, some are scavengers and very few are parasitic. Many beetles are regarded as major pests of agricultural plants and stored products. They attack all parts of living plants as well as processed fibres, grains and wood products. Bark beetles can cause great harm to trees, some species can infest foodstuffs, whilst chafers and leaf beetles cam cause serious harm to crop. In contrast to this many beetles are beneficial and play an important role in nutrient recycling, i.e., returning organic matter through multitrophic interaction which contribute to soil fertility such as scavengers and wood boring beetles. Predatory species such as lady beetles are important biological control agents of aphids and scale insects.

Order Coleoptera is grouped into 4 suborders 16 superfamily and 128 families (Borrer & Delong, 2005). Some of the families are Scarabaeidae, Carabidae, Coccinellidae, Buprestidae etc. The Scarab beetles (family Scarabaeidae) under superfamily Scarabaeoidea comprise a species rich group and a conspicuous component of the beetle fauna of the world. These Scrab beetle (Scarabaeidae: Scarabaeinae) consist of species collectively called true dung beetles. Most of the beetles of this subfamily feed exclusively on dung. Adults of these beetles are noticeable due to their relatively large size, bright color, often elaborate ornamentation and interesting life histories. Life histories of scarab beetles are incredibly diverse and includes adults that feed on dung carries fungi, vegetations, pollen, fruits, compost of roots. On the other hand, some scarab beetles live in the nests of ants, in the nest of termites, or in the nest of rodents and birds.

The superfamily Scarabaeoidea considered as a monophyletic group (Lawrence & Britlon, 1991), includes 12 families, 43 subfamilies, 118 tribes and 94 subtribes recognized as valid taxa (Smith, 2006). The family Scarabaeidae consist of 6 subfamilies, and are very large and the most diverse and widely distributed group of beetles, and according to some estimates about 27,800 species of this family are known worldwide (Chandra & Gupta, 2013). Within the family Scarabaeidae, the two subfamilies Aphodiinae and Scarabaeinae include approximately 6,850 species worldwide, and the subfamilies Orphninae, Melolonthinae, Dynastinae, Rutelinae, Cetoniinae, Trichiinae and Valginae includes approximately 20,950 species (Ratcliffe & Jameson, 2001).

The great diversity and ability to adapt, along with being a key component in the working of biosphere, make these beetles one of the most abundant and successful of all insects.

2. MATERIALS AND METHODS

Study Area (Aligarh Muslim University):

Aligarh Muslim University is funded by the central Government of India. It was established by Sir Syed Ahmad Khan as Musalmanan-e-Hind in 1875–78. It was designed to train Muslims for government service in India and to prepare them for advance training in British Universities. The Mohammedan Anglo-Oriental College became Aligarh Muslim University in 1920. The main campus occupies over 467.6 hectares (1,115 acres) in Aligarh city, Uttar Pradesh.

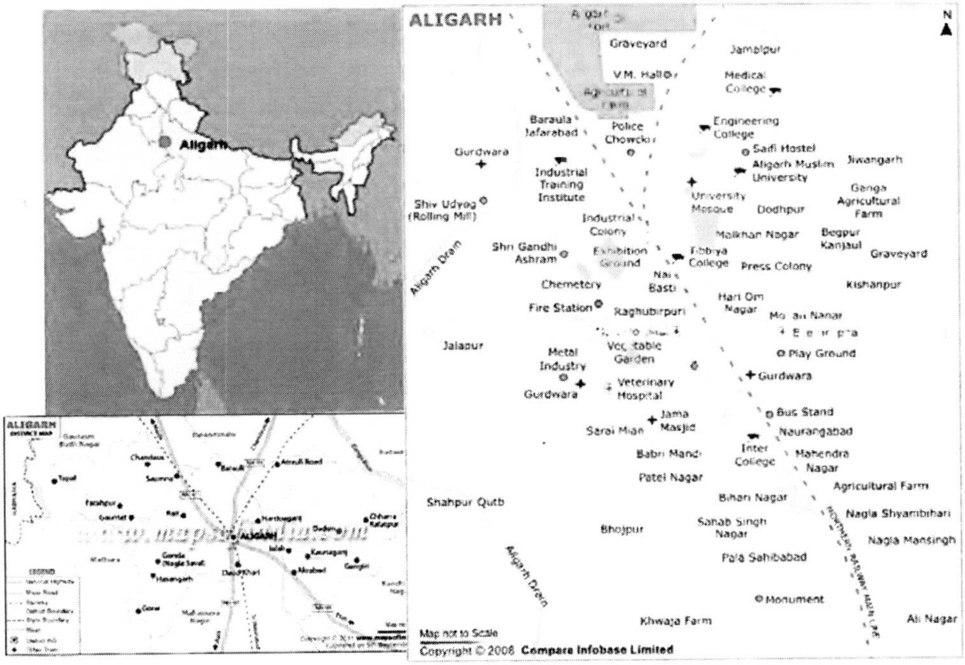

Collection site

Collection, Preservation and Identification:

A faunistic survey of was conducted in and around Aligarh Muslim University campus during 2013–14 for the investigation of the dung beetle fauna. The dung beetles were collected from dung pats of dogs, cows and human during day with the help of forceps. After that they were killed using benzene vapors in killing bottle. They were further processed for permanent preservation and labeled properly. The pinned specimens were identified with the help of the literature (Arrow, 1931; Janssens, 1940; Balthasar, 1963a, 1963b). The identified specimens were deposited in the museum of Zoology Department, Aligarh Muslim University, Aligarh.

Historical Review:

The important taxonomic works on the Oriental fauna are rather few. In the series of 'Fauna of British India' on various groups of the family Scarabaeidae, Arrow studied subfamily Scarabaeinae (Coprinae) of Laparosticti in vol. III (1931) and dealt 354 species in total. Later, Balthasar (1963a, 1963b) wrote two monographs on Scarabaeinae of the Palaearctic and Oriental regions. Further studies on the scarab beetles of India have been carried out by Chandra (1986, 1999), Gupta (1986), Kabakov (2006) and Krikken (2009). Mittal (1993) published an annotated list of the scarab fauna of Western Uttar Pradesh and recorded 151 species of these beetles belonging to 41 genera and 12 subfamilies. Recently, Mittal (2005) also gave the diversity and conservation status of 136 species of Laparosticti Scarabaeidae belonging to 26 genera and 8 subfamilies from North India. Subsequently Chatterjee (2010) reported 44 species belonging to 16 genera and 3 subfamilies while studying the scarab beetle fauna of Uttarakhand. Chandra et al. (2012) published an updated checklist of the scarab beetles of Uttarakhand comprising 167 species belonging to 52 genera, 21 tribes, 9 subfamilies and 3 families of superfamily Scarabaeoidea. The Comprehensive research on the scarab beetle diversity of central India was made by Chandra (2000, 2003), Chandra & Ahirwar (2007), Chandra & Singh (2010) and Chandra & Gupta (2011; 2012 a, b, c). Chandra (2000) published an inventory of 96 species of scarabaeid beetles from Madhya Pradesh. Later, concentrating on the scarab diversity of central India, Chandra & Ahirwar (2007) published a comprehensive account of scarab beetles of Madhya Pradesh and Chhattisgarh and recorded 124 species/subspecies belonging to 45 genera in 11 subfamilies. Recently, Chandra & Singh (2010) reported 22 species of scarab beetles belonging to 11 genera and six subfamilies from Achanakmar Wildlife Sanctuary, Chhattisgarh. Chandra & Gupta (2011b) recorded two genus *Onthophagus Latreille* from Madhya Pradesh, India for the first time. Chandra & Gupta (2012) published new distributional record of five species of Onthophagus from Central India. Chandra & Gupta (2012b) reported five species of genus Onthophagus Latreille, from Madhya Pradesh for the first time from India, and gave their description of external male genitalia.

3. RESULTS AND DISCUSSION

Taxonomy and Diversity of Dung Beetles:

Family: Scarabaeidae

Scrab Beetles:

The members of the family Scarabaeidae vary greatly in size, color and habits. They are heavy-bodied, oval or elongate, usually convex beetles, with tarsi 5-segmented (rarely, the front tarsi are absent), and the antennae 8- to 11-segmented and lamellate. The last three (or rarely more) of the antennal segments ate expanded into plate-like structures that may be spread apart or united to form a compact terminal club. The front tibiae are more or less dilated, with the outer edge toothed or scalloped. The scarabs vary greatly considerably in habits. They are mainly are dung feeders, but also feed on decomposing plant materials, carrions, some live in the nest or burrows of vertebrates, or in the nests of ants or termites, a few feed on fungi, many feed on plant materials such as grasses, foliage, fruits and flowers, and some of these are serious pests of lawns, golf greens, or various agricultural crops.

Subfamily Scarabaeinae:

Dung Beetles and Tumble Bugs:

These beetles are robust, 5–30 mm in length, and feed on dung. Most of them are dull black, but some are metallic green in color. The tumble bugs *(Canthon and Deltochilum)* are black, measuring about an inch in length, the hind tibiae rather slender, and there are no horns on the head or pronotum. The members of other genera in this subfamily have the middle and hind tibiae swollen at the tip. In *Phanaeus,* which is usually a little less than an inch in length, the body is brilliant green with the pronotum golden, and the males have a long horn on the top of the head. Other beetles in this subfamily range in length from 4–9 mm. the tumble bugs are usually common in pastures and are interesting insects to watch. They chew off a piece of dung, work it into a ball, and roll this ball a considerable distance; they usually work in pairs, one pushing and the other pulling the ball with their hind legs. The ball is then buried in the soil, and the eggs are laid in the ball. The larvae are thus assured a food supply, and the location of the ball provides protection.

(i) Genus Digitonthophagus Balthasar, 1959:

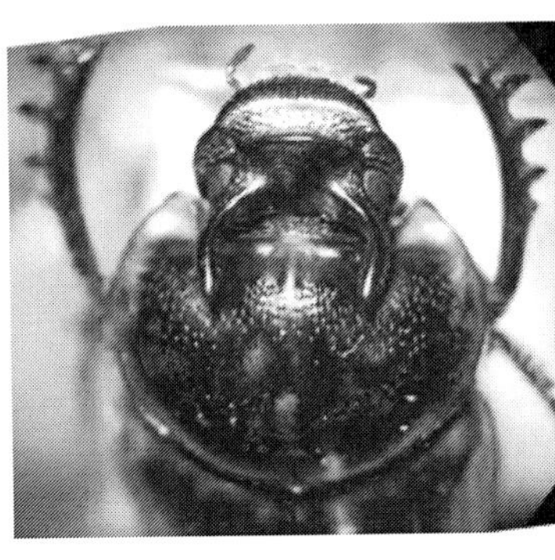

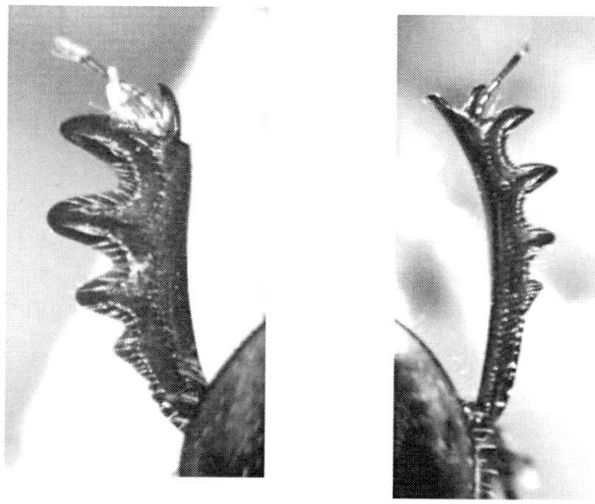

Figs 1–6. *Digitonthophagus gazella:* 1–3 female: 1, body; 2, head with thorax; 3, protibia. 4–6 male: 4, body; 5, head with thorax; 6, pro tibia.

Diagnosis:

Medium to large sized; yellowish brown; pronotum green sometimes red with some granules; head and pronotum with metallic lustre; lateral sides of pronotum yellow; elytra yellow; protibiae elongate, slender, clypeus rounded at front; vertex with a pair of backwardly directed horns in males at posterior region of head, sometimes a tubercle present in between horns; in females, head with two strong and straight carina or sometimes with two horns like males.

***Digitonthophagus gazella* (*Fabricius, 1787*):**

(Figs 1–6)

Scarabaeus gazella Fabricius, 1787, I: 377.

Onthophagus catta, Arrow, 1931, III: 230, fig. 23 (♂).

Onthophagus (Digitonthophagus) gazella, Balthasar, 1963b, II: 365.

Digitonthophagus gazella, Löbl, Krell, Ziani, and Král, 2006, 3: 161.

Diagnosis:

Length: 8–13mm; Breadth: 5–8mm. Testaceous yellow, with the head, pronotum (except a narrow pale margin at the sides and base), the median part of the metasternum, the front tibiae, the greater part of the middle and hind tibiae, and a large oval spot upon the lower surface of the fore posterior femora, greenish- or coppery-black. The edges of the prothorax, elytra, and all the segments of the body and legs are also dark. Broadly oval and convex, smooth and moderately shining, with a thin clothing of yellow setae upon the legs and lower surface. The pronotum with sparse granules in median part, replaced behind by fine scattered punctures, and is smooth and unpunctured round the circumference. The elytra finely striate, with the intervals flat and unpunctured. The pygidium bears an angulate basal carina and a few scattered punctures. The metasternum bears only a few setigerous punctures.

In male, the clypeus is rather feebly rugulose and shining. The vertex bears a pair of slender horns, curving outward and inclined a little backward, a little flattened at the base and not united. The front of the pronotum is very smooth and shining. The front tibia is elongate, slender and curved, its inner extremity produced into a long spur and its external teeth rather widely separated.

(ii) Genus Gymnopleurus Illiger, 1803:

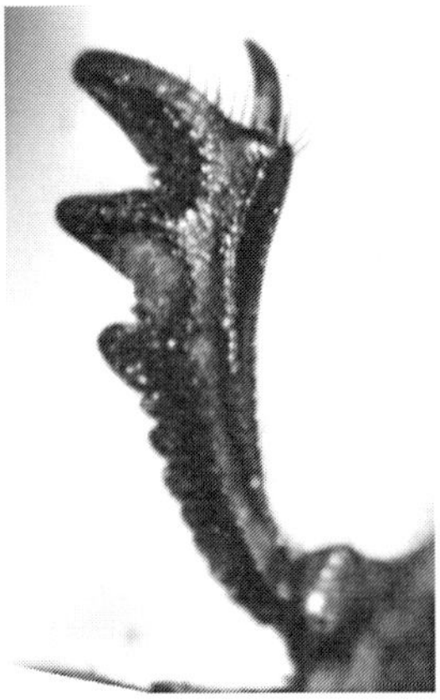

Figs 7–12. ***Gymnopleurus* (*Metagymnopleurus*) *gemmatus:*** **7–9 female: 7, body; 8, head with thorax; 9, pro tibia. 10–12 male: 10, body; 11, head with thorax; 12, pro tibia.**

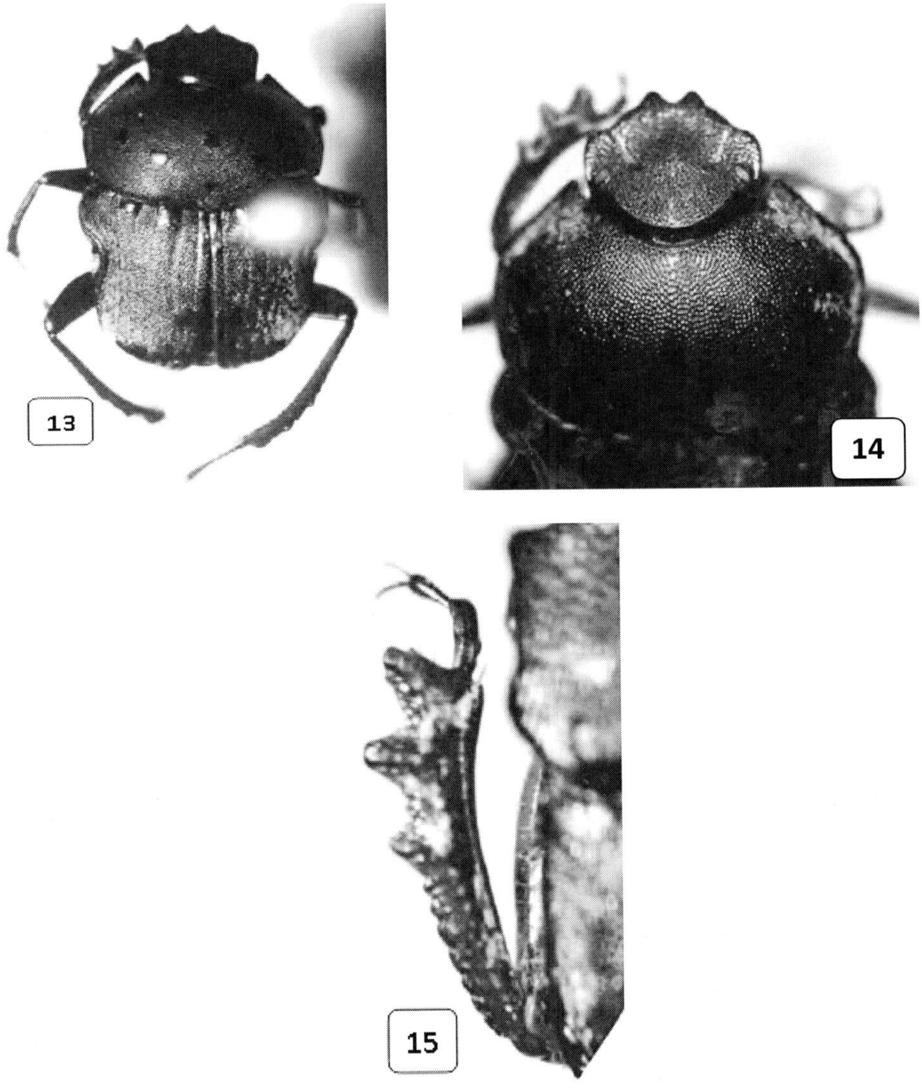

Figs 13–15. *Gymnopleurus* (*Metagymnopleurus*) *parvus:* 13, body; 14, head with thorax; 15, pro tibia.

Diagnosis:

Broadly oval, convex and depressed; antennae 9 segmented with 3rd joint as long as 4th and 5th together, 6th one extremely short; clypeus bi or quadridentate at front; pronotum moderately convex, strongly punctate or granular with shining spots, smooth or hairy, front angles acutely produced, lateral sides rounded and carinate; lateral side of elytra deeply excised behind shoulder exposing metasternal epimera; elytra either finely striate, or sometimes hairy with shining spots or sometimes smooth and strongly confluently punctate; protibiae tridentate, remaining surface strongly serrate; profemora hollowed at anteriorly bearing a sharp a tooth beyond middle and sharply carinate, tooth sometimes absent; outer and inner edge of meta and meso tibiae serrated or toothed.

(iii) *Gymnopleurus (Metagymnopleurus) gemmatus Harold, 1871:*

(Figs 7–12)

Gymnopleurus gemmatus, Harold, 1871, Col. Hefte, VIII: 117.

Gymnopleurus gemmatus, Arrow, 1931, Faun. Brit. India, (Lamellicornia: Coprinae), III: 53, pl. III, Fig. 1.

Gymnopleurus gemmatus, Janssens, 1940, Verh. Kon. Nat. Mus. Belg., 2 (18): 62.

Gymnopleurus (Gymnopleurus) gemmatus, Balthasar, 1963, Mon. Scarab. Aphod. Palae. Ori Reg., (Coleoptera: Lamellicornia), Coprinae, I: 204, pl. VIII, Fig. 2.

Gymnopleurus (Metagymnopleurus) gemmatus, Kabakov, 2006, Beet. subf. Scarab. Faun. Russ. adj. Coun.: 79. *Scarabaeus granulatus,* Fabricius (nec Olivier), 1792, Ent. Syst.: 65.

Diagnosis:

Black with upper surface clothed with minute grey setae, leaving many shining bare patches upon the pronotum. Body oval, not very convex, legs slender. Front tibia armed with three strong teeth, occupying the terminal half of the outer edge and finely serrate in the upper half, the middle and hind tibia rather feebly toothed externally. Front femora with flattened anterior edge and is not toothed. Clypeus produced into two blunt lobes in front. Pronotum granular, except upon the smooth elevated patches. Sides are straight in front, strongly rounded behind, front angles acutely produced, hind angles bluntly prominent, the base feebly rounded, with a slight oblique depression on each side of the middle. Elytra are irregularly rugose, finely striate, very deeply excised at the sides. Sides of sternum and abdomen are clothed with recumbent grey setae, abdomen is rounded laterally at the base. Pygidium is closely punctured, setose at the base and bare at the apex. In male the front tibia is longer and narrower than that of female, a little bent before the end, feebly serrate internally and armed with shorter teeth externally. Terminal spur bent at the end and rather blunt. In female the front tibial spur is slender, curved and very acute.

(iv) *Gymnopleurus (Metagymnopleurus) parvus MacLeay, 1821:*

(Figs 13–15)

Gymnopleurus parvus, MacLeay, 1821, Horae Ent., I, 2: 517.

Gymnopleurus parvus, Arrow, 1931, Faun. Brit. India, (Lamellicornia: Coprinae), III: 52, pl. III, Fig. 2.

Gymnopleurus parvus, Janssens, 1940, Ver. Kon. Nat. Mus. Bel., 2 (18): 62, pl. II, Fig. 10.

Gymnopleurus (Gymnopleurus) parvus, Balthasar, 1963, Mon. Scarab. Aphod. Palae. Ori. Reg., (Coleoptera: Lamellicornia), Coprinae, I: 202.

Gymnopleurus (Metagymnopleurus) parvus, Kabakov, 2006, Beet. subf. Scarab. Faun. Russ. adj. Coun.: 79.

Diagnosis:

Dark, coppery or greenish-black, sometimes steely blue, upper surface with dense minute greyish setae, leaving a few shining denuded spots. Oval body, moderately convex with slender legs. Front tibia armed with three strong teeth, occupying terminal half of the outer edge, basal half finely serrate. Front femur flat at its anterior edge, not toothed. Middle and hind tibia coarsely toothed. Clypeus produced into two lobes in front. Pronotum highly convex with sides strongly rounded, front angles acutely produced, hind angles very obtuse, base gently rounded with a minute impression on each side of the middle. Elytra finely densely granular, finely striate, lateral margins strongly excised behind shoulders. Sides of abdomen rounded at the base. In male, spur of the front tibia is flat, truncate at the end with sharp angles. In female, the spur of front tibia is long slender and sharply pointed.

(v) ***Genus* Onitis *Fabricius, 1798:***

Figs 16–18 *Onitis philemon:* 16, body; 17, head with thorax; 18, pro tibia.

Diagnosis:

Oblong with legs stout; meso and meta tibiae strongly dilated; antennae 9 segmented; head with a clypeo frontal suture either complete or interrupted by a tubercle; clypeus entire or feeble excised in middle at front margin; pronotum with two basal impressions on each side of middle at base, sparsely or closely or strongly punctate or entirely smooth; scutellum visible; elytra with a strong lateral carina, finely striate; pygidium smooth or densely hairy.

(vi) *Onitis philemon (Fabricius, 1801):*

(Figs 16–18)

Onitis philemon Fabricius, 1801. Syst. Eleut. I: 30.

Onitis philemon, Arrow, 1931. Faun. Brit. India, (Lamellicornia : Coprinae), 3: 393–394.

Onitis philemon, Balthasar, 1963. Mon. der Scarabaeidae und Aphodiidae der Palaeark. und Orientalis. Region (Coleoptera: Lamellicornia), Coprinae, II: 41.

Diagnosis:

Length: 14–19 mm Breadth: 7.5–10.5 mm. Green, coppery or bronzy-black, moderately shining. Oval, not very elongate nor very convex. The head is rugulose, with the ocular lobes rather smooth, the clypeus parabolic, an extremely feeble e margination of the front edge in the middle, and separated from the forehead by an interrupted curved carina, with a short transverse carina just before it and a conical tubercle just behind it. Pronotum is fairly strongly closely but unevenly punctured, without well-marked median line. The base is obtusely angulate in the middle. The elytra are moderately strongly striate and the intervals very finely and sparsely punctured. The 1st, 3rd, and 5th intervals are a little narrower and usually a little more elevated than the 2nd and 4th. The pygidium is very feebly and sparsely punctured. The metasternal shield is smooth and feebly punctured, except near the anterior angles, where it is thinly hairy, and the sides of the metasternum are densely granulate and hairy. In male, the clypeus is granulate. The front legs are elongate, the tibia slender, strongly curved in front, armed with four feeble teeth externally, finely serrate internally in the basal half and strongly produced at the apex. The middle femur has a sharp tooth near the end of the lower edge, and sometimes a rounded lobe in the middle, and the middle tibia is slender at the base and abruptly dilated near the middle. The hind trochanter is a little toothed beneath. In female, the clypeus is closely transversely rugose. The front tibia is broad and armed with four strong teeth and an articulated terminal spur.

(vii) *Genus Onthophagus Latreille, 1802:*

(Figures 19–33)

Diagnosis:

Medium to small sized; oval, flat, and convex; entirely smooth or clothed closely or sparsely with setae on dorsal or ventral surface; clypeus variable in shape, rounded or bilobed or acuminate in front margin; pronotum closely or sparsely punctate or rugose or granulate with some smooth area, front angles acute or blunt, base rounded; scutellum not visible; mesometasternal suture nearly straight; metasternal shield broad with anterior part flat, sloping or sometimes vertical in middle and occasionally forming a short compressed process; abdomen short, 6th sternite extremely short in males; protibiae tri or quadridentate; meso and meta tibiae dilating greatly from base to extremity with terminal margin nearly straight, but sometimes trilobed; antennae 9-segmented but sometimes 8-segmented, basal segment serrated in some; sexual dimorphism strongly developed; some dung feeders and some carrion feeders.

(viii) Onthophagus (Onthophagus) mopsus (Fabricius,1792):

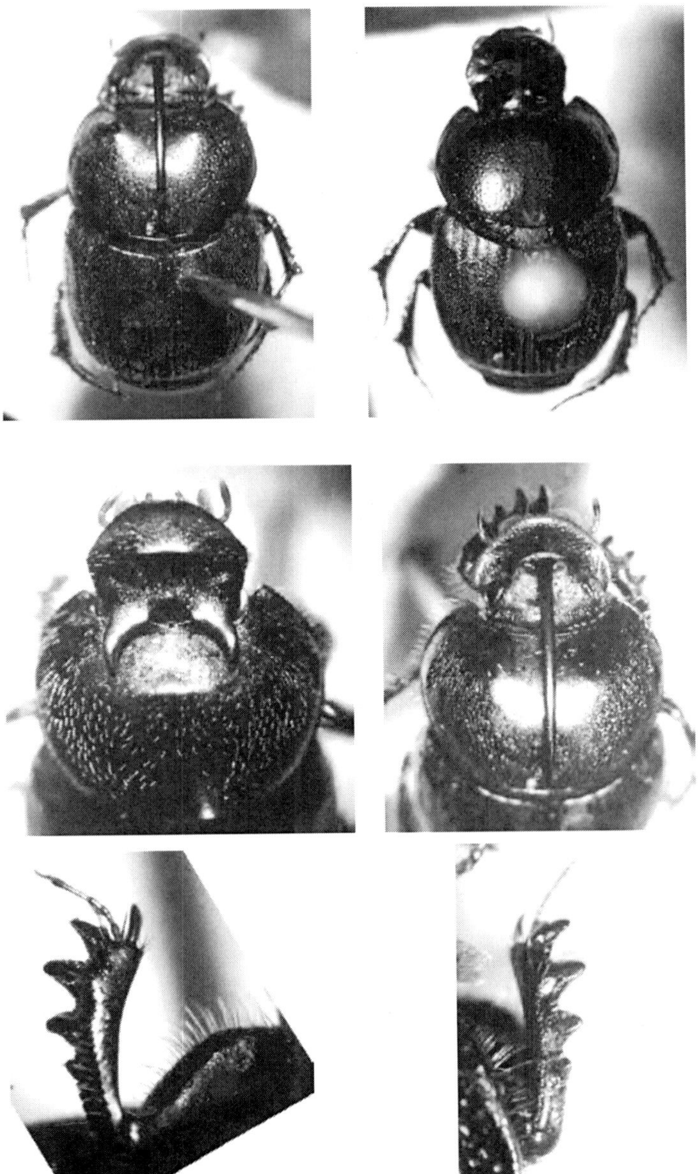

Figs 19–24 *Onthophagus* (*Onthophagus*) sp. 19–21 O. mopsus male: 19, body; 20, head with thorax; 21, pro tibia. 22–24 O. cervus male: 22, body; 23, head with thorax; 24, pro tibia.

(Figs 19–21)
Scarabaeus mopsus Fabricius, 1792: 58.
Onthophagus mopsus Arrow, 1931: 328-329.
Onthophagus (Onthophagus) mopsus Balthasar, 1963: 441.

Diagnosis:

Length 7.5 mm, maximum width 5.4 mm. Oval body, not very convex, blackish brown with a slight metallic lustre but not very shining. Dorsally, clothed with fairly closely and very minute greyish setae. Head bears an extremely long and slender horn, arising in front of eyes and curving backward beyond hind margin of pronotum. In middle, pronotum is smooth and unpunctured and a little flattened, with a strong tubercle on each side.

(ix) *Onthophagus (Onthophagus) cervus (Fabricius, 1798):*

(Figs 22–24)
Copris cervus Fabricius, 1798. Ent. Syst. Suppl.: 31.
Onthophagus cervus, Arrow, 1931. Faun. Brit. India, (Lamellicornia : Coprinae), 3: 348–349.
Onthophagus (Onthophagus) cervus, 1963. Balthasar, Mon. der Scarabaeidae und Aphodiidae der Palaeark.und Orientalis. Region (Coleoptera : Lamellicornia), Coprinae, II: 307.

Diagnosis:

Length: 6–7 mm Breadth: 3.5–4.5 mm. Shining black, with the head and pronotum, and also the elytra and lower surface, coppery or green and the elytra decorated with orange, forming a broad external margin, extending almost to the suture posteriorly, but generally irregular, and only reaching the 4th elytral interval basally, the lateral margin sometimes interrupted on the 7th interval or wanting in the middle. The femora, antennae, mouth-organs, pygidium, and sides of the abdomen are also usually yellow, and the upper surface is thinly clothed with minute yellow setae. Oval and moderately convex. The head is not very broad, the ocular lobes rounded at the sides and not projecting much in front of the eyes. The clypeus is a little produced, its margin strongly reflexed and straight in the middle. The pronotum is fairly strongly, not very closely but unevenly punctured with smaller punctures intermixed. The front angles are rather sharp, the lateral margins straight in front, sinuate behind, and the base rounded. The elytra are finely striate, the intervals slightly convex and rather finely but unevenly punctured. The pygidium is punctured like the pronotum with intermixed large and small punctures. The metasternal shield bears only a very few fine punctures, and the sides of the metasternum bear larger but not very numerous punctures. In male, the clypeus is a little produced, but truncate and not pointed, very shining, lightly punctured with intermixed large and small punctures, and separated from the forehead by a not very strong carina. A pair of horns arise behind the eyes, sloping backward, broad at the base but not united, each bent at a right angle internally just beyond the base, curved outward and rapidly narrowed. The pronotum is rather flat, smooth, and shining in its anterior part. In female, the clypeus is strongly and closely punctured and not shining. There is a slight prominence in the middle of the anterior part of the pronotum just behind the front margin.

(x) ***Onthophagus (Onthophagus) quadridentatus (Fabricius, 1798):***

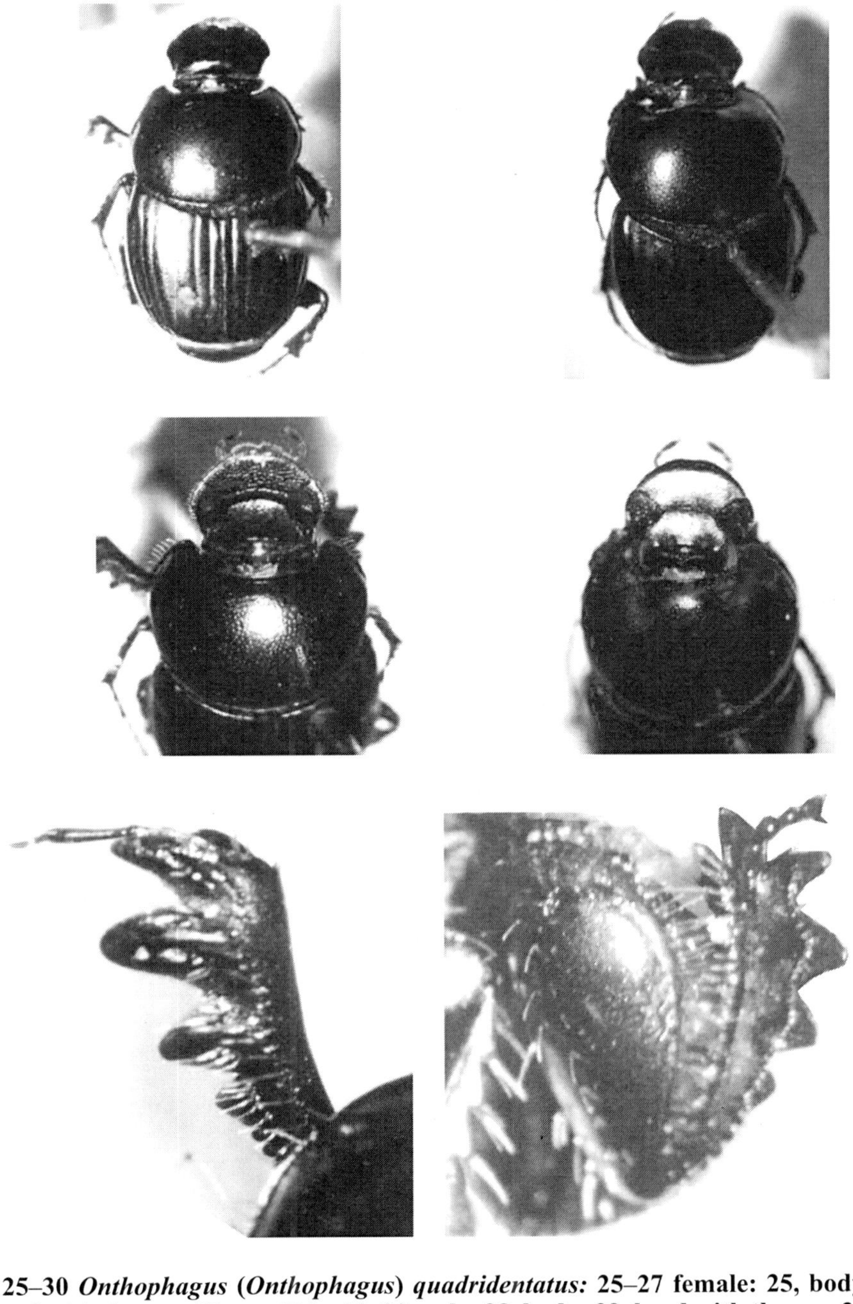

Figs 25–30 *Onthophagus* (*Onthophagus*) *quadridentatus:* 25–27 female: 25, body; 26, head with thorax; 27, pro tibia. 28–30 male: 28, body; 29, head with thorax; 30, pro tibia.

Copris quadridentatus Fabricius, 1798. Ent. Syst. Suppl. 34

Onthophagus quadridentatus, Arrow, 1931. Faun. Brit.India, (Lamellicornia : Coprinae), 3: 304–305.

Onthophagus (Onthophagus) quadridentatus, Balthasar, 1963. Mon. der Scarabaeidae und Aphodiidae der Palaeark. und Orientalis. Region (Coleoptera :Lamellicornia), Coprinae, II: 493–494.

Diagnosis:

Length: 6–8 mm Breadth: 4–5 mm. Black, smooth and shining, with the elytra a little less shining, the antennae and mouth-organs yellow, and the tarsi reddish. The upper surface is without hairs or setae, and the lower surface bears only very scanty yellow hairs. Broadly oval and moderately convex. The head is broad, the eyes are not very small seen from above, the ocular lobes are strongly rounded in front of the eyes, and the clypeal margin is evenly rounded and strongly reflexed. The pronotum is finely and not closely punctured. The front angles are not very sharp, the lateral margins are nearly straight in front, feebly sinuate behind, and the base is gently rounded and finely margined. The elytra are strongly striate. The metasternal shield is finely and sparingly punctured, and the sides of the metasternum a little more strongly. In male, the head is very smooth and shining, with very fine and sparse punctures, the clypeus separated from the forehead by a slight rounded carina, and the hind margin of the head is produced backward, forming a pair of short horns curving upward and outward and slightly approximating at the tips. Within and near the base of each horn is a short upward directed spine-like tooth. The anterior part of the pronotum is a little flattened on each side of the middle. In female, the clypeus is transversely rugose and separated by a strong, nearly straight carina from the strongly punctured forehead. The pronotum is usually rather more strongly punctured than that of the male.

(xi) *Onthophagus (Onthophagus) spinifex (Fabricius, 1781):*

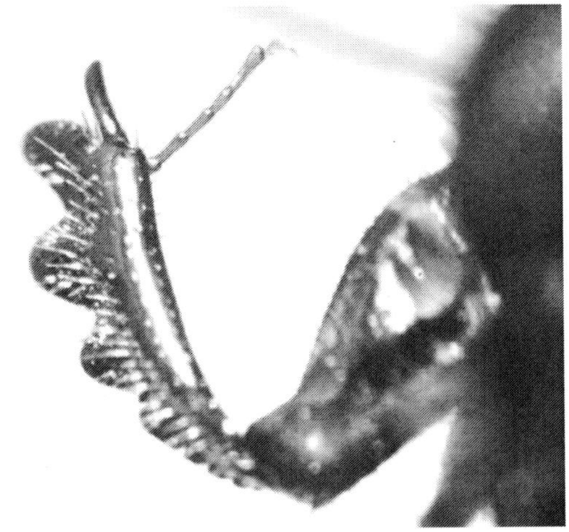

Figs 31–33 *Onthophagus (Onthophagus) spinifex:* 31, body; 32, head with thorax; 33, pro tibia.

Scarabaeus spinifex Fabricius, 1781. Spoc. Ins. i, 29.

Scarabaeus spinifer Oliver, 1789. Ent. i, 3, 148, pI. 12, fig. 112.

Scarabaeus aeneus Fabricius, 1781. Spec. Ins. i, 34; Oliv. torn. cit,. 131, pI. 14, figs 128, a & b.

Copris spinifex Fabricius, 1801. Syst. EI. i, 49.

Onthophagus reflexicornis Redtenbacher, 1867. Roise d. Novara, 57.

Diagnosis:

Length: 7.5–10 mm Breadth: 4.5–6 mm. Dark blue, dark metallic or golden-green, with the legs and lower surface almost black, the antennae, mouth-organs, and tarsi reddish, and clothed above and beneath with yellowish hairs and setae. Broadly oval and largely convex. The head is rather broad, nearly semicircular, with the clypeus coarsely transversely rugose and separated from the forehead by an arcuate carina. The pronotum is moderately closely studded with small granules at the sides (in the male or upon almost the whole surface in female). The elytra are finely striate, the intervals flat, not shining, and finely an irregularly granular. The pygidium is rather strongly and closely punctured, with a fairly close clothing of recumbent hairs. The metasternal shield is rather strongly and evenly, but not closely, punctured, and the sides of the metasternum are fairly closely granular. In male, the head bears a long slender horn, arising between the eyes, a little flattened in its basal part, curving gently backward and sometimes extending beyond the middle of the body. The pronotum is very smooth and shining in the middle and very slightly longitudinally grooved. The lateral granules change to asperate punctures nearer the middle, and these may change to simple punctures near the middle and base. In male, hornless phase, the cephalic horn is replaced by a transverse carina. The pronotum has a short steep declivity at the front margin antl bears a pair of small tubercles rather close together at the top of the declivity. The granules of the sides change into asperate punctures upon the disc, which is not smooth except near the middle of the base. The pronotum has a steep, smooth anterior declivity in the middle and a smooth hollow on each side, with a slightly elevated prominence between the two hollows, a little depressed in the middle. The entire surface, except the front margin and lateral hollows, is covered with granules.

4. CONCLUSION

The result of the study is based on examination of 40 specimens that included eight species within four genera *(Digitonthophagus, Gymnopleurus, Onitis, Onthophagus)* of dung beetle tribe Scarabaeinae. The material examined represented Aligarh Muslim University localities. Most of the specimens were collected during November and December, 2013. All the specimens were collected during day time.

Biodiversity surveys provide fundamental information needed for conservation planning, protected area justification and design, and development of management plans (Spector & Forsyth, 1998). Thus, here an attempt is made to explore dung beetle diversity in the A.M.U. Campus so that this may be utilized in biodiversity conservation and management plans of the university area. The occurrence and abundance of coprophagous beetles in a given locality is mainly affected by a variety of biotic and abiotic factors including fauna, flora, solar radiation, temperature, soil type, soil pH, rainfall, and most significantly the supply of dung for food (Fincher et al., 1970). There are several studies from tropical and temperature systems that indicate a severe impact of local and regional-scale anthropogenic habitat changes on patterns of dung beetle abundance and species diversity (Nicholas et al., 2008). The global modification, fragmentation and loss of tropical forest habitat are reported to lead to high local extinction rates across forest restricted dung beetle communities (Nicholas et al., 2008). Thus, they satisfy all of the criteria of an ideal bio-indicator for investigating impacts of anthropogenic disturbances to ecosystems.

To conclude, species collected in this study represent patterns of diversity, richness, abundance, distribution of functional groups, altitudinal and habitat variations in dung beetle assemblage. Using diversity measure agricultural lands found to be more diverse than mixed forests in the sanctuary. The present study tries to give a brief accoung on the diversity of one of area from the A.M.U. campus so that these results may be utilized in future to assess diversity and conservation problems from the region. Further studies will include standardize additional methods and collections.

ACKNOWLEDGEMENT

The authors thank the Chairman, Department of Zoology, Aligarh Muslim University, Aligarh, for providing research facilities. They are grateful to Dr. Devanshu Gupta, Scientist 'C', Zoological Survey of India, Kolkata for his help in collection and identification of the specimens and for providing literature and diagnosis of the species.

REFERENCES

Arrow, G.J. (1931). The Fauna of British India including Ceylon and Burma. Col. Lamell. III, (Coprinae). Taylor & Francis, London, 121–428, 61 fig., 19 pls., map.

Balthasar, V. (1963a). Monographie der Scarabaeidae und Aphodiidae der Palaearktischen und Orientalischen Region. (Coleoptera: Lamellicornia), Verlag der Tschechoslowakischen Akademie der Wissenschaften Prag, I, 1–391, 137 figs., 24 pls.

Balthasar, V. (1963b). Monographie der Scarabaeidae und Aphodiidae der Palaearktischen und Orientalischen Region. (Coleoptera:Lamellicornia), Verlag der Tschechoslowakischen Akademie der Wissenschaften Prag, II, 1–627, 226 figs., 16 pls.

Borror, D.J., DeLong, D.M. & Triplehorn, C.A. (2005). An Introduction to the Study of Insects. Philadelphia: Saunders College Pub, 1981. Print.

Chandra, K. (1986). Taxonomic studies on Pleurostict Scarabaeidae (Coleoptera) of North-West India. PhD Thesis. Faculty of Science, Kurukshetra University, Kurukshetra,

251pp, plate 28, map 14.

Chandra, K. (1999). Annotated check-list of Aphodiinae (Scarabaeidae : Coleoptera) from India, Rec. Zool. Surv. India, 97 (Part-2) 87–108.

Chandra, K., (2000). Inventory of Scarabaeid Beetles (Coleoptera) from Madhya Pradesh, Print J., 15(11): 359–353. 21.

Chandra, K., (2003). Insect Biodiversity in Madhya Pradesh and Chhattisgarh, Advancements in Insect Biodiversity (edited by Rajiv Gupta), Agrobios (India), 37–52 & 325–337.

Chandra, K., Ahirwar, S.C. (2007). Insecta : Coleoptera: Scarabaeidae, Fauna of Madhya Pradesh (including Chhattisgarh), State Fauna Series, 15 (Part-1): 273–300.

Chandra, K. & Gupta, D. (2012a). Diversity and relative abundance of Pleurostict Scarabaeidae (Coleoptera) in Achanakmar-Amarkantak Biosphere Reserve, Central India, World Journal of Zoology, 7(2): 147–154; http://dx.doi.org/10.5829/idosi.wjz.2012.7.2.63229.

Chandra, K. & Gupta, D. (2012b). New distributional record of five species of *Onthophagus* (Coleoptera: Scarabaeidae: Scarabaeinae) from Central India. Scholarly Journal of Agricultural Science, 2(1): 8–12. Chandra, K. & Gupta, D. (2012c). First report on five species of genus *Onthophagus Latreille,* 1802 (Coleoptera, Scarabaeidae) from Madhya Pradesh, India, and their description of external male genitalia. Biodiversity Journal, 3(2): 99–106.

Chandra, K & Gupta, D (2011a). Study of Scarabaeid Beetles (Coleoptera) of Veerangana Durgavati Wildlife Sanctuary, Damoh, Madhya Pradesh, India, Deccan Current Science, 05: 272–278.

Chandra, K & Gupta, D (2011b). Two New Records of genus *Onthophagus Latreille,* 1802 (Coleoptera : Scarabaeidae) from Madhya Pradesh, India. Uttar Pradesh Journal of Zoology, 31(2), 253–255.

Chandra, K. & Gupta, D. (2011c). New record of two species of *Caccobius* from Madhya Pradesh (Coleoptera: Scarabaeidae), 13(3): 117.

Chandra, K. & Gupta, D. (2013). Studies on Tribe Gymnopleurini (Coleoptera: Scarabaeidae: Scarabaeinae) from Madhya Pradesh and Chhattisgarh with A Checklist from India. Proceeding of National Academy of Sciences, India, Section B Biological Sciences. 10.1007/s40011-013-0229z.

Chandra, K. & Singh, S.P. (2010). Scarabaeid beetles (Coleoptera) of Achanakmar Wildlife Sanctuary, Chhattisgarh. National Journal of Life Sciences, 7(1): 71–74.

Chandra, K., Gupta, D., Uniyal, V.P., Bharadwaj, M. & Sanyal, A.K. (2012). Studies on Scarabaeid Beetles (Coleoptera) of Govind Wildlife Sanctuary, Garhwal, Uttarakhand, India. Biological Forum — An International Journal, 4(1): 48–54.

Chatterjee, S.K. (2010). Insect: Coleoptera: Scarabaeidae: (Cetoniinae, Dyanstinae, and Rutelinae), Fauna of Uttarakhand, State, fauna Series, 18 (Part-2): 311–383.

Fabricius, J.C. (1781). Species insectorum, exhibentes eorum differentias specificas, synonyma auctorum, loca natalia, metamorphosin, adiectis observationibus, descriptionibus. Tom.I, Carol.Ernest.Bohnii, Hamburgi et Kilonii, VIII: 1–552.

Fabricius, J.C. (1787) Mantissa Insectorum. Hafniae, 348 pp.

Fabricius J.C., 1792. Entomologia systematica emendata et aucta, secundum classes, ordines, genera, species adiectis, synonymis, locis, observationibus, descriptionibus. Tom. I. Hafniae: Christ. Gottl. Proft., xx+330 pp.

Fabricius, J.C. (1798) Supplementum Entomologiae Systematicae. Hafniae, 572 pp.

Fabricius, J.C. (1801). Systema Eleutheratorum secundum ordines, genera, species: adiectis synonymis, locis, observationibus, descriptionibus. Tomus I, Impensis bibliopoli academici novi, Kiliae: 1–506.

Fincher, G. T., Stewart, T. B. & Davis, R. (1970). Attraction of coprophagous beetles to feces of various animals. Journal of Parasitology, 56: 378–383.

Gupta, A.K. (1986). Studies on Dung Beetles (Coleoptera: Scarabaeidae: Laparosticti) of North-West India. PhD Thesis. Faculty of Science, Kurukshetra University, Kurukshetra, 251pp+24pls.

Harold, E.von. (1871). Berichtigungen und Zusätze zum Catalog Coleopterorum Synonym and Systematic, Coleopterological Booklets. Munich 8: 117–123.

Illiger, J.K.W. (1803). Verzeichniss der in Portugall einheimische Käfer Erste Leiferung. Magazin für Insektenkunde, 2: 186–258.

Janssens, A. (1940). Monographie des Gymnopleurides. Verhandelingen Koninklijk Natuurhistorisch Museum Belgie. Brussel, 2(18): 1–74.

Kabakov, O.N. (2006). Beetles of the subfamily Scarabaeinae of the fauna of Russia and adjacent countries. Moscow, Scientific publications, KMK: 1–374.

Krikken, J. (2009). Drepanocerine dung beetles: a group overview, with description of new taxa (Coleoptera: Scarabaeidae: Scarabaeinae). Haroldius, 4: 1–30.

Latreille, PA (1802). Hist. Nat. Gen. Part. Crust. Insec., volume 3. F. Dufart, Paris, France, 3(1–467): 139–159.

Lawrence, J.F. & Britton, E.B. (1991). Coleoptera, p 553–683. In: CSIRO The insects of Australia. A textbook for students and research workers, 2nd edn, vol 2. Melbourne University Press, Carlton, p 1137.

Löbl, I., Krell, F.T., Ziani, S. & Kral, D. (2006). Scarabaeidae: Scarabaeinae: Onthophagini, pp. 159–176. In I. Löbl & A. Semetana (ed.), Catalogue of Palaearctic Coleoptera – volume 3. Apollo Books, Stenstrup, 690 pp.

MacLeay, W.S. (1821). Horae entomologica: or essays on the annulose animals, S. Bagster, London 1(2): 161–524 (490–524).

Mittal, I.C. (1993). Natural manuring and soil conditioning by dung beetles. Tropical Ecology, 34(2): 150–159.

Mittal, I.C. (2005). Diversity and Conservation Status of Dung Beetles (Laparosticti: Scarabaeidae: Coleoptera) in North India, Bulletin of the National Institute of Ecology, 15: 43–51.

Nichols, E., Spector, S., Louzada, J., Larsen, T., Amezquita, S. & Favila, M. E. (2008). Ecological functions and ecosystem services provided by Scarabaeinae dung beetles. Biological Conservation, 141: 1461–1474.

Olivier, A.G. (1789). Entomologie, ou histoire naturelle des insectes, avec leurs caractères génériques et spécifiques, leur description, leur synonymie, et leur figure enluminée. Coléoptères, Baudouin, Paris, 1: 1–190.

Ratcliffe, B.C. & Jameson, M.L. (2001). Scarabaeoidea: Scarabaeoid beetles (= Lamellicornia). (URL:http://www-museum.unl.edu/research/entomology/Guide/Scarabaeoidea/Scarabaeoidea-pages/Scarabaeoidea-Overview/ScarabaeoideaO.html). In: Ratcliffe, B.C. & M.L.

Jameson (eds.). Generic Guide to New World Scarab Beetles (URL: http://www-museum.unl.edu/ research/entomology/ Guide/index4.htm). Accessed on: 01.06.2014.

Redtenbacher, L. (1867). Reise der Osterreichischen Fregatte Novara um die Erde in den Jahren 1857-58-59 unter der Befehlen des Commodore B.von Wullerstorf-Urbair.

Zoologie II part. Coleopteren, Wien, (1868): 1–249.

Smith, B.T.A. (2006). A Review of the Family-Group Names for the Superfamily Scarabaeoidea (Coleoptera) with Corrections to Nomenclature and a Current Classification, Coleopterists Society Monograph Number 5:144–204.

Spector, S. & Forsyth, A.B. (1998). Indicator taxa in the vanishing tropics, pp. 181–210 in A. Blamford and G. Mace (eds.), Conservation in a changing world. Zoological Society of London, London.

Zhang, Z.Q. (2011). An outline of higher-level of classification and survey of taxonomic richness. Zootaxa. 3148: 1–237.

❐

Chapter-30

BIODIVERSITY AND CONSERVATION OF ANNELIDS AT DARJEELING, WEST BENGAL, INDIA

Md. Nurul Hasan* and C.K. Mandal

Zoological Survey of India, FPS Building, Kolkata-700016
**Email of Corresponding Author: nhasan.mld@gmail.com*; mandalsucker@gmail.com*

ABSTRACT

Himalayan region is rich in biodiversity and biodiversity needs conservation. Loss of important species like earthworm from the terrestrial biome leads to the loss of environmental properties. More than 57 species of Earthworms are found and the number of new species are increasing day by day at Darjeeling. Recently one earthworm species Octochaetona jahanarai has been collected from Mirik, Darjeeling (Latitude 26.8908° North and Longitude 88.1825° East) West Bengal India. Which is new to science. This species of earthworm has an elongated robust body. This species belongs to Family Octochaetidae and genus Octochaetona. It has 16 species of Octochaetona genus in India. The total length of the body is 8.2cm. Breadth 3 mm. anal pore is 0.3 mm in diameter. Mouth is 1 mm in diameter. Total number of body rings is 157. Male pores open between the rings 18. Female pore is obscure. Clitellar region reddish brown, large and balloon shaped. This is the remarkable character of the species. Neck is also distinguished character of the earthworm. Colour is brown in living condition. Colour fades up in preserved state. Lumbricine setal arrangement is observed in the body.

Keywords: Octochaetona jahanarai sp. nov. Clitellata, Octochaetidae, Mirik lake side, Darjeeling district, West Bengal, India.

1. INTRODUCTION

Earthworms occur in diversified habitats frommanure, compost, litter, humus, kitchen drainage, forest land, grassland, agricultural land, plant nursery, etc. They are omnivorous, but most of them derive nutrition from dead organic matter. The beneficial effect of earthworms in increasing soil fertility has been documented since the time of Darwin (1881). Because of the vast number of species available in India and also because of the economic value, understanding of the distribution, abundance and ecological condition of the species has become very important. Octochaetidae family has 26 genera in India (Julka, 1988).Halder (1998), Bandyopadhyay et. al (2008), Goswami and Mandal (2014 and 2015) Mandaland Mitra (2008), Mandal and Halder (2004), Mandal et. al (2006 and 2008), Paliwal and Mandal (2008), Sinha (2009), Julka et.al (2009) Hasan et, al. (2018a, 2018b) Indian authors and Baird (1869), Beddard (1883, 1891), Perrier (1872), Stephenson (1916, 1917, 1923), Michaelsen (1891, 1900) foreign author worked on earthworms of which are valuable for the future workers of taxonomy. There are more than 6000 species of earthworms in the world, where as in India 561.

2. MATERIALS AND METHODS

After making the collection, earthworms were sorted out and cleaned. Then the cleaned earthworms were placed in a tray with a small quantity of water and were slowly killed by anaesthetizing with 70% alcohol by adding drop by drop at frequent intervals. Earthworms usually die in an extended condition by this process. Just after death, the Earthworm were kept in 70% alcohol for permanent preservation. This new species of earthworm *Octochaetona jahanarai* has been collected from Mirik lake side, Darjeeling district (Latitude 26.8908^0 North and Longitude 88.1825^0 East), West Bengal, India. The new species is compared with the material of other species available in hand and information on other species is taken from literature.

PLATE 1.

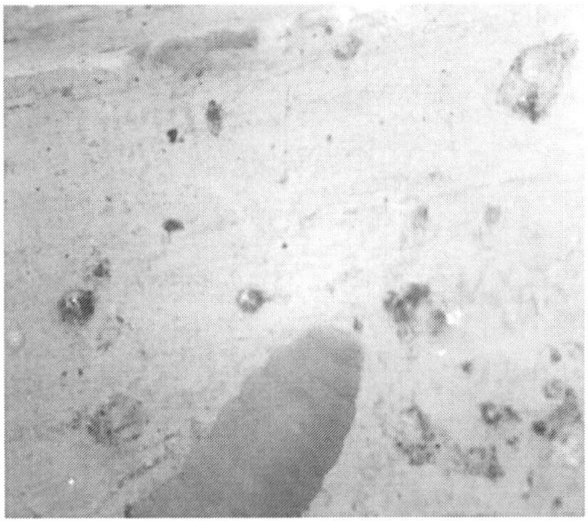

Fig.1. Prostomium of *Octochaetona jahanarai*

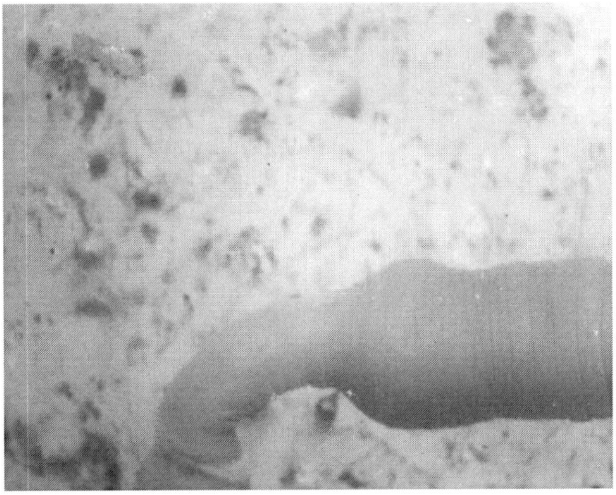

Fig.2. Balloon shaped clitellum of *Octochaetona jahanarai*

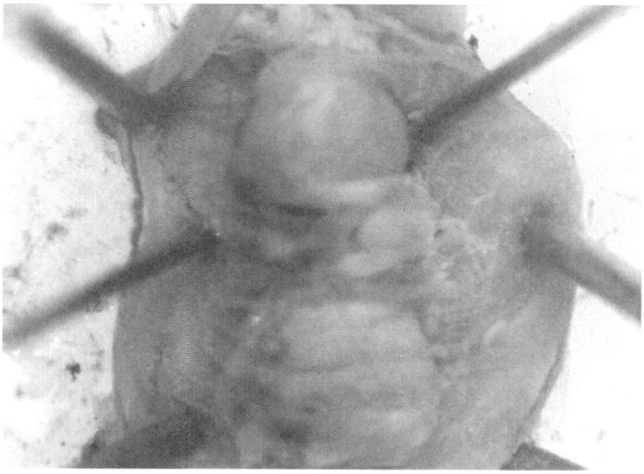

Fig.3. Stomach of *Octochaetona jahanarai*

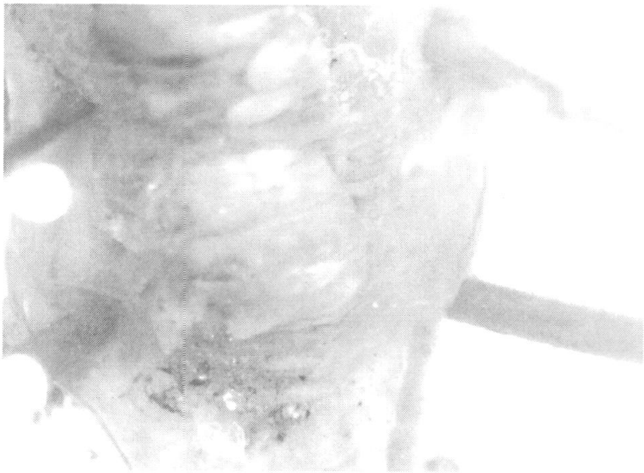

Fig.4. Seminal vesicle of *Octochaetona jahanarai*

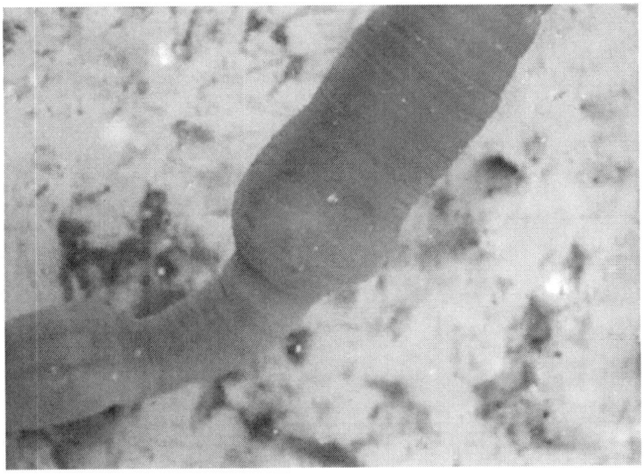

Fig.5. Neck structure of *Octochaetona jahanarai*

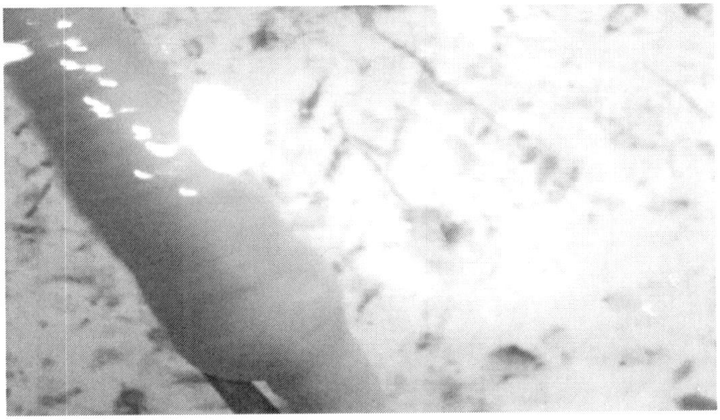

Fig.6. Male pore of *Octochaetona jahanarai*

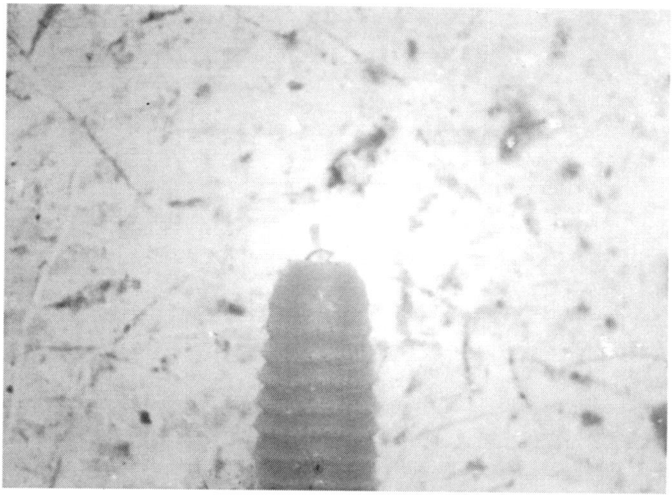

Fig.7. Showing segmentation anal pore of *Octochaetona jahanarai*

3. HOLOTYPE

Z.S.I. Reg. No. An5735/1, 1 ex., Coll. Md. Nurul Hasan, 10.ix.2016, (Latitude 26.8908^0 North and Longitude 88.1825^0 East), Mirik, Darjeeling, West Bengal, India.

Diagnosis:

Body (Plate 1)

The total length of the body is 8.2cm. Breadth 3 mm. anal pore is 0.3 mm in diameter. Mouth is 1 mm in diameter. Total number of body rings is 157. Male pores open between the rings 18. Female pore obscure. Clitellar region reddish brown large and balloon shaped. This is the remarkable character of the species. Neck is also distinguished character of the earthworm. Colour is brown in living condition. Colour fades up in preserved state. Lumbricine setal arrangement in the body.

Habit & Habitat:

The species of earthworms live in the bush. It lives in the shadow place and beneath the half feet below of the soil. In the month of June -July due to rain it comes out from the soil in search of dry habitat, then collection becomes easy. Breeding season is July -October. It embraced to each other from opposite end at the time of breeding. Male pore is below the female pore which is opposite in leeches.

Distribution:

West Bengal, India

Etymology:

The specific name is proposed according to the name of the one of the author's mother who collected the specimen from Darjeeling.

Remarks:

The present described species *Octochaetona jahanarai* is somewhat similar to *Octochaetona surensis* in body shape. It is distinguished from *Eutyphoeus* in its colour, number of rings in the body (Plate 1). All the species of Octochaetidae family are different from the new species of the earthworm.

Comparative species:

Octochaetona surensis: An 3960/1, 20 ex., Balasore, Odisha.An 4078/1, 18ex. Goalpara, Birbhum, West Bengal.

***Octochaetona paliensis*:** An 4011, 1 ex., Bhadrak, Odisha.An 4792/1, 6ex. Mahavura college, Karnataka.

Octochaetona beatrix: An 4533/1, 5 ex. Biswanathpur, Kalahandi, Orissa. An 4589/1, 5 ex. Baroda, Gujarat.

***Octochaetona fermosi*:** An 4593/1, 4 ex. Lahore, Pakisthan. An 4686/1, 1ex. Gangtok, Sikkim.

***Octochaetona rosea*:** An 4790/1, 8ex.MahavuraCollege, Karnataka.

***Octochaetona berkudensis*:** An 4820/1, 1ex. Odisha, India. An 4825/1, 9ex. Odisha India.

Material examined: 1 ex. Mirik, Darjeeling, West Bengal India. (Latitude 26.8908^0 North and Longitude 88.1825^0 East).

Inference:

If we keep up the conservation of any area it will be convenient to new species and faunal biodiversity.

ACKNOWLEDGEMENT

We are grateful to the Director, Zoological Survey of India, for lucid suggestions during the study of the Earthworm, and my special thanks to Dr. Ch. Satyanarayana O/C General Non-Chordata Section and Sri S. S. Mishra, Scientist-C for their inspiration and deep gratitude to Dr. L.K. Sing, Scientist-D and the O/C Spirit Building for his valuable suggestions.

REFERENCES

Baird, W. 1869. Description of a new species of earthworm (Megascolex diffringens) found in North Wales. Proceedings of the zoological society of London (1869): 108(Part-30: 40-43.

Bandyopadhyay, P.K., Mandal C.K. and Mitra, A.K. 2008. Earthworms of North 24 Parganas Rec. Zool. Surv. India: 108(Part-3): 21-25.

Beddard, F.E. 1883. Note on some earthworms from India. Annals and Magazine of Natural History, (5) 12: 213-224.

Beddard, F.E. 1891. The classification and distribution of earthworms. Proceedings of the Royal Physical Society Edignborough, 10; 235-290.

Darwin, C. 1881. The formation of vegetable mould through the action of worms with obser-vations on their habits. Murry, London, 298 pp.

Goswami, Rinku and Mandal, C.K. 2014. Earthworm of South 24 Parganas with ecological notes, Rec. Zool. Surv. india.114 (part-4):681-684.

Goswami, Rinku and Mandal, C.K. 2015. A study on Earthworm Population and Diversity with Special reference to physicochemical Parameters in Different habitats of

South 24 Parganas District in West Bengal. Rec. Zool. Surv. india.115 (part-1):31-38.

Halder, K.R. 1998. Annelida: Oligochaeta: earthworms, State Fauna Series3: Fauna of West Bengal, part 10: 17-93.

Haldar, K.R. Dhani. S. And Mandal, C.K. 2004. On some earthworm of present in unnamed Collections of ZSI. Rec.Zool.Surv. India.107 (part-3):79-93, 2007.

Julka, J.M. 1988. Fauna of India & the adjacent Countries: *Megadrile Oligochaeta* (Earthworms); Haplotaxida: Lumbricina: Megascolecoidea: Octochaetidae, 400 p.p. Zoological Survey of India, Calcutta.

Julka, J.M., Paliwal, R. and Kathireswari, P. 2009. Biodiversity of Indian Earthworms-overview. In Proceedings of Indo-Us Workshop on Vermitechnology in Human Welfare, Edwards, C.A., Jayaraj, R. and Jayaraj, I.A. (Eds) RohiniAchagam, Coimbatore, India. 36-56.

Mandal, C.K. & Haldar, K.R. 2004. Earthworm of Uttarakhand. State Fauna Series, 18(Part-3),181-191.

Mandal, C.K., Misra, A. and Mitra. S. 2006. A report on earthworm species of Tarakeswar and Adjoining area of Hooghly district, west Bengal with some ecological notes. Rec.Zool.Surv. India. 106 (part-3).123-130.

Mandal, C.K., Misra. A. Roy, G.C. and Biswas's, 2008. Record and abundance earthworm Ballavpur wildlife sanctuary, Birbhum, west Bengal. Rec.Zool.Surv.India.108 (part-3):97-108.

Hasan N., Das Ujjal and Mandal C.K. 2018a. New record of an earthworm Perionyx fulvus from Malda and Howrah District West Bengal, India. International Journal of Advanced Research in Basic Engineering Science and Technology (IJARBEST). Vol. 4: Issue 2; 13-16.

Hasan N., Das, Ujjal and Mandal C.K. 2018b. New record of an earthworm Octochaetona surensis from Kolkata, West Bengal, India. International Journal of Advanced Research in Basic Engineering Science and Technology (IJARBEST). Vol. 4: Issue 2; 1-4.

Michaelsen, W. 1891.Oligochaeta des Naturhistorischen Museums in hamburg. IV. Jahrbuch der Hamburgischen Wissenschaftlichen anstalten, Hamburg, 8: 3-42.

Michaelsen, W. 1900. Oligochaeta. Liferung 10. Berlin: R. Friedlander, I-XXIX +1-575.

Perrier, P. 1872. Recherches pour server α' l'histoire des lombriciensterrestres. Nauvelles archives de Museum d'HistoireNaturelle, Paris, 8: 5-198.

Paliwal R. and Mandal, C.K. 2008. Earthworm fauna of Museum Tank. Rec. Zool. Surv. India, Occ. Paper No. 333. 46-48.

Sinha, R.K. 2009. Earthworms: the miracle of nature (Charles Darwin's unheralded soldiers of mankind & farmer's friend'). Environmentalist, 29: 339-340.

Stephenson, J. 1916. On a collection of Oligochaeta belonging to the Indian Museum. Records of the Indian Museum, 12: 299-354.

Stephenson, J. 1917, Perionyx. gravelyi. Rec. Indian Museum. 13: 378.

Stephenson, J. 1923. The Fauna of British India including Ceylon and Mayanmar. Taylor & Francis, 518 pp.

❒

Chapter-31

ECOLOGICAL INFLUENCE ON THE FORENSICALLY IMPORTANT FLIES OF THE FAMILY CALLIPHORIDA

Reema Sonker and Kalpana Singh*
Department of Zoology, University of Lucknow, Lucknow-226 007
**Email of Corresponding Author: drkalpanasingh@gmail.com*

ABSTRACT

Forensic entomology is a modern branch of applied science deals with collecting and analyzing insect evidence found over and around the carcass to aid in forensic investigations. One of the most important applications of forensic entomology is the determination of postmortem interval (PMI). Flies of family Calliphoridae are very important in this respect as these flies are often the first carrion insect to relocate a carcass for feeding and breeding purpose. The behaviour and life cycle of various species of this family are used in the determination of post-mortem interval. This article explored the various environmental factors affecting the arrival and colonization of Calliphoridae flies over a carcass. Generally, the Calliphoridae fly's attendance over a carcass is affected by temperature, rainfall, ambient light intensity, wind speed, climate, the season of the year, and activities of other insect fauna present in the area.

For the estimation of post-mortem interval that is based on the thermally dependent growth pattern of different instars of blowflies, consideration of these environmental factors on the development and ecological behaviour of these flies are very important. Environmental conditions of the area directly affect the phenomenon of cold tolerance, migration growth and diapauses of Calliphorid fly species. These factors not only have direct effects on the growth and development of these flies but also have marked influence on the predation, parasitism, and likelihood of resource depletion thus ultimately affects the arrival pattern and population size variation of different species of Calliphorid flies.

Keywords: Forensic entomology, Postmortem interval, Calliphoridae, Colonization, Factors.

1. INTRODUCTION

Insects of some specific groups colonize the dead remains in a predictable sequence (Early & Goff, 1986). This sequence of insect visitation on dead remains termed as heterotrophic succession. It occurs in a predictable pattern for a given set of circumstances (Keh, 1985; Schoenly & Reid, 1987) and correlated with stages of decomposition (Anderson & Vanlaerhoven, 1996; Payne, 1965; Rodriguez & Bass, 1983; Reed 1958; Megnin, 1894). These insects are attracted to dead remain by decay odour and feed on carcass as well as on necrophagus insects (Smith, 1986). A number of biochemical and physical changes in decomposing carcass make it a rich source of food for a number of arthropods with time (Anderson & Cervenka, 2002; Anderson, 2010). These arthropod species exploit these resources in the way that their activities make it attractive for certain groups and less attractive for some other groups (Anderson, 2010). The analysis of this sequence of insect visitation on the different stages of decomposition of carrion can be used in determination of postmortem interval (Anderson & Vanlaerhoven, 1996).

The three groups of terrestrial invertebrates, Hymenoptera Coleoptera and Diptera form the majority of these characteristic assemblages of carrion fauna (Smith, 1986). Generally, species of Diptera dominate the early stages of decomposition while Coleoptera species occurs in advanced and dry stages (Wolff et al., 2001).

The cadaveric faunal succession in tropical region initiated by Calliphorid flies followed by flies of family Muscidae and Sarcophagidae (Greenberg, 1991; Smith; 1986; Anderson & Vanlaerhoven, 1996; Goff, 1993). Next in this queue are specimens of various families of Coleoptera (Smith, 1986; Catts & Haskell, 1990; Goff, 1993).

Flies of family commonly known as blow fly constitute the primary carrion fauna (Greenberg 1991; Byrd and Castner, 2010). This is very diverse family of Diptera represented by more than 1000 species in different region of world (Smith 1986; Merritt & De Jong, 2015). These flies are forensically very important as both adult and larvae of flies of this family are associated with different stages of decomposition (Greenberg, 1991; Villet, 2011; Whitworth, 2006). Flies of family Calliphoridae are the first carrion insects in outdoor faunal succession (Smith, 1986; Anderson &Vanlaerhoven, 1996; Payne, 1965; Vanlaerhoven & Anderson, 1999) requires, protein for the proper development of ovary and oocyte (Avancini, 1988)and also a suitable and secure breeding ground for oviposition and development of their offspring (Greenberg, 1991). It has been reported that Calliphoridae flies can relocate a dead body within minutes of death and oviposit on it within hours (Anderson & Vanlaerhoven, 1996; Payne, 1965; Shean et al., 1993; Vanlaerhoven & Anderson, 1999) but in presence of blood and other body fluids oviposition take place very soon (Nuorteva, 1977; Holt et al., 1979).

This host location of Calliphorid flies is triggered by sulphur-rich volatile compounds produced by microbial activities during decomposition (Ashworth & Wall, 1994). Their antennae have highly specialized sense organs which are stimulated by odor and gases of these volatile compounds. However the oviposition in Calliphorid fly is related to physiological changes in female, pheromones & moisture (Collatz & Hoeger, 1980), but the ammonia rich compounds of decomposition stimulate the oviposition in Calliphoridae fly. These volatile compounds produced in decomposition of dead remains, received by the olfactory receptors of the ovipositor of adult female fly (Catts & Haskell, 1990; Browne, 1965).These flies show the group oviposition resulting in the formation of egg masses. The pheromones produced by previously arrived flies helps another fly in site selection (Holt et al., 1979; Ashworth & Wall, 1994). Sometime these egg masses consist of several different fly species (Greenberg, 1991).This group oviposition behavior reduced the effects of predation (Greenberg, 1991) and also astrategy to avoid the vulnerable effects of low temperature. Earlier oviposition in Calliphorid fly in the presence of blood or other body fluid is explained by (Browne, 1962) that, blow flies are very sensitive to water and the tarsal contact with blood increased the oviposition rate as water is major constituent of blood (80%).

The specific species of family Calliphoridae involved in this heterotrophic succession are affected by the various variables of the postmortem site or locations such as such as habitat, season and climate, and manner of death (Anderson, 2001).

These factors also include geographical region, temperature, relative humidity, rain fall, whether or not carcass exposed to sun or covered or wrapped, vegetation and types of soil (Anderson, 2010). These factors affect the species composition, richness, abundance, and diversity of carrion flies in a particular geographical region (Azwandi & Ahmad, 2009; Gennard, 2012; Gunn, 2008; Heo et al., 2011; Nazni et al., 2011; Omar et al., 2003)

These factors not only have direct effects on the growth and development of these flies but also have marked influence on the predation, parasitism, and likelihood of resource depletion thus ultimately affects the arrival pattern and population size variation of different species of Calliphorid flies.

2. EFFECTS OF TEMPERATURE

Temperature is one of the most significant factors predict the activities and behaviour of a number of species of Calliphoridae (Cilek & Schreiber, 1996). Daily Variation in amount of the temperature of the day is a well-recognized factor affects the blow fly's attendance and their life cycle on the carcasses (Byrd & Butler, 1996).

In summer season adult female activity on the carcasses begins after dawn (Baldridge et al., 2006) However, after sunset there is an abrupt drop-off in their activities (Payne, 1965). But in case of artificial lighting nocturnal activities of Calliphoridae flies has been recorded by (Greenberg, 1990). In winter season these flies active on the ephemeral resources at midday regardless of atmospheric morning temperature (Payne, 1965).

Temperature not only affect the blow fly attendance on the carcasses but also it is one of the most important variables influencing the life cycle of Calliphoridae flies as the rate of development of larvae is directly depends on temperature. Increases in atmospheric temperature generally reduce the time of development of maggots of flies (Campobasso et al., 2001). Aggregations of large masses of fly's larvae also generate heat by their activities and metabolism. This heat increases the temperature of microsere so the rate of development of larvae (Campobasso et al., 2001). The heat generated by the large masses of maggots is directly related to the maggot mass density and weight of carcass. The amount of heat absorbed or dissipated from the maggot masses has a significant effect on the development of the fly's larvae and on the decomposition rate of the carcass (Goodbrod & Goff, 1990).

3. EFFECTS OF RELATIVE HUMIDITY

Insects are able to choose an environment that is suitable in terms of humidity this is the way to control water loss (Harrison et al., 2012). Insects prefer high humid habitats for longer life span and greater fecundity (Ouedraogoa et al., 1996). Temperature and humidity are two important variable of environment which is positively correlated with oviposition (Tomberlin & Sheppard, 2002). Flies of family Calliphoridae feed on patchy resources such as decomposing carcass and often lay eggs in masses (Cruikshak & Wall, 2002) this is very important adaptation of these flies to protect their eggs from predation and better survival values of their offspring in adverse environment. In extreme temperature and humidity conditions egg clustering prevent the egg mortality (Stamp, 1980). Decrease in humidity resulting in dehydration of eggs (Norhisham et al., 2013). Loss in water increase the development period of egg stage and induce dormancy in some species (Zrubek & Woods, 2006). Eggs of Calliphoridae flies also face high rate of mortality in high humid condition due to excessive moisture (Guarneriet al.,2002; Norhisham et al., 2013). Those species of flies which are able to tolerate a wider range of humidity have greater egg hatching ability then other species. The successful development of egg and hatching of larvae from the egg depend on the water loss. The low rate of survival of these flies at egg stage is due to desiccation (Davies, 1947).

4. EFFECTS OF PHOTOPERIOD

Photoperiod is a crucial exogenous variable which play important role in development of Calliphoridae flies. Gomes et al., (2006) have observed an altered behaviour in blow fly larvae in artificial light condition. The larvae of *Chrysomya megacephala* burried to deep pupation site when exposed to twenty-four-hour light in comparison to12:12 light/ dark cycle (Gomes et al., 2006). Induction of diapause in the life of blow fly is also an important effect of Photoperiod. The larvae of Calliphora vicina enter a post feeding stage just before pupation. This diapause is a photoperiodically regulated trait exhibited by cold adapted species of Calliphoridae (Saunders et al., 1986). The larvae of female flies which are exposed to long day hours develop into next generation fly without the induction of diapause, but the eggs laid by those female flies which are exposed to autumnal short days develops into larvae which enter into post feeding diapause stage (Saunders, 1987). This photoperiodically induced diapause is fully expressed if these competent larvae exposed to very low temperature (Saunders et al., 1986). Nabity et al.,(2007) has observed that cyclic light resulting in the faster development of larvae of Phormia regina with decreasing temperature.

5. EFFECTS OF WIND SPEED

Generally, speed of wind has mixed effects on the flight activity of Diptera. But according to Blackwell, (1997) and Mohr et al., (2011) host seeking capability of flies does depressed by speed of wind. But studies of Vogt & Starick, (1985) and Vogt et al., (1985) on *Chrysomya rufifacies* indicated that wind has not any effect on the Calliphorid flies though they have not any data about the speed of wind. According to Ashworth and Wall, (1994) wind has positive effects on blowfly attendance over a carcass as volatile compound released by the decomposing remains spread by wind in the atmosphere and helps the flies in relocation of carcass.

6. EFFECTS OF SEASON

The abundance and species composition of Calliphorid flies over a carcass is highly influenced by the seasons of the year. The decrease in temperature during winter and in rainy season resulting in the lower abundance of these flies on the carcasses or dead remains while high temperature of spring and summer favors the higher abundance of these flies on the ephemeral sources. (Gruner et al., 2007; Benbow et al., 2013; Brundage et al., 2011).

Weidner et al., (2015) have observed that species composition of Calliphoridae flies shows variation in different season of the year. The occurrence of these species on the dead remains can predict about the season of death of the victim. Four species of genus *Calliphora* the *C. livida, C. vomitoria, C. vicina,* and *C. cadaverina* strongly indicate the spring season of the year. The flies of this genus were highly abundant in winter, spring and in fall (Watson and Carlton, 2005; Mohr and Tomberlin, 2014; Matuszewski et al., 2010; Weidner et al., 2015). Hall, (1948) and Greenberg, (1973) stated that species of genus *Lucilia* are strongly indicators of summer season in USA as these species were highly abundant in June through August. Brundage et al., (2011) has observed that *Phormia regina* is highly abundant in different region of California and adult specimen of this Calliphorid fly was collected during all four seasons of the year. Similarly, *Lucilia sericata* was a dominant species in summer and fall season in New Jersey (Anderson and Vanlaerhoven, 1996).

7. EFFECTS OF NUTRITION

Nutrient quality is also a very important variable affect the growth and development of Calliphoridae larvae on the carcass. The feeding substrate or type of tissues fed to maggots of Calliphoridae flies has a direct relationship with their rate of development and body size of adult fly (Kaneshrajah & Turner, 2004).

A number of research studies indicate that rate of development of these flies increase with better quality of nutrients (Weissmann & Podmanická, 1970; Kaneshrajah &Turner, 2004).The experimental data of the Sharpe & Hu, (1980) reveal the fact that if the Calliphorid larvae were reared at high protein content diet they show increase rate of development but growth rate was highest at high carbohydrate content diet. Clark et al., (2006) studied the effect of different type of tissues fed to blowflies on the rate of development and concluded that porcine tissue results in faster development in comparison to bovine tissue.

Kaneshrajah & Turner, (2004) performed a comparative analysis on the effect of different types of feeding tissues on the rate of development of Calliphoridae flies and observed that larvae feeds on the lung tissues have twice the development rate in comparison to those feeding on the brain, heart, kidney and liver tissues.

8. EFFECTS OF NOCTURNAL OVIPOSITION

Generally, Calliphorid flies are heliotropic active in day hours and rest at night. Although these flies not oviposit in night but there is some exception reported by Greenberg, (1990). He has recorded the nocturnal oviposition in *Calliphora vicina, Lucilia sericata* and in *Phormia terrae-novae* under artificial light. But nocturnal oviposition has not been reported in many other researches studies in other regions (Byrd & Butler, 1997; Haskell et al., 1997).

9. EFFECTS OF SIZE OR TYPE OF CARCASS

Size or type of dead remains is also a very important variable which affects the Calliphorid fly's attendance on the carcass. Denno and Cothran, (1975) have observed that *Lucilia sericata* and *Phormia regina* shows preference for small sized carcass. But according to, Kuusela & Hanski (1982) species composition of Calliphoridae flies does not affected by the size of carrion but large sized carcass show the highest abundance of these flies.

10. BIOTIC FACTORS

Not only abiotic factors fly of family Calliphoridae are also affected by a number of biotic variables. The intraspecific and interspecific species interactions on the decomposing remains are responsible for competition, mutualism and commensalism among the individuals of different species on the carcasses. Interaction of different species on the dead remains resulted in the formation of discrete ephemeral communities (Kneidel 1985; Hanski, 1987).

The high density of different instars of Calliphoridae flies on the carrion often results in competition for food. This scramble competition due to limited food resources lead to decrease in growth rate and high rate of mortality (Davies 1999; Ireland and Turner, 2006; Vanlaerhoven, 2015). The overcrowding of larvae on the carcass resulted in the undersized larvae, pupae and adults (Saunders and Bee, 1995).The increase internal carcass temperature due to this larval aggregation generate large amount of heat so rate of development of larvae increases (Turner & Howard, 1992; Slone & Gruner, 2007). Decrease rate of development increase the chances of parasitism or predation for maggots

(Nijhout, 2003). Aside from competition positive effects of intraspecific and interspecific species interactions also exist on the carcass.

The coexistence of different species of Calliphoridae on the carrion also affects the longevity and survival rate of particular species of Calliphoridae. The primary fauna of carrion flies modifies the habitat and make it suitable for the next coming species (Hanski, 1987; Hanski and Kuusela, 1977). This behaviour of commensalism has been reported for those species of Calliphoridae that coexist on dead remains. The greater survival rates and large sized adults of *Phormia regina* has been reported from the carrion if this species share the feeding substrate with *Lucilia sericata* (Rosati, 2014).This facilitation of feeding substrate for *Phormia regina* is due to the metabolic activities of *Lucilia sericata*. This carrion feeding species released salivary enzymes which allow the feeding tissues more suitable and efficient for the *Phormia regina* (Charabidze et al., 2011). Beside the commensalism flies of family Calliphoridae adopted many different mechanisms for their coexistence on same resource and manage the feeding substrate in better way (Atkinson and Shorrocks, 1981; Kneidel, 1984; Hanski, 1987; Ives, 1991).One of these mechanisms is resource partitioning which reduces the species interactions among different species on the same resource. According to aggregation model of coexistence the different competing species can coexist on the same resource sonly when they adopt an irregular pattern for their aggregation (Atkinson and Shorrocks, 1984). Flies of family Calliphoridae also show the spatial preferences in different habitats and environment (Fiene et al., 2014). Baseline studies on Calliphoridae flies have documented differences in species composition of these flies in different habitats (Anderson, 2010).Temporal availability also influences the species composition of blow flies and their extent of interaction on the ephemeral resources (Shorrocks & Bingley, 1994).

11. CONCLUSION

This review article demonstrates that there are differences in Calliphoridae fly development due to influence of a number of abiotic and biotic factors, which have a profound effect on the community dynamics as these factors, such as temperature, humidity and the presence of other species on the carrion fluctuate over a spatial and temporal scale.

ACKNOWLEDGMENT

Authors are grateful to Prof. Omkar, Head, Department of Zoology, University of Lucknow, Lucknow for providing the necessary laboratory facilities.

REFERENCES

Anderson, G.S. (2001). Insect succession on carrion and its relationship to determining time of death. In Byrd JH, Castner JL (Eds.) Forensic Entomology: The Utility of Arthropods in Legal Investigations. CRC Press, Florida, 143-176.

Anderson, G.S. (2010). Factors that influence insect succession on carrion. In Byrd, J. & Castner, J. (Eds.), Forensic Entomology: The Utility of Arthropods in Legal Investigations. Boca Raton, Florida, USA. CRC Press. 2: 201-250.

Anderson, G.S. & Cervenka. (2002). Insects associated with the Body: Their Use and Analyses. In Haglund, W. & Sorg, M. (Eds.), Advances in Forensic Taphonomy:Method, Theory and Archaeological Perspectives.Boca Raton, Florida, USA. CRC Press, 173-200.

Anderson, G.S. & Vanlaerhoven, S.L. (1996). Initial studies on insect succession on carrion in South Western British Columbia. Journal of Forensic Sciences, 41 (4): 617-625.

Ashworth, J.R. & Wall, R. (1994). Responses of the sheep blowflies *Lucilia sericata* and *L. cuprina* to odour and the development of semiochemical baits. Medical Veterinary Entomology, 8:303-309.

Atkinson, W.D. and B. Shorrocks. (1981). Competition on a divided and ephemeralresource: a simulation model. Journal of Animal Ecology 50: 461-471.

Avancini, R.M.P. (1988).The influence of non-protein diet on ovarian development in *Chrysomya putoria.* (Diptera: Calliphoridae) Revista Brasilia De Entomologia, 32 (2):103-105.

Azwandi, A. & Abu Hassan, Ahmad (2009). A preliminary study of the decomposition and dipteran associated with exposed carcasses in an oil palm plantation in Bandar Baharu, Kedah, Malaysia. Tropical Biomedicine,26:1-10.

Baldridge, R. S., S. G. Wallace, and R. Kirkpatrick. (2006). Investigation of nocturnal oviposition by necrophilous flies in central Texas. Journal of Forensic Science 51:125-126.

Barton Browne, L. (1965). An analysis of the ovipositional responses of the blow fly *Lucilia cuprina* to ammonium carbonate and indole. Journal of Insect Physiology, 11(8): 1131-1143.

Barton Browne, L. (1962). The relationship between oviposition in the blow fly *Lucilia cuprina* and the presence of water. Journal of Insect Physiology, 8(4): 383-390.

Benbow, M.E., Pechal, J.L. and R.M. Mohr. (2013). Community and landscape ecology of carrion. In: M.E. Benbow, J.K. Tomberlin and A.M Tarone (eds.), Carrionecology, evolution, and their applications. CRC, Boca Raton, FL. 151-185.

Blackwell, A. (1997). Diel flight periodicity of the biting midge Culicoides impunctatus and the effects of meteorological conditions. Med. Vet. Entomol. 11: 361-367.

Brundage, A., Bros, S. & Honda, J.Y. (2011). Seasonal and habitat abundance and distribution of some forensically important blow fies (Diptera: Calliphoridae) in Central California. Forensic Science International 212:115–120.

Byrd, J. H. & Butler, J. F. (1996). Effects of temperature on *Cochliomyia macellaria* (Diptera: Calliphoridae) development. Journal of Medical Entomology, 33(6):901-905.

Byrd, J. H. & Butler, J. F. (1997). Effects of temperature on *Chrysomya rufifacies* (Diptera: Calliphoridae) development. Journal of Medical Entomology, 34(3):353-358.

Byrd, J.H. & Castner, J. (2010). Insects of Forensic Importance. In J. Byrd, & J. Castner (Eds.), Forensic Entomology: The Utility of Arthropods in Legal Investigations. Boca Raton, Florida, USA, CRC Press,2:39-126.

Campobasso, C. P., Di Vella, G. & Introna, F. (2001). Factors affecting decomposition and Diptera colonization. Forensic Science International, 120: 18-27.

Catts, E.P. & Haskell, N.H. (1990). Entomology and Death: A Procedural Guide. Joyce's Print Shop, South Carolina. (ed. 1st): 1-182.

Charabidze, D., Bourel, B. and D. Gosset. (2011). Larval-mass effect: characterisation of heat emission by necrophageous blowflies (Diptera: Calliphoridae) larval aggregates. Forensic Science International, 211: 61–66.

Cruickshank, I. and R. Wall. (2002). Aggregation and habitat use by Lucilia blowflies (Diptera: Calliphoridae) in pasture. Bulletin of Entomological Research 92: 153-158.

Cilek, J. E., and E. T. Schreiber. (1996). Diel host-seeking activity of Chrysops celatus (Diptera: Tabanidae) in North-West Florida. Fla. Entomol, 79: 520-525.

Collatz, K.G. & Hoeger, U. (1980). Age-related changes in the body composition of mated and unmated blow flies, Phormia terraenovae. Experimental Gerontology, 15(5): 433-441.

Clark, K. Evans, L. Wall, R. (2006). Growth rates of the blowfly, *Lucilia sericata*, on different body tissues, Forensic Science International, 156: 145–149.

Davies, L. (1947). Laboratory studies on the egg of the blowfly *Lucilia sericata* (MG). Journal of Experimental Biology, 23: 71.

Denno, R.F. and Cothran, W.R. (1975). Niche relationships of a guild of necrophagous flies. Annals of the Entomological Society of America, 68: 741-754.

Early, M. and Goff, M. (1986). Arthropod succession patterns in exposed carrion on the island of O'ahu, Hawaiian Islands, USA. Journal of Medical Entomology, 23(5): 520-531.

Fiene, J.G., Sword, G.A., VanLaerhoven, S.L. and A.M. Tarone. 2014. The role of spatial aggregation in forensic entomology. Journal of Medical Entomology, 51: 1-9.

Goff, M. L. (1993). Estimation of postmortem interval using arthropod development and successional patterns. Forensic Science Review, 5(2): 81-94.

Gomes L, Gomes G, Oliviera H, Sanches M and Zuben CV (2006). Influence of photoperiod on body weight and depth of burrowing in larvae of *Chrysomya megacephala* (Fabricius) (Diptera, Calliphoridae) and implications for forensic entomology.Revista Brasileira de Entomologia 50(1):76–79.

Goodbrod, J.R. & Goff, M.L. (1990). Effects of larval population density on rates of development and interactions between two species of *Chrysomya* (Diptera: Calliphoridae) in laboratory culture. Journal of Medical Entomology, 27(3):338-343.

Greenberg, B. (1973). Flies and disease, vol. 2. Biology and disease transmission. Princeton University Press, Princeton, NJ.

Greenberg, B. (1990 b). Nocturnal oviposition behavior of blow flies (Diptera: Calliphoridae). Journal of Medical Entomology, 27(5):807-810.

Greenberg, B. (1991). Flies as forensic indicators. Journal of Medical Entomology, 28(5): 565-577.

Gennard, D. (2012). Forensic entomology: An introduction. Wiley, 272.

Gruner, S.V., Slone, D.H. & Capinera, J.L. (2007). Forensically important Calliphoridae (Diptera) associated with pig carrion in rural North-Central Florida. Journal of Medical Entomology, 44:509-515.

Guarneri, A.A., Lazzari, C., Diotaiuti, L., and M.G. Lorenzo. (2002). The effect of relative humidity on the behavior and development of *Triatoma brasiliensis*. Physiological Entomology 27: 142-147.

Gunn, A. (2008). Essential forensic biology. John Wiley and Sons Ltd, England.

Hanski, I. (1987). Carrion fly community dynamics: patchiness, seasonality and coexistence. Ecological Entomology, 12: 257-266.

Hanski, I. and Kuusela, S. (1977). An experiment on competition and diversity in the carrion fly community. Annales Entomologici Fennici, 43: 108-115.

Hall, D.G. (1948). Blow flies of North-America, Indiana: Lay Fayette, Thomas Say Foundation.

Haskell, N. H., Hall, R.D., Cervenka, V.J. & Clark, M.A. (1997). On the body: insects' life stage presence and their postmortem artifacts. In: Haglund, W.D. And Sorg, M. H. (Eds.). Forensic Taphonomy, The Postmortem Fate of Human Remains. CRC Press, Boca Raton, FL. 415-448.

Heo Chong Chin, Kurahashi, H. & Abdullah, M. (2011).Opportunistic Insects Associated with Pig Carrions in Malaysia. Sains Malaysiana, 40(6): 601–604.

Holt, G.G., Adams, T.S. & Sundet, W.D. (1979). Attraction and ovipositional response of screwworms, *Cochliomyia hominivorax* (Diptera: Calliphoridae) to stimulated bovine wounds. Journal of Medical Entomology, 16:248-253.

Harrison, J.F., Woods, H.A. and S.P. Roberts (eds). (2012). Ecological and environmental physiology of insects. Oxford University Press: Oxford and New York. pp. 372.

Ives, A.R. (1991). Aggregation and coexistence in carrion fly community. Ecological Monographs 61(1): 75-94.

Ireland, S. and B. Turner. (2006). The effects of larval crowding and food type on the size and development of the blowfly, *Calliphora vomitoria*. Forensic ScienceInternational, 159: 175-181.

Kaneshrajah, G.B., Turner, B. (2004). *Calliphora vicina* larvae grow at different rates on different body tissues, Int. J. Legal Med. 118 242–244.

Keh, B. (1985). Scope and applications of forensic entomology. Annual Review of Entomology, 30: 137-154.

Kneidel, K.A. (1984). Influence of carcass taxon and size on species composition of carrion breeding diptera. American Midland Naturalist, 111: 57-63.

Kneidel, K. A. (1985). Competition and disturbance in communities of carrion-breeding Diptera. Journal of Animal Ecology, 53: 849-865.

Kuusela, S. & Hanski, I. (1982). The structure of carrion fly communities: The size and type of carrion. Holarctic Ecology, 5: 337-338.

Matuszewski, S., Bajerlein, D., Konwerski, S. & Szpila, K. (2010). Insect Succession and Carrion Decomposition in Selected Forests of Central Europe. Part 2: Composition and Residency Patterns of Carrion Fauna. Forensic Science International, 195(1-3):42-51.

Mégnin, P. (1894). La faune des cadavres. Encyclopédie Scientifique des Aide-Memoire. Gauthier-Villars et Fils, Paris.

Merritt, R.W. and G.D. De Jong. (2015). Arthropod communities in terrestrial environments. In: M.E. Benbow, J.K. Tomberlin and A.M Tarone (eds.), Carrionecology, evolution, and their applications. CRC, Boca Raton, FL. 65-91.

Mohr, R. M., Mullens, B. A. & Gerry, A. C. (2011). Diel patterns of female host seeking, male swarming, and sugar feeding in Fannia conspicua (Diptera: Muscidae) in southern California. J. Med. Entomol. 48: 188-195.

Mohr, R. M. & Tomberlin, J.K. (2014). Environmental Factors Affecting Early Carcass Attendance by Four Species of Blow Flies (Diptera: Calliphoridae) in Texas.J. Med. Entomol. 51(3): 702-708.

Nabity, P., Higley, L. G. & Heng-Moss, T. (2007). Light-induced variability in development of forensically important blow fly Phormia regina (Diptera: Calliphoridae). Journal of medical entomology, 44: 351–358.

Nazni, W.A., Lee, H.L., Chen, C.D., Heo, C.C., Abdullah, A.G., Wan-Norjuliana W.M., Chew, W.K., John, J., Rosli, H. & Sofian, M.A. (2011). Comparative insect fauna succession on indoor and outdoor monkey carrions in a semi-forested area in Malaysia. Asian Pacific Journal of Tropical Biomedicine, Supplement: S232-S2238.

Nijhout, H.F. 2003. The control of body size in insects. Developmental Biology, 261: 1-9.

Norhisham, A.R., Abood, F., Rita, M. and K.R. Hakeem. 2013. Effect of humidity on egg hatchability and reproductive biology of the bamboo borer (*Dinoderus minutes* Fabricius). SpringerPlus 2:9.

Nuorteva, P. (1977). Sarcosaprophagous insects as forensic indicators. In Forensic

Medicine: A Study in Trauma and Environmental Hazards. Tedeschi, C.G., Eckert, W.G. and Tedeschi, L.G. (eds). Saunders, Philadelphia. 2:1072-1095.

Omar, B., Marwi, M. A., Ahmad, A., Zuha, R. M. and Jeffery, J. (2003). Synanthropic index of flies (Diptera: Musciade and Calliphoridae) collected at several locations in Kualalumpur and Gombak, Malaysia. Tropical Biomedicine, 20(1): 77-82.

Ouedraogoa, P.A., Soua, S., Sanona, A., Mongea, J.P., Huignarda, J., Trana, B., and P.F. Credland. (1996). Influence of temperature and humidity on populations of *Callosobruchus maculatus* (Coleoptera: Bruchidae) and its parasitoid *Dinamus basalis* (Pteromalidae) in two climatic zones of Burkina Faso. Bulletin of Entomological Research 86: 695-702.

Payne, J. A. (1965). A summer carrion study of the baby pig Sus scrofa Linnaeus. Ecology, 46 (5): 592-602.

Reed, H. B. (1958). A study of dog carcass communities in Tennessee, with special reference to the insects. American Midland Naturalist, 59 (1): 213-245.

Rodriguez, W. C. & Bass, W. M. (1983). Insect activity and its relationship to decay rates of human cadavers in East Tennessee. Journal of Forensic Sciences, 28 (2): 423-432.

Rosati, J.Y. (2014). Spatial and temporal variability in the carrion insect community: using blow flies (Family: Calliphoridae) as a model system to study coexistence mechanisms at multiple scales (Doctoral Dissertation). Retrieved from scholar.uwindsor.ca

Saunders D (1987). maternal influence on the incidence and duration of larval diapauses in *Calliphora vicina*. Physiol Entomol 12:331–338.

Saunders DS, Macpherson JN and Cairncross KD (1986) Maternal and larval effects of photoperiod on the induction of larval diapause in two species of fly, *Calliphora vicina* and *Lucilia sericata*. Exp Biol 46(1):51–58.

Saunders, D.S. and A. Bee. 1995. Effects of larval crowding on size and fecundity of the blow fly, *Calliphora vicina* (Diptera: Calliphoridae). European Journal of Entomology 92: 615-622.

Schoenly, K. & Reid, W. (1987). Dynamics of heterotrophic succession in carrion arthropod assemblages: discrete series or a continuum of change? Oecologia, 73:192-202.

Sharpe, P. and Hu, L. (1980) Reaction kinetics of nutrition dependent poikilotherm development. J theor Biol 82:317–333.

Shean, B. S., Messinger, B. A. & Papworth, M. (1993). Observations of differential decomposition on sun exposed v. shaded pig carrion in Coastal Washington State. Journal of Forensic Sciences, 38 (4): 938-949.

Shorrocks, B. and Bingley M. (1994). Priority effects and species coexistence: experiments with fungal-breeding Drosophila. Journal of Animal Ecology 63: 799-806.

Slone, D.H. and S.V. Gruner. (2007). Thermoregulation in larval aggregations of carrion breeding blow flies (Diptera: Calliphoridae). Journal of Medical Entomology 44: 516-523.

Smith, K. (1986). A Manual of Forensic Entomology. London: The Trustees of British Museum of Natural History, London and Comstock Publishing Associates, New York 1-207.

Stamp, N.E. (1980). Egg deposition patterns in butterflies: why do some species cluster their eggs rather than deposit them singly? American Naturalist 115: 367-380.

Tomberlin, J.K. and Sheppard, D.C. (2002). Factors influencing mating and oviposition of black soldier flies (Diptera: Stratiomyidae) in a colony. Journal of Entomological Science, 37: 345-352.

Turner, B., and T. Howard. (1992). Metabolic heat generation in Dipteran larval aggregations: a consideration for forensic entomology. Medical andVeterinary Entomology 6:179-181.

VanLaerhoven, S.L. (2015). Modeling species interactions within carrion food webs. In: M.E. Benbow, J.K. Tomberlin and A.M Tarone (eds), Carrion ecology, evolution, and their applications. CRC, Boca Raton, FL. 231-245.

VanLaerhoven, S.L., and G.S. Anderson. (1999). Insect succession on buried carrion in two biogeoclimatic zones of British Columbia. Journal of Forensic Sciences, 44:32-43.

Villet, M.H. (2011). African carrion ecosystems and their insect communities in relation to Forensic entomology. Pest Technology 5: 1-15.

Vogt, W. G., and N. T. Starick. (1985). A wind-oriented fly trap for quantitative sampling of adult Musca vetustissima Walker. J. Aus. Entomol. Soc. 24: 223-227.

Vogt, W. G., T. L. Woodburn, R. Morton, and B. A. Ellem. (1985). The influence of weather and time of day on trap catches of males and females of *Lucilia cuprina* (Wiedemann) (Diptera, Calliphoridae). Bull. Entomol. Res. 75: 315-319.

Watson, E.J. and Carlton, C.E. (2005). Insect succession and decomposition of wildlife carcasses during fall and winter in Louisiana. J Med Entomol, 42(2):193–203.

Weissmann, L. and Podmanická, D. (1970). [Influence of nutrition and breeding temperature on the lenght of larval development of Scotia segetum Den. et Schiff]. Biologia (Bratisl) 25(8):537–545.

Weidner, L.M. Jennings, D.E. Tomberlin, J.K. and hamilton, G.C. (2015). Seasonal and Geographic Variation in Biodiversity of Forensically Important Blow Flies (Diptera: Calliphoridae) in New Jersey, USA. J. Med. Entomol. 1–10.

Whitworth, T. (2006). Keys to the genera and species of blow flies (Diptera: Calliphoridae) of America north of Mexico. Proc. Entomol. Soc. Wash. 108: 689-725.

Wolff, M., Uribe, A., Ortiz, A. and Duque, P. (2001). A preliminary of study forensic entomology in Medellin, Columbia. Forensic Sci. Int. 120: 53-59.

Zrubek, B. and Woods, H.A. (2006). Insect eggs exert rapid control over an oxygen water tradeoff. Proceedings of the Royal Society B 273: 831-834.

❐❐❐